GENETICS AND CONSERVATION
OF RARE PLANTS

Genetics AND Conservation OF Rare Plants

Edited by

DONALD A. FALK
KENT E. HOLSINGER

CENTER FOR PLANT CONSERVATION

New York   Oxford
OXFORD UNIVERSITY PRESS
1991

Oxford University Press

Oxford   New York   Toronto
Delhi   Bombay   Calcutta   Madras   Karachi
Petaling Jaya   Singapore   Hong Kong   Tokyo
Nairobi   Dar es Salaam   Cape Town
Melbourne   Auckland

and associated companies in
Berlin   Ibadan

Library of Congress Cataloging-in-Publication Data
Genetics and conservation of rare plants /
edited by Donald A. Falk, Kent E. Holsinger.
 p. cm.   Includes bibliographical references and index.
 ISBN 0-19-506429-1
 1. Rare plants—Genetics. 2. Plant conservation.
 3. Plant populations. I. Falk, Donald A. II. Holsinger, Kent E.
QK86.A1G45   1991
631.5'23—dc20   91-10735

9 8 7 6 5 4 3 2 1

Printed in the United States of America
on acid-free paper

6-25-92

Foreword

As an aspiring young botanist growing up in San Francisco, I was always thrilled at the discovery of botanical novelties, those rare, unusual plants that one came upon unexpectedly and that often turned out to be new or otherwise interesting. For example, my rediscovery of two very local endemics on the serpentine soils of the Presidio, *Arctostaphylos hookeri* subsp. *ravenii* (now endowed as part of the National Collection of the Center for Plant Conservation) and *Clarkia franciscana,* was of particular enjoyment to me as a young teen-ager. My initial study of *Clarkia* helped to steer me toward a research career that has been heavily focused on the plant family Onagraceae, and helped to give me an early appreciation of the interest and importance of rare plants.

Although California, with its thousands of local endemics, was a uniquely suitable place to make those discoveries, and many other botanists have shared that appreciation, until recently we have not appreciated much about plant population biology. A deeper appreciation of the reproductive, ecological, and evolutionary significance of genetic variability is of fundamental importance in plant conservation. How do we deal with the limited variability that is often present in the diminished populations of rare plants? What if the last few individuals differ remarkably from one another or represent hybrids with other species? If only one genetically self-incompatible individual should survive, what strategies should be undertaken to try to save it? In our attempt to solve problems of these kinds, the detailed exploration of plant genetics, population biology, and evolution that this book represents is of great significance, and clearly long overdue.

As we enter the final decade of the twentieth century, enormous pressures are being exerted on the world's capacity to support us in a sustainable fashion. These pressures are nowhere more evident than in California, scene of my childhood discoveries and home to more than 600,000 additional people every year. Fields, woods, and meadows are all-too-rapidly giving way to housing developments and freeways, and the choice localities where evolutionary novelties have appeared and still flourish locally are being lost with every passing year. A similar fate is befalling species and habitats in Hawaii, Florida, Puerto Rico, and many other regions. Our human future will be secure only when we are able to attain stability in our numbers and in the way we use the planet's resources. At that time, the kinds of plants, animals, fungi, and microorganisms that remain available to enjoy, to study, and to use will depend in large extent on the actions that we take now.

Around the world, plants are disappearing even more rapidly than they are in the United States. The Center for Plant Conservation model, with a network of cooperating botanical gardens, each dedicated to the preservation of a particular set of species native to its region in the gardens and in seed banks, seems to have wide applicability. It would be highly advantageous to replicate this model in various nations around the world, and thus to help stem the loss of plant species that poses such a serious problem for us all. The genetic principles that are the subject of this book would need to be applied generally as the plants were brought into cultivation, stored in seed banks, and then reintroduced into suitable habitats.

This volume, providing as it does a panorama of some of the most important tools that we have available for conserving plant species, will assist us greatly in that task. The principles explored here are in general of equal importance for the conservation of other organisms, but have been organized for the first time around the priorities of plant conservation. We still have a great deal to learn, and in a field evolving as rapidly as conservation biology, we feel certain that there will be continued deepening of our understanding of the biology of plants. This application to conservation is thus one of the most important contributions of scientific research to sustaining the planet we live on. Meanwhile, this volume gives us additional hope and, more than that, some of the tools that we badly need to help us conserve our biotic heritage.

PETER H. RAVEN
Missouri Botanical Garden
St. Louis

Preface

To conserve biological diversity, conservation programs must be guided by the biology of the species or systems that they seek to preserve. This is, in fact, the basic axiom of conservation biology. For this reason, the scientific research community must provide the best available data and models to guide program decisions and allocations of resources. Scientific research is not a luxury; it is the foundation of good conservation. We cannot conserve what we do not understand.

The practice of conservation, however, takes place in a context of limited resources. Consequently, practitioners must frequently extrapolate and improvise, making pragmatic decisions based on incomplete understanding of the biology of the species or systems they wish to protect. As a result, conventional wisdom and speculation are more often the guide than are hard data and good theory. This is especially true for the conservation of plants, which has been hampered by the reliance on borrowed and extrapolated models from animal biology and by the paucity of research into the biology of rare plant species.

Both the population dynamics and the genetics of rare species are key areas of research in biological conservation. These fields are no longer simply of theoretical interest to research biologists. Genetics and demography are increasingly recognized as crucial to success in long-term management of any species. Thus programs to conserve and manage endangered plant species will benefit significantly from emerging models in plant genetics and population biology, models that provide a basis for biologically sound strategies to manage populations of endangered species on an efficient and sustainable basis. This increased scientific understanding is particularly important in the development of integrated conservation strategies, combining population and species management in species-specific combinations of on-site (*in situ*) and off-site (*ex situ*) conservation techniques.

This volume is intended to present current scientific understanding in areas of genetics and population biology that are critical for management and conservation of endangered plant species. To our knowledge, this is the first attempt in book form to treat issues of the genetics and population biology of rare plants systematically. The structure and logic of the book should be noted: we begin with several chapters examining the basic population biology, ecology, and genetics of rare plants. Successive sections discuss patterns of distribution of genetic variation, cost-effective methods of capturing this variation in conservation programs, and the design of large-scale strategies for conserving diversity. We hope this work will offer insight into the biology of some of the rarest life forms on earth. Such knowledge is, we believe, both an end in itself and the key to successful conservation.

This book grew from papers originally presented at a Conference on the Genetics and Conservation of Rare Plants coordinated by the Center for Plant Conservation and held at the Missouri Botanical Garden in St. Louis in March 1989. The intent of the meeting was to bring together leading research scientists and conservation practitioners and to summarize current research in population biology and genetics with relevance to conservation programs and policies for rare plant species. Since rare plant species are still poorly represented in the published literature, the Center sought to establish a scientific basis for conservation. Over the course of the meeting, the partitioning and distribution of genetic variation within species, sampling strategies designed to capture a representative sample of that variation, and long-term strategies for maintenance of the genetic diversity present in the sample emerged as core issues. We paid special attention to practical problems, such as finding ways to help field collectors in selecting the optimal number of populations to be sampled (or protected, in the case of habitat conservation). Similarly, we examined research that correlates genetic diversity with life history, demography, and ecological factors for their potential predictive value. We also wanted to help the conservation community understand the factors that affect the dynamics of small populations that are characteristic of many rare species (e.g., genetic drift, inbreeding, and environmental stochasticity).

After the meetings, we solicited additional papers in several key areas of research that were not represented at the conference, with the intention of making this book a well-rounded statement on the biology of rare plants. We asked each author not only to summarize current research, but also to provide a commentary on the tangible implications of his or her work for conservation. These sections are intended to form a bridge between research and application—and to serve as the basis for the summary sections and guidelines at the end of the volume.

RELEVANT ISSUES

Studies of rare plant species are important not only because they address the immediate practical concerns of species conservation, but also because they contribute directly to a better understanding of the ecological and evolutionary processes that are fundamental to all of life's diversity. When rare species are poorly understood, common ones often will be also. For example, it was common until recently to assume that most endangered plant species are rare because they are poor competitors. Many studies have shown that this conclusion is in error. Rare plant species are often specialists for a rare microhabitat, and their demographic and competitive characteristics may be identical to those of their more widespread relatives. Partly because of such studies, the roles of disturbance-mediated competition, recruitment limitation, local interaction, and microhabitat specialization in plant population and community dynamics are much more widely recognized now than they were ten years ago.

Studies of rare species not only deepen our understanding of common ones, but also offer insight on questions of fundamental importance in ecology and evolutionary biology, questions that are directly relevant to the design of appropriate management strategies for rare and endangered plant species. For example: What is the relationship between electrophoretic variation, DNA polymorphism, and genetic diversity at loci that are important in evolutionary responses to environmental change? What factors are primarily responsible for determining the distribution and abundance of genetic variation in plant populations? What evolutionary and ecological consequences do occasional or periodic fluctuations in population size have? What is the relationship between niche width and genetic diversity? Are the dynamics of plant populations determined more by stochastic environmental variation or by deterministic competitive interactions among species? Is the distributional range of a species more often the result of historical patterns of dispersal or of physiological limits to growth and reproduction?

Answers to these and similar questions are required for the intelligent management of rare species. Thus the distinction between basic and applied research begins to disappear where rare species are concerned. Research directed toward improving the chances of a rare species' survival will also contribute to the resolution of basic issues in biology, and any research intended to clarify those issues will contribute to improved management strategies for rare species. This interpenetration of basic and applied research on rare species is evident in many chapters in this book. It is especially evident in the four chapters that form the first part of this volume.

PART I

The chapters in Part I focus on the biology and genetics of rare plant species. They provide the biological foundation on which specific conservation strategies can be built. Many rare plant species are endangered in part because their populations are small. Barrett and Kohn discuss the ecological reasons why some populations are small, but their focus is on the genetic consequences of life in small populations. Their survey of both the theoretical and the empirical literature shows that the effect of drastic reductions in populations size can be seen even in populations that now number in the tens of thousands, whether the population bottleneck was a result of a natural catastrophe or the colonization of new geographical areas. They also provide some simple techniques to measure the effect on individual reproduction and survival of continued reproduction in small populations.

While Barrett and Kohn examine the genetic consequences of small population size, Huenneke focuses on the ecological role of genetic diversity. She points out that the loss of genetic diversity is sometimes a direct threat to the continued viability of plant populations. More frequently, however, the loss of genetic diversity is a *consequence* of population failure rather than its cause. Those endangered species known to have suffered recent changes in distribution and abundance are more likely to be threatened by the loss of genetic diversity than are those that have always been rare.

Animal conservationists have commonly employed demographic techniques to the management of endangered species, but plant conservationists have concentrated primarily on the conservation of habitat. Menges shows how demographic techniques used by animal conservationists can be modified for the management of rare plant populations. As for most animal populations that have been analyzed, it appears that ecological factors, specifically environmental stochasticity, pose a greater threat to the continued existence of most small populations than do genetic factors.

Although much of what we know about the conservation biology of plants is based on the analysis of temperate-zone species, most of the species that face immediate extinction are found in the tropics. In addition, the characteristics of rare species in the tropics may be very different from those of typical rare species in the temperate zones. Because of the extraordinary species diversity characteristic of tropical forests, well over half the tree species found there are rare in the sense that they occur in very low population densities. As Bawa and Ashton point out, these low population densities pose formidable problems for conservationists, not the least of which is that the conservation of rare tropical forest plants, especially trees, is likely to require the protection of areas hundreds of square kilometers in size.

PART II

The chapters in Part I of this book describe the broad biological principles that are fundamental to the design of any conservation program for rare plants; those in Part II discuss the principles that underlie one important aspect of the development of conservation programs: the conservation of genetic diversity. Hamrick, Godt, Murawski, and Loveless begin this section by analyzing the effect of a variety of life-history characteristics (geographical range, life form, mode of reproduction, breeding system, seed dispersal mechanism, and successional status) on the distribution and abundance of genetic variation in plant populations. Several broad patterns are discernible, such as the tendency to greater between-population diversification in self-fertilizing species than in outcrossers and to greater total genetic diversity in widespread species than in endemics. They also observe that these broad patterns obscure considerable differences among species with similar life-history characteristics. These differences presumably reflect, to a large extent, the vagaries of evolutionary history. As Karron observes, history has a major impact on genetic diversity in rare plant species. He shows that rare species in one genus are often more genetically diverse than common ones in another genus. Even within a single genus, geographical distribution may provide only a crude guide to the amount of genetic diversity present within a species.

To conserve genetic diversity in rare plants, we must know more than the pattern of genetic differentiation among populations. We must design sampling schemes that will capture representative samples of the genetic diversity present in each species, whether in samples for off-site collections or in natural populations that are to be protected. As Brown and Briggs point out, capturing all the diversity present is an impossible, and perhaps undesirable, goal. Drawing on their experience with the rare flora of southeastern Australia, they point out that most rare plants, at least in temperate zones, are found in very few populations. Although the design of sampling or protection schemes for the conservation of genetic diversity in crop plants and their wild relatives is an enormous task, the observation that most rare plants exist in only a few populations results in simple guidelines for the selection of populations to sample and protect.

PART III

Having explored the general considerations that shape conservation collections, we turn to their management and evaluation. Schaal, Leverich, and Rogstad begin by summarizing methods available to conservation biologists for assessing the genetic

variation in a sample. Morphological variation is widely used to provide a crude estimate of genetic differentiation among individuals in a species, but distinguishing between genetic and purely phenotypic differences requires an enormous effort. Allozyme electrophoresis, probably the most widely used tool among population
geneticists today, offers a deeper insight into genetic variation and has now been used
to study hundreds of species. But allozyme studies can detect variation only at genes
encoding soluble enzymes, which may not always be representative of the entire genome. DNA techniques, which permit direct examination of the genome, avoid many
drawbacks of the other methods. Precisely because of their ability to examine the entire
genome, however, these methods are prone to other difficulties. Sequences that are
noncoding, highly repetitive, or invariant may be surveyed along with those that manifest significant variation among individuals, leading to difficulties in interpretation.
The laboratory techniques are also expensive and not yet widely available, making
their application to conservation problems limited at present. In the long run, DNA
analysis will provide an enormously powerful tool to understand the phylogeny and
population genetics of plants, both for basic science and for conservation.

Many conservation collections will be maintained as stored seed, which permits a
broader genetic representation to be included than would be possible with growing
plants. Eberhart, Roos, and Towill discuss methods for maintaining seed collections,
including both conventional and cryogenic techniques. While not all species have
seeds amenable to low-temperature storage, this technology offers a cost-effective way
to capture much of the variability in an off-site collection. Ongoing procedures to
monitor the viability of seed collections must be an integral part of any such program,
especially for rare wild species whose characteristics in seed storage are not yet well
known.

PART IV

The last major section of this book represents a synthesis of the preceding papers, and
an interpretation of their meaning for conservation. It is here that *Genetics and Conservation of Rare Plants* reaches its essential objective: the application of theory and
research to models for conserving diversity. Millar and Libby begin with a strategy for
conserving the diversity in widespread species, for as they point out, such species are
usually more genetically diverse than rare endemics. Moreover, the methods used for
widespread plants can be adapted to species with narrow ranges. They describe a five-
point program that integrates on-site and off-site approaches, employing techniques
of reintroduction and habitat management. Rieseberg follows with three case studies
of rare plants in *Cercocarpus* and *Helianthus* where hybridization is a significant part
of the species' biology. Many rare plants are thought to be of recent hybrid origin, but
anthropogenic hybridization with widespread species is a significant threat to the genetic integrity of some rare plants. As Rieseberg shows, rare plants are especially
vulnerable to genetic assimilation, outbreeding depression, and the loss of adaptive
ecotypic variation that often results.

Plant conservationists are often painfully aware of the seeming head start of animal
biologists in addressing genetic issues. Zoos, for example, have been involved in captive-breeding and reintroduction programs for years, an activity that is still relatively

new to botanical gardens. Templeton offers many useful insights from the perspective of animal breeding and genetic conservation, discussing techniques that will be applicable to plant programs. One issue that emerges as particularly significant is that of the genetic content of the initial sample: if a collection is eventually used as the core of a reintroduction program, the initial samples act as analogues of a founder population. If every species reintroduction effort were treated as an empirical test of the founder effect, complete with controls and baseline measurements of diversity, we would very quickly accumulate a far deeper understanding of the behavior of small, recently established populations than we now have.

This section concludes with two synthetic papers expressive of the overall intent of this book, drawing together the presentation of scientific evidence and reasoning toward conservation strategies. Holsinger and Gottlieb begin by summarizing the scientific insights of preceding chapters into the biology of rare plants, paying special attention to what is known about the population genetics of rare species. They then offer a set of recommendations for biologically sound conservation programs, including an assessment of priority species, the significance of interpopulation diversity, the adaptive significance of rare alleles, and the integration of on-site and off-site approaches. As they observe, there is far from a consensus on every relevant point, and many of these points, such as the degree of genetic difference among populations and the adaptive significance of this variation, have profound implications for conservation programs. Falk extends this reasoning by joining models in two basic areas that inevitably shape genetic conservation programs: population genetics and the economics of establishing and maintaining collections. No conservation program can do everything, and the resources available to any program are inevitably limited. Thus resource managers face a series of difficult choices about the allocation of scarce resources among an abundance of needs. The model offered here, joining biological and economic theory, is a preliminary sketch of what must eventually become a major area of theory in conservation biology.

APPENDIX

Readers whose primary interest lies in the application of theory to actual conservation practice will find the most immediate help in the Appendix. The "Genetic Sampling Guidelines for Conservation Collections of Endangered Plants," developed by the Center for Plant Conservation for use in its own rare plant program, are intended to provide concrete guidance to those whose intent is to capture a representative sample of the genetic diversity in target species as they establish living collections of rare plants. Plant species (even the rare ones) are so variable in genetic structure, ecology, and reproductive biology as to defy broad generalization. This, in turn, creates problems for the conservationist who seeks to manage or collect a species whose biology is known only anecdotally, or not at all. The "Guidelines" wrestle with this demon, in an attempt to offer practical guidance to the designer and field collector of a conservation collection, based on observations about population genetics found elsewhere in the volume. We believe that these "Guidelines" are the most detailed statement of their kind to date and hope that they will find broad application.

BIBLIOGRAPHY

Locating information on any particular aspect of plant conservation biology is a difficult task. Papers dealing with plant conservation might appear in any one of a hundred or more professional journals, in conference proceedings, or in government agency reports. In preparing their chapters for this book, the contributors located and read most of this literature before synthesizing it into chapters that, collectively, answer many questions that face plant conservationists. As a result, the integrated bibliography, which consists of some 850 individual references, is the most extensive compilation of research and writing on rare and endangered plants that has ever been published, and we expect it to be an invaluable resource for conservationists, ecologists, evolutionists, and geneticists for many years to come.

This book is the culmination of several years' effort by many people, including many whose contributions are in danger of remaining invisible. Editors and authors may strut and fret their hour upon the stage, but someone has to feed the cast and keep the curtain from falling on their heads. To all those who contributed ideas, time, and creativity, we tip our hats.

It is customary to leave mention of funding sources to the very end. We must observe, however, that unless someone had been willing to take a chance and commit resources to this idea, the project would never have progressed beyond the idea state. Our thanks to the Pew Charitable Trusts, the John D. and Catherine T. MacArthur Foundation, and the National Science Foundation for helping turn our ideas into reality.

From its inception, this project had the good fortune to be steered by a committee possessed of great insight, prestige, and scientific judgment. We thank Peter Ashton, Robert Cook, Leslie Gottlieb, William Libby, Barbara Schaal, Mike Soulé, and Alan Templeton for giving the boat a push in the right direction. Frank Thibodeau was also instrumental in developing the early designs for this project.

The Science Advisory Council of the Center for Plant Conservation helped sharpen the focus on the application of science to the practice of conservation. Thanks to the members of that group, who continue to give unselfishly of their time and expertise in guiding the Center's programs: John Fay, Bob DeFilipps, Bob Jenkins, Art Kruckeberg, Dick Lighty, Gren Lucas, Bruce MacBryde, Bob Mohlenbrock, Larry Morse, Norton Nickerson, and Barbara Schaal.

At the conference itself, Kamal Bawa, Hugh Bollinger, Vernon Heywood, Doug Schemske, and Norm Weeden presented papers, moderated sessions, and helped lead discussions. We are also immensely grateful to the staff of the Missouri Botanical Garden—especially Peter Raven, Marshall Crosby, and George Rogers—for opening the magnificent facilities of that institution for our use.

Many individuals offered the ideas, creativity, experience, advice (both solicited and unsolicited), time, and encouragement of themselves and their institutions. For this we especially thank Jim Affolter, Marlin Bowles, Bob Breunig, Bill Brumback, Marcy Demauro, Tom Elias, Peggy Fiedler, Tim Flynn, Rob Gardner, Ed Guerrant, Subodh Jain, Dan Janzen, Bob Jenkins, Charlotte Jones-Roe, Julie Kierstead, Steven King, Hal Koopowitz, Carol Lippincott, Jim Locklear, Tom Lovejoy, Dave Michener,

Orlando Mistretta, Hal Mooney, Gary Nabhan, Rob Nicholson, Bob Ornduff, Brian Parsons, Bruce Pavlik, Richard Primack, George Rogers, John Schwegman, Jon Shaw, Cal Sperling, Susan Wallace, Peter White, Del Wiens, and Robert Wyatt.

Without the support and involvement of the staff of the Center for Plant Conservation, this project would never have gotten off the ground. Along with members of the University of Connecticut, these people have worked long and hard behind the scenes: Linda McMahan, Mike O'Neal, Peggy Olwell, Nicole Roth, Russell Stafford, and Kerry Walter. The editors also acknowledge their eternal debt to Grace Padberg, Linda DeBruyn and Jennifer Klein, who carried the entire logistical load from conception to publication, including preparation of manuscripts and tending the frayed nerves of the organizers. The index might never have been done without the patient help provided by Roberta Mason.

We cannot imagine an editor more possessed of vision, wit, patience, and scientific judgment than William Curtis of Oxford University Press. He and Irene Pavitt have been a joy to work with, and there should be no mistake that the quality of the final product shows their steady hands.

Finally, we dedicate this project to our mothers and fathers, for it is their nature and nurture that set us on this road in the first place.

St. Louis, Mo.   D.A.F.
Storrs, Conn.   K.E.H.
February 1991

Contents

Contributors, xvii

I POPULATION BIOLOGY AND GENETICS OF RARE SPECIES

1. Genetic and Evolutionary Consequences of Small Population Size
    in Plants: Implications for Conservation, 3
    Spencer C. H. Barrett and Joshua R. Kohn

2. Ecological Implications of Genetic Variation
    in Plant Populations, 31
    Laura Foster Huenneke

3. The Application of Minimum Viable Population Theory
    to Plants, 45
    Eric S. Menges

4. Conservation of Rare Trees in Tropical Rain Forests:
    A Genetic Perspective, 62
    K. S. Bawa and P. S. Ashton

II DISTRIBUTION AND SAMPLING OF GENETIC VARIATION

5. Correlations Between Species Traits and Allozyme Diversity:
    Implications for Conservation Biology, 75
    J. L. Hamrick, M. J. W. Godt, D. A. Murawski, and M. D. Loveless

6. Patterns of Genetic Variation and Breeding Systems
    in Rare Plant Species, 87
    Jeffrey D. Karron

7. Sampling Strategies for Genetic Variation in *Ex Situ* Collections
    of Endangered Plant Species, 99
    A. H. D. Brown and J. D. Briggs

III MANAGEMENT AND ASSESSMENT OF OFF-SITE COLLECTIONS

8. Comparison of Methods for Assessing Genetic Variation in Plant
    Conservation Biology, 123
    Barbara A. Schaal, Wesley J. Leverich, and Steven H. Rogstad

9.   Strategies for Long-Term Management of
     Germplasm Collections, 135
     S. A. Eberhart, E. E. Roos, and L. E. Towill

IV CONSERVATION STRATEGIES FOR GENETIC DIVERSITY

10. Strategies for Conserving Clinal, Ecotypic, and Disjunct Population
     Diversity in Widespread Species, 149
     Constance I. Millar and William J. Libby

11. Hybridization in Rare Plants: Insights from Case Studies in
     Cercocarpus and *Helianthus*, 171
     Loren H. Rieseberg

12. Off-Site Breeding of Animals and Implications
     for Plant Conservation Strategies, 182
     Alan R. Templeton

13. Conservation of Rare and Endangered Plants:
     Principles and Prospects, 195
     Kent E. Holsinger and L. D. Gottlieb

14. Joining Biological and Economic Models for Conserving
     Plant Genetic Diversity, 209
     Donald A. Falk

     Appendix
     Genetic Sampling Guidelines for Conservation Collections
     of Endangered Plants, 225
     Center for Plant Conservation

Bibliography, 239
Index, 277

Contributors

P. S. ASHTON
*Department of Organismic
    and Evolutionary Biology
Harvard University
Cambridge, Massachusetts 02138*

SPENCER C. H. BARRETT
*Department of Botany
University of Toronto
Toronto, Ontario M5S 3B2
Canada*

K. S. BAWA
*Department of Biology
University of Massachusetts
Boston, Massachusetts 02125*

J. D. BRIGGS
*CSIRO Division of Plant Industry
GPO Box 1600
Canberra
A.C.T. 2601
Australia*

A. H. D. BROWN
*CSIRO Division of Plant Industry
GPO Box 1600
Canberra
A.C.T. 2601
Australia*

S. A. EBERHART
*Plant Germplasm Research
U.S. Department of Agriculture
Forest Service
Northern Plains Area National Seed
    Storage Laboratory
Fort Collins, Colorado 80523*

DONALD A. FALK
*Center for Plant Conservation
Missouri Botanical Garden
P.O. Box 299
St. Louis, Missouri 63166*

M. J. W. GODT
*Department of Biology
University of Georgia
Athens, Georgia 30602*

L. D. GOTTLIEB
*Department of Genetics
University of California
Davis, California 95616*

J. L. HAMRICK
*Department of Botany
University of Georgia
Athens, Georgia 30602*

KENT E. HOLSINGER
*Department of Ecology
    and Evolutionary Biology
University of Connecticut
Storrs, Connecticut 06268*

LAURA FOSTER HUENNEKE
*Department of Biology
New Mexico State University
P.O. Box 30001
Department 3AF
Las Cruces, New Mexico 88003*

JEFFREY D. KARRON
*University of Wisconsin
Department of Biological Sciences
P.O. Box 413
Milwaukee, Wisconsin 53201*

JOSHUA R. KOHN
Department of Botany
University of Toronto
Toronto, Ontario M5S 3B2
Canada

WESLEY J. LEVERICH
Department of Biology
St. Louis University
St. Louis, Missouri 63103

WILLIAM J. LIBBY
Departments of Genetics and Forestry
  and Resource Management
University of California
Berkeley, California 94720

M. D. LOVELESS
Department of Biology
College of Wooster
Wooster, Ohio 44691

ERIC S. MENGES
Archbold Biological Station
P.O. Box 2057
Lake Placid, Florida 33852

CONSTANCE I. MILLAR
U.S. Department of Agriculture
Forest Service
Pacific Southwest Forest and Range
  Experiment Station
Box 245
Berkeley, California 94701

D. A. MURAWSKI
Department of Biology
University of Georgia
Athens, Georgia 30602

LOREN H. RIESEBERG
Rancho Santa Ana Botanic Garden
1500 North College Ave.
Claremont, California 91711

STEVEN H. ROGSTAD
Department of Biology
Washington University
St. Louis, Missouri 63130

E. E. ROOS
Plant Germplasm Research
U.S. Department of Agriculture
Forest Service
Northern Plains Area National Seed
  Storage Laboratory
Fort Collins, Colorado 80523

BARBARA A. SCHAAL
Department of Biology
Washington University
St. Louis, Missouri 63130

ALAN R. TEMPLETON
Department of Biology
Washington University
St. Louis, Missouri 63130

L. E. TOWILL
Plant Germplasm Research
U.S. Department of Agriculture
Forest Service
Northern Plains Area National Seed
  Storage Laboratory
Fort Collins, Colorado 80523

I

POPULATION BIOLOGY AND GENETICS OF RARE SPECIES

1

Genetic and Evolutionary Consequences of Small Population Size in Plants: Implications for Conservation

SPENCER C. H. BARRETT and JOSHUA R. KOHN

Plant population sizes vary in space and time both within and among species. This variability is the result of complex interactions among the life-history features of populations, local environmental conditions, and the historical ecology of particular species. Populations range in size from many thousands, as in some forest trees and savannah grasses, to small colonies of a few plants in rare orchids and certain parasitic plants. Considering the sessile habit of plants and the ease with which they can be counted, data on population sizes are surprisingly sparse in the botanical literature.

Several features of the population biology of plants may account for the limited census data on population size, in comparison with many animal groups. Accurate estimates can be complicated by subterranean seed banks and because some plants exhibit prolific clonal growth (Harper 1977). These features lead to problems in defining what constitutes a plant population (Crawford 1984). Whereas the ramet may be the unit of interest to ecologists, the genet (or genotype) is of more significance to geneticists and evolutionary biologists. Determination of the number of genets in a population requires many genetic markers, and the number is therefore difficult to obtain for most clonal species (Silander 1985a).

The significance of the size of populations to their breeding structure, genetics, and evolutionary dynamics was first recognized by Wright (1931, 1938, 1946) and is today a central focus of concern in conservation biology. One of the goals of many conservation programs, in addition to habitat preservation, is to maintain existing levels of genetic variation in species that are rare or threatened (Frankel and Soulé 1981; Simberloff 1988). Loss of variation is thought to reduce the ability of populations to adapt to changing environments and increase their susceptibility to pest and disease pressures (Fisher 1930; Hamilton 1982; Beardmore 1983).

The total genetic variation maintained within a species can be partitioned in a hierarchical manner, according to the way it is distributed among regions, populations, and individuals within populations. Four evolutionary forces—mutation, natural selection, migration, and random genetic drift—interacting with an organism's recombination system, account for the manner in which variation is distributed among levels in the hierarchy. The relative importance of these factors differs among species and ecological groups. Where populations are small and isolated from one another, genetic drift will have a dominant influence on population genetic structure. Populations in which drift predominates lose variation more readily than do populations in which drift is not a factor, and are thus particularly prone to local extinction. Conservation practices require an understanding of the effects of stochastic forces on genetic variability, since populations conserved in reserves, and samples maintained in zoos and botanical gardens, are usually restricted in size (Berry 1971; Franklin 1980; Lande and Barrow-clough 1987).

Another consequence of small population size is the occurrence of inbreeding or mating between close relatives. The loss of fitness upon inbreeding is known as *in-breeding depression*. Although the genetic mechanisms underlying inbreeding depression and its converse, *heterosis*, are still not completely understood (see below), it is generally agreed that inbreeding reduces reproductive performance in most animals, particularly mammals (Ralls et al. 1986, 1988). These effects can ultimately lead to the loss of valuable genetic stocks if intense inbreeding is practiced. Because of the occurrence of inbreeding depression in small populations of captive animals, a variety of breeding protocols have been developed in recent years to overcome its debilitating effects (Chesser et al. 1980; Templeton and Read 1983).

The effects of inbreeding on fitness in plants are likely to be a good deal more variable than in animals because plants have such diverse reproductive systems and population structures (Charlesworth and Charlesworth 1987). Generalizations on the harmful effects of inbreeding, and recommendations on the number of individuals required to maintain viable populations, may need to be modified from those available for most animals because many plant species regularly inbreed. In some cases it may be more important to determine whether *outbreeding depression*, the fitness decline that can result from hybridization, occurs (Templeton 1986). Information on outbreeding depression in plants will be of particular importance to *ex situ* conservation strategies if they involve controlled breeding of samples collected from different populations.

A burgeoning literature on conservation biology of rare and endangered species has developed during the past decade, with several volumes devoted solely to biological aspects of rare plant conservation (Simmons et al. 1976; Synge 1981; Bramwell et al. 1987). In contrast to works on animal conservation, the botanical literature concerned with nondomesticated species has largely neglected genetic considerations in devising strategies for rare plant conservation, focusing instead on ecological and demographic issues. Here we attempt to redress the balance by examining the genetic and evolutionary consequences of small population sizes in wild plants and their implications for conservation biology. We consider the problem from both theoretical and empirical standpoints, using data from natural plant populations wherever possible, and attempt to distinguish those features of plants that make generalizations from the animal literature unsound

We begin by reviewing some of the genetic consequences of small population size in plants, paying particular attention to the role of genetic bottlenecks in reducing diversity. This is followed by a consideration of mating patterns in small populations and an assessment of the significance of inbreeding depression and heterosis to fitness. We conclude by discussing the implications of what is known about the genetics of small populations for *in situ* and *ex situ* conservation practices and suggest empirical studies that could be profitably undertaken on rare plants both to enhance the chances for success of the conservation effort and to increase understanding of the role of population genetic structure for species survival.

ECOLOGICAL CAUSES OF SMALL POPULATION SIZE

Despite the paucity of data on population size variation in plants, natural history observations indicate that populations of some species are large, continuous, and stable, some are regularly small and sparsely distributed, whereas others fluctuate dramatically from season to season. These differences are usually associated with contrasting life-history strategies of individual species. However, population sizes can also vary considerably within species, particularly in those subject to frequent cycles of colonization and extinction. Figure 1.1 illustrates the range of population sizes that occur in *Eichhornia paniculata,* a colonizer of ephemeral aquatic habitats in northeastern Brazil that are often short-lived as a result of drought, floods, and human disturbance. There is a large variance in population size within each of the two years that populations were sampled. In addition, 32% and 51% of populations sampled in 1987 and 1988, respectively, contained fewer than 100 individuals. This suggests that populations of *E. paniculata* are likely to be especially subject to genetic drift. Indeed, evidence is available to support this suggestion (Barrett et al. 1989).

Populations may be small for a variety of reasons. The major reasons, which have been summarized by Harper (1977), are as follows.

1. The available sites are few and separated by distances beyond a species' normal dispersal ability.
2. The carrying capacity of the site is low.
3. The habitability of the site is of short duration because of successional displacement.
4. Colonization is in its early stages, and full exploitation of the site has not occurred.

To these may be added the diverse factors associated with environmental catastrophes, including fires, grazing, droughts, floods, and insect and disease outbreaks. Because of the variety of ecological factors that limit population size, patterns of genetic variation in small populations that originate in contrasting ways are likely to differ. As we shall see, populations that remain small for long periods are likely to be considerably less diverse than those that have only recently become small.

It is often assumed that rare and endangered species inevitably occur in small populations that are geographically isolated from one another. The ecological and evolutionary processes that give rise to rarity are, however, sufficiently complex that we cannot assume that all rare plants exhibit these patterns (Kruckeberg and Rabinowitz 1985). Emphasizing the ecological heterogeneity of rare species, Rabinowitz (1981) constructed a classification of rarity based on local population size, habitat specificity, and geographical range. For each situation two categories (large vs. small, wide vs. narrow) were erected. Since species with large population sizes, wide ranges, and broad habitat preferences are unlikely to be rare, the classification scheme resulted in

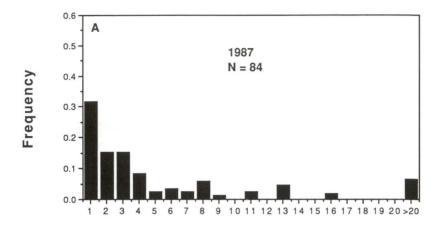

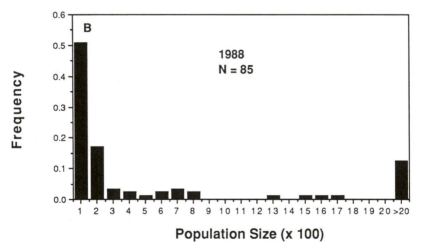

Figure 1.1. Distributions of population size in *Eichhornia paniculata* from northeastern Brazil during 1987 and 1988. Sample sizes in A and B refer to the number of populations censused in each year. (B. C. Husband and S. C. H. Barrett unpubl. data)

seven forms of rarity, only four of which were characterized by small local populations. On theoretical grounds, these differences in the spatial distribution and size of populations are likely to result in contrasting patterns of population genetic structure. Although several rare plants surveyed for genetic variation appear to be genetically depauperate because of small population size effects (see below), it may be premature to assume that this is a universal feature of rare species (Stebbins 1980; Griggs and Jain 1983; Nickrent and Wiens 1989).

The distribution and amounts of genetic diversity within and among populations of rare plants are likely to depend on whether a species has always been rare or whether it has recently become so as a result of human influences. Species that occur naturally in sparsely distributed, small populations may possess genetic systems adjusted to close inbreeding, as well as adaptations that offset the disadvantage of rarity—for instance, during mating (e.g., bird pollination in *Eucalyptus stoatei* [Hopper and Moran 1981]). In contrast, species that have experienced severe reductions in population numbers owing to habitat destruction or grazing (e.g., *Gossypium mustelinum*, [Pickersgill et al. 1976]) may be more susceptible to genetic stresses imposed by small population size. These considerations highlight the importance of considering the ecology of rare species and the diverse ways that rarity can arise (e.g., Stebbins and Major 1965; Wiens et al. 1989). These differences will have important implications for population genetic structure, the sampling of genetic material, and the specific types of breeding programs that are planned.

GENETIC CONSEQUENCES OF SMALL POPULATION SIZE

Although some rare plants maintain large population sizes, a majority occur in small populations, often with decreasing numbers (Ashton 1987). It is therefore important to assess whether significant genetic deterioration can arise through sudden or gradual decreases in numbers. Harmful genetic effects may result from inbreeding in small populations (see below). We concentrate in this section, however, on how genetic variation is affected by small population size. We consider how stochastic forces such as bottlenecks, founder events, and genetic drift influence different classes of genetic variation, and assess how important these processes are likely to be in wild populations of plants and those that are to be conserved. Before we do this, however, it is necessary to clarify concepts of population size since, from the genetical perspective, natural populations often behave as if they are smaller than a direct count of individuals would suggest.

Effective Population Size

Population sizes of relevance to conservation genetics may not be equivalent to the number of individuals in a population. The reason is that the size of the breeding population is rarely equal to the total population size (N) of most models concerned with the genetics of finite populations. Factors that cause this disparity include temporal fluctuations in numbers, nonrandom mating, differential fertility, unequal sex ratios, age and size structure, and gene flow among populations (Kimura and Crow 1963; Crawford 1984; Heywood 1986). These factors violate the assumption that each of the N individuals has an equal probability of contributing gametes to the next generation. A more useful parameter, first introduced by Wright (1931), is N_e, the effective population size. This represents the size of an idealized population in which individuals contribute equally to the gamete pool and have the same variation in allele frequencies and levels of inbreeding as the observed population. The effective size can be estimated from the actual number of breeding individuals, when factors such as mating system, sex ratio, and variation in fertility are known. Alternatively, N_e can be inferred from the between-year variance of allele frequencies at neutral loci (Kimura and Crow 1963; Crow and Denniston 1988). Estimates of N_e for a variety of animal groups are summarized by Crawford (1984), who draws attention to the lack of data from plant populations. The few examples that are available suggest that effective number for populations of plant species may be considerably smaller than the census number (Jain and Rai 1974; MacKay 1981).

Genetic Bottlenecks and Founder Events
Theoretical Considerations

A *bottleneck* is a sharp reduction in the number of individuals of a species in a particular place or time. If the restriction in number is accompanied by significant loss of genetic diversity, the term *genetic bottleneck* is often used. *Population bottlenecks* usually refer to a collapse in the number of individuals within a single population as a result of some environmental catastrophe. Population bottlenecks may or may not be associated with genetic bottlenecks; this depends on a variety of demographic and genetic factors that are discussed below. Bottlenecks are often associated with colonizing events, if one or a few individuals establish populations in previously unoccupied territory. Irrespective of whether bottlenecks occur within a single population or through the founding of new populations, they usually involve sampling error; small samples are rarely representative of the populations from which they are drawn. The theoretical analysis of bottlenecks is of importance for the development of nonequilibrium models in population genetics as well as in studies of speciation (Carson 1968; Barton and Charlesworth 1984; Carson and Templeton 1984). In addition, the study of bottlenecks can have practical applications for genetic resource conservation in cultivated plants and animals as well as endangered species (Maruyama and Fuerst 1984).

A genetic bottleneck is a single event in time. When populations remain small for any extended period, the sampling effects become cumulative. This gives rise to random changes in gene frequency because of the sampling of gametes from generation to generation. This process is referred to as *random genetic drift*. In large populations, on average, only small random changes in gene frequency occur as a result of drift; however, where population sizes are small (e.g., <100), gene frequencies can undergo large fluctuations in different generations, leading to loss of alleles. Therefore, large populations should maintain higher levels of genetic variability than do small populations (Wright 1931; Kimura and Crow 1964). Owing to the paucity of population size data for plants, support for the theory is limited, although what data are available are in accord with the prediction that small populations are more likely to lose polymorphism at neutral loci through stochastic forces (Figure 1.2).

The term *founder effect* was coined by Mayr (1963) to refer to "the establishment of a new population by a few original founders . . . which carry only a small fraction of the total genetic variation of the parental population." Random processes of this type are central to the theories of speciation proposed by Mayr (1963). He invoked founder events as the mechanisms by which genetic revolutions, resulting in reproductive isolation, are triggered. As originally conceived by Mayr, the founder principle was thought to reduce genetic variation affecting both quantitative traits and levels of heterozygosity. However, as pointed out by Lewontin (1965), and more recently reviewed by Lande (1980), measures of genetic variation that weigh genotypes by their frequencies, such as heterozygosity or quantitative genetic variation, are not particu-

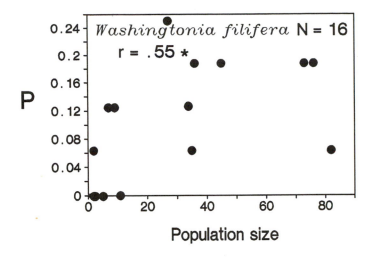

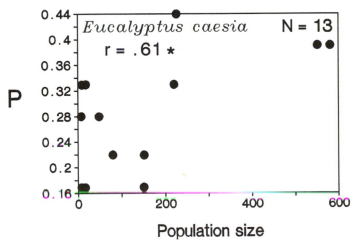

Figure 1.2. The relationship between population size and the proportion of isozyme loci that are polymorphic (P) in *Washingtonia filifera* (McClenaghan and Beauchamp 1986) and *Eucalyptus caesia* (Moran and Hopper 1983). Sample sizes (*N*) refer to the number of populations examined. In neither species was there a significant relationship between heterozygosity and population size. Both species are rare and maintain small population sizes. A total of 16 and 18 loci were screened in *W. filifera* and *E. caesia*, respectively.

larly sensitive to sampling error because only rare alleles tend to be lost each generation. Thus founder events in outcrossing organisms can be less effective at reducing genetic variability than is generally believed. Refinement of this point was later made by Nei et al. (1975) in their theoretical analysis of the effects of bottlenecks on genetic variation.

The loss of genetic variability concomitant with a bottleneck has both qualitative and quantitative aspects. A reduction in the number of alleles, especially rare ones, is much greater than the loss of heterozygosity and genetic variance per se (Figure 1.2). Nei et al. (1975) used a model in which a severe size reduction occurred suddenly in a population that was in equilibrium between mutation and genetic drift. They found that the amount of reduction in average heterozygosity per locus depends not only on the size of the bottleneck, but also on the subsequent rate of population growth. If population sizes increase rapidly, reduction in average heterozygosity is minimal, even if the number of founders is small. In contrast, the loss in the average number of alleles per locus is profoundly affected by bottleneck size, but little by the rate of population growth following the bottleneck. This difference occurs mainly because genetic drift eliminates many low-frequency alleles. However, the average number of alleles per locus increases faster than average heterozygosity when large population sizes are restored.

The situation modeled by Nei et al. (1975) is likely to apply most often to exotic pests and weeds, whose populations expand rapidly following introduction to a new region (Elton 1958; Parsons 1983). More frequently, however, wild populations experience periods during which size fluctuates. In the case of rare species, populations may go through repeated bottlenecks, or, where plants or animals are brought into captivity, a single bottleneck may be followed by a protracted period of small population size. Several of these more complex situations have been examined analytically by Denniston (1978), Watterson (1984), and Maruyama and Fuerst (1984, 1985a, 1985b). In general, their results confirm that the loss of alleles greatly exceeds reductions in heterozygosity, when populations remain small following a bottleneck, although protracted periods of small population size will result in a significant loss of variation.

The effects of stochastic loss of genetic variation from small managed populations have been considered by Lacy (1987). Using computer simulation, he investigated the interacting effects of genetic drift, mutation, immigration, population subdivision, and various forms of selection on the loss of heterozygosity over 100 generations in populations ranging in size from 20 to 500. He found that genetic drift was the predominant force controlling reductions in genetic variation. Mutation had no noticeable effect on populations of the size typical for many species managed in nature reserves, zoos, or botanical gardens. Immigration from a large source population can retard, halt, or even reverse the loss of genetic variation, even with only one or a few migrants per generation. He also found that unless selection is considerably stronger than is commonly observed in nature (but see Endler 1986), it is ineffective in counteracting genetic drift when population size is 100 or fewer. Subdivided populations rapidly lose variation from within each subpopulation but retain variation across subpopulations better than a single panmictic population does (although see Fuerst and Maruyama 1986). Lacey's results, some of which are illustrated in Figure 1.3, indicate that conservation biologists should be most concerned about the effects of genetic drift in reducing genetic variation in managed populations. The regular introduction of immigrants and the division of managed populations into smaller breeding groups that interchange enough migrants to prevent unacceptably high levels of inbreeding are two strategies that would ameliorate the loss of variability due to drift.

Most theoretical work on the stochastic loss of variation through bottlenecks and founder events has considered diploid, outbreeding organisms with separate sexes. This genetic system is common among many animal groups, but is unrepresentative

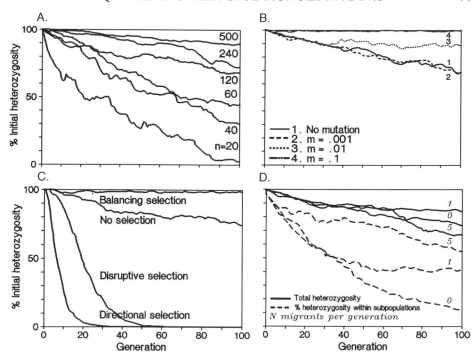

Figure 1.3. The percent of initial heterozygosity retained in populations with time (in generations) for (A) population sizes ranging from 20 to 500; (B) populations with equal forward and backward mutation rates of 0, 10^{-1}, 10^{-2}, or 10^{-3} per generation; (C) populations subjected to balancing selection, no selection, disruptive selection, and directional selection (relative fitnesses of 0.8 : 1.0 : 0.8; 1.0 : 0.8 : 1.0; and 1.0 : 0.8 : 0.6, for balancing, disruptive, and directional selection, respectively); (D) populations (solid lines) and five subpopulations (dashed lines) with an average of 0, 1, and 5 inter-subpopulation migrants per generation. Individual trajectories in each figure represent the average of 25 simulated populations. Except where indicated, all figures are based on simulations involving a population size of 120 individuals. (After Lacy 1987)

of most plant species where polyploidy, hermaphroditism, and varying degrees of self-fertilization commonly occur. These features of plant genetic systems modify in different ways the significance of stochastic influences on patterns of genetic variation within populations. For example, in autopolyploids with polysomic inheritance, population size restrictions will have less of an effect on the loss of genetic variability because high levels of segregational heterozygosity are maintained in populations (Haldane 1930; Mayo 1971). In allopolyploid species with high levels of gene duplication, loss of heterozygosity is unimportant for loci that exhibit "fixed" heterozygosity. Similarly, the diverse array of mating systems in plants will vary in the extent to which they are influenced by bottlenecks and genetic drift. For instance, in selfing species severe population restriction followed by population growth might not affect heterozygosity, but could cause severe loss of allelic variation. Theoretical work could usefully explore the effect of various mating systems and types of polyploidy on the stochastic loss of genetic variation from small populations of plants. Approaches of this type have been employed to investigate the influence of genetic drift and nonequilibrium conditions on the maintenance of variation at loci controlling the mating-system polymorphism tristyly (Heuch 1980; Morgan and Barrett 1988), and at the self-incompatibility locus in species with multiallelic homomorphic systems (Imrie et al. 1972).

Empirical Evidence

Examination of the patterns of genetic variation at isozyme loci in different groups of colonizing animals has provided some evidence for the role of bottlenecks in reducing variation (Taylor and Gorman 1975; Huettel et al. 1980; Richardson et al. 1980; Bryant et al. 1981; Berlocher 1984; Baker and Moeed 1987; Menken 1987). Although there are few comparable studies in plants, there are good reasons to believe that bottlenecks and founder events play a more significant role in governing patterns of variability, particularly in colonizing species (Barrett and Richardson 1986). One reason is that many cosmopolitan plants reproduce either through clonal propagation or by predominant self-fertilization. With uniparental reproduction and limited gene flow from the source region, genetically uniform population systems are likely in the introduced range. Examples of genetic bottlenecks in colonizing plants have recently been discussed by Barrett and Husband (1989), and isozyme surveys that have revealed extensive areas of genetic uniformity in cosmopolitan taxa are reviewed in Barrett and Shore (1989).

Not all plant isozyme studies that have revealed genetically depauperate populations involve cosmopolitan species with uniparental reproductive systems. Karron (1987a), in a literature survey, compared levels of genetic diversity at isozyme loci in 11 pairs of congeneric species that were either rare or common. Diversity was reduced in the rare species, irrespective of the reproductive system, perhaps as a consequence of bottlenecks associated with declining population numbers. Several authors have invoked bottlenecks during the evolutionary history of a species as an explanation to account for the low levels of genetic variation maintained in contemporary populations. Examples involve rare and widespread species and both inbreeders and outbreeders (Table 1.1). In most cases, restriction of populations to small refugia, owing to climatic changes associated with Pleistocene glacial events, are postulated as the causal factor resulting in genetic bottlenecks.

Inferences on the role of historical bottlenecks in reducing genetic variation can be strengthened if genetic data are available for congeneric species with similar life histories and reproductive systems. In *Pinus* spp., virtually all taxa investigated, except the locally restricted *P. torreyana* and more widespread *P. resinosa,* exhibit high levels of genetic diversity (Ledig 1986a). Although fewer data are available for Australian *Acacia* spp., several species of the genus that occupy environments similar to that of the genetically depauperate *A. mangium* possess higher levels of genetic diversity. Moran et al. (1989) suggest that these taxa were able to maintain wider distributions during interglacial floods because of broader ecological tolerance. In contrast, several geographical surveys of isozyme variability in different European and North American taxa of *Salicornia* have revealed very low levels of variability within each taxon (Jefferies and Gottlieb 1982; Wolff and Jefferies 1987a, 1987b). It is possible that destruction of coastal environments, as a result of glacial advance during the Pleistocene, affected all taxa of *Salicornia* throughout the north temperate zone. Alternatively, low variation may be unrelated to bottleneck effects and instead is associated with the predominantly selfing habit of these species and adaptation to the highly specialized salt marsh environment with its high predictability and low biotic diversity.

These examples illustrate some of the difficulties that are associated with inferring the past occurrence of genetic bottlenecks. Comparisons among related taxa can provide indirect evidence on the role of bottlenecks; however, simple correlation analysis frequently confounds differences in evolutionary history among related species with contemporary ecological factors that also affect population genetic structure. Although intraspecific studies reduce these problems, difficulties can still arise when genetic differentiation among conspecific populations occur in traits that directly influence recombination (e.g., Glover and Barrett 1987). Clearly, the major problem in evaluating the role of past genetic bottlenecks in reducing variability is that so little is known about the evolutionary history of most plant species (but see Cwynar and MacDonald 1987). Recent migration events can provide more convincing evidence of genetic bottlenecks, particularly if the source region is known (e.g., *Echinochloa microstachya* in Australia [S. C. H. Barrett and A. H. D. Brown unpubl. data]) or the number of individuals and time of introduction have been documented (e.g., *Sarracenia purpurea* [Schwaegerle and Schaal 1979]). However, to understand fully the genetic consequences of finite population size effects in plants, experimental approaches are required. These can involve the monitoring of artificially established colonies of known size and genetic composition (Martins and Jain 1979; D. W. Schemske unpubl. data) or the investigation of natural colonization processes (Harding and Mankinen 1972; Jain and Martins 1979; Barrett and Husband 1989).

One of the few experimental studies in plants explicitly designed to investigate the effect of restrictions in population size on genetic variation was conducted by Polans and Allard (1989) using field collected genotypes of *Lolium multiflorum,* an outbreeding grass. Artificial populations composed of one, two, four, or eight plants were established for three successive generations, and changes in allele frequencies and levels of heterozygosity were monitored at three isozyme loci. The results obtained were consistent with expectations for neutral loci subject to genetic drift. The severity in loss of genetic variation increased with smaller population sizes. Restrictions in population size also resulted in significant phenotypic effects in four of six quantitative traits that were studied. Smaller populations flowered more slowly, and were shorter with fewer tillers and seeds, than populations of larger size (Figure 1.4). This indicates the potential ill effects of restrictions in population size on traits that are likely to affect fitness.

When population sizes of experimental populations of *L. multiflorum* were increased, some of the detrimental effects were still discernible. However, the expression of quantitative traits recovered to levels similar to those in the original population, after a single generation of random mating, in composite populations formed by mixing genotypes from the small inbred populations. Recovery from the debilitating effects of drift presumably results from reassociation of favorable genetic combinations in mixtures of inbred lines.

The study by Polans and Allard (1989) considered the effects of small population size on both isozyme loci and quantitative characters. This is important because species that demonstrate low variation at isozyme loci owing to bottleneck effects may not be devoid of variation at other gene loci (e.g., Moran et al. 1981; Giles 1983; Warwick et al. 1987). In fact, theoretical work by Lande (1976, 1977, 1980) suggests that if

Table 1.1. Plant Taxa in Which Genetic Bottlenecks Arising from Founder Events and Drift May Account for the Low Levels of Isozyme Polymorphism

Taxon	Current distribution	Life form	Breeding system	Features of bottleneck	Source
Ancient bottlenecks					
Acacia mangium	Widespread Australasia	Tree	Outcrosser	Interglacial flooding in tropical Australasia	Moran et al. 1989
Chrysoplenium iowense	Narrow endemic Iowa	Herb	Outcrosser	Pleistocene glaciation, small refugia	Schwartz 1985
Clarkia franciscana	Narrow endemic N. California	Herb	Selfer	Bottleneck via a rapid speciation event?	Gottlieb 1973a
Eucalyptus caesia	Narrow endemic W. Australia	Tree	Outcrosser	Small isolated populations on granite rocks	Moran and Hopper 1983
Howellia aquatilis	Narrow endemic N.W. U.S.	Herb	Selfer	Pleistocene glaciation, small refugia	Lesica et al. 1988
Pedicularis furbishiae	Narrow endemic Maine	Herb	Outcrosser	Postglacial colonization and frequent extinctions	Waller et al. 1987
Pinus resinosa	Widespread E. U.S.	Tree	Outcrosser	Pleistocene glaciation, small refugia	Fowler and Morris 1977
Pinus torreyana	Narrow endemic S. California	Tree	Outcrosser	Climate change during xerothermic period	Ledig and Conkle 1983
Salicornia europaea	Widespread N. temperate zone	Herb	Selfer	Interglacial flooding of salt marshes	Wolff and Jefferies 1987a, 1987b
Stephanomeria malheurensis	Narrow endemic Oregon	Herb	Selfer	Bottleneck via a rapid speciation event	Gottlieb 1977b
Typha spp.	Widespread N. America	Herb	Clonal	Pleistocene glaciation, small refugia?	Sharitz et al. 1980
Washingtonia filifera	Narrow endemic S.W. U.S.	Palm	Outcrosser	Colonization from refugial populations	McClenaghan and Beauchamp 1986

Species	Distribution		Breeding system	Bottleneck event	Reference
Eichhornia paniculata[a]	Widespread disjunct N. Brazil and Caribbean	Herb	Variable	Bottleneck during colonization of Jamaica	Glover and Barrett 1987
Turnera ulmifolia[a]	Widespread neotropics	Herb	Outcrosser	Bottleneck during colonization of Caribbean	Barrett and Shore 1989
Recent bottlenecks					
Avena barbata	Cosmopolitan	Herb	Selfer	Bottleneck during introduction to California	Clegg and Allard 1972
Bromus tectorum	Europe N. America	Herb	Selfer	Bottleneck during introduction to N. America	S. Novak, unpubl. data
Chondrilla junceum	Mediterranean Australia	Herb	Apomict	Bottleneck during introduction to Australia	Burdon et al. 1980
Echinochloa spp.	Cosmopolitan	Herb	Selfer	Bottleneck during introduction to Australia	S. C. H. Barrett and A. H. D. Brown unpubl. data
Eichhornia crassipes	New and Old World tropics	Herb	Clonal	Bottleneck during introduction to Old World	S. C. H. Barrett unpubl. data
Sarracenia purpurea	N. America	Herb	Outcrosser	Bottleneck during introduction to Ohio reserve	Schwaegerle and Schaal 1979
Xanthium spp.	New World Australia	Herb	Selfer	Bottleneck during introduction to Australia	Moran and Marshall 1978

Note: Ancient bottlenecks refer to events that occurred prior to historic times; except where indicated, these are manifested at the species level. Recent bottlenecks refer to events occurring within historic times and are evident at the population level.
[a]Population-level bottleneck.

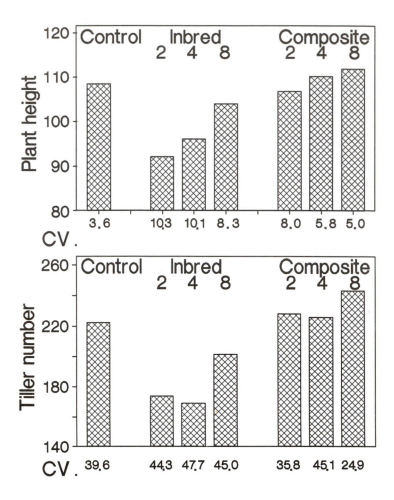

Figure 1.4. The effect of experimental bottlenecks on plant height (cm) and tiller number in *Lolium multiflorum*. Experimental populations of restricted size (2, 4, and 8) were maintained for three generations and then grown in a common garden with a control treatment and composite mixture of genotypes from the inbred treatments. The mean and coefficient of variation for each treatment are illustrated. (After Polans and Allard 1989)

populations expand rapidly following a bottleneck, the high mutability of polygenic characters can generate sufficient variability for rapid evolution into new adaptive zones. Rates of restoration of genetic variability following a bottleneck are orders of magnitude faster for quantitative characters than for heterozygosity at isozyme loci. Spontaneous mutation in metrical traits produces additive genetic variation on the order of 10^{-3} times the environmental variance (σ_e^2) per generation, whereas the usual single-locus rates for major mutants are about 10^{-6} per generation. The high mutation rate of metrical traits follows because many genes contribute to their expression. Because of the high mutability of polygenic characters, even small isolated populations that have lost most of their genetic variability through a bottleneck can generate sufficient variation for evolutionary change to occur.

When genetic variation in quantitative characters is the result of additive effects, variation should decrease in proportion to $1/(2N_e)$ after a bottleneck of N_e individuals (Lande 1980). However, if nonadditive effects, such as epistasis and dominance, contribute to genetic variation, then the effects of restrictions in the size of populations will not be a simple function of N. Recent theoretical work by Goodnight (1987, 1988) indicates that additive genetic variance can increase, at least temporarily after a bottleneck, as inbreeding converts nonadditive genetic variation from the donor population to additive genetic variation in the derived population. Few studies, other than those on houseflies (e.g., Bryant et al. 1986a, 1986b), have assessed these theories, and none have been undertaken on plants. Since so little work has been conducted on the quantitative genetics of metrical traits in wild plant populations (reviewed in Mitchell-Olds and Rutledge 1986), it is difficult to evaluate how large the contribution of nonadditive gene effects is likely to be for most quantitative traits.

MATING IN SMALL POPULATIONS

The frequency and intensity of inbreeding are often far greater in plants than in most animal groups. A variety of factors account for this difference, including the sessile habit of plants, restricted gene dispersal through pollen and seeds, small population sizes, and high levels of self-fertilization and sib-mating. These factors not only influence the levels and organization of genetic variation in plant populations, but also reduce effective population size, and hence increase inbreeding.

The level of inbreeding (F) in a population increases over time at a rate dependent on the effective population size (N_e such that $\Delta F = 1/(2N_e)$ per generation (Wright 1922; Falconer 1981). Therefore, populations become inbred more rapidly when they are of small size (Figure 1.5). For very small populations, the influence of size on inbreeding may overwhelm effects brought about by the system of mating alone. Nevertheless, mating patterns are prime determinants of the levels of inbreeding in both large and small populations.

In this section, we begin by discussing the measurement of inbreeding and its relationship to the mating system of plants. We then consider how inbreeding influences offspring fitness by reviewing both theoretical models and empirical data on inbreeding depression in wild plant populations. Finally, we review evidence for the occurrence of outbreeding depression in plant populations and its relationship to mating systems.

Types of Inbreeding in Plants

Plants display a wide variety of breeding systems that differ in their influence on mating patterns and population genetic structure (Richards 1986). The mating parameter with the largest influence on genetic structure is the *selfing rate* (*s*), the proportion of matings that result from self-fertilization. Although self-incompatible and dioecious species are prevented from self-fertilization, inbreeding may still occur through matings between related individuals. Biparental inbreeding is particularly likely in plants, since nearby individuals are often closely related, owing to restricted gene dispersal. The selfing rate of self-compatible species varies, depending on floral biology and

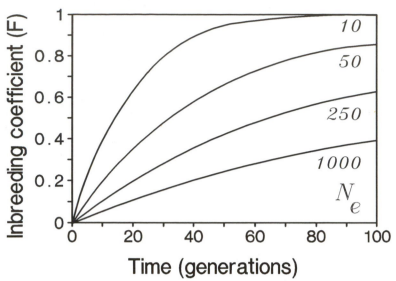

Figure 1.5. Changes in the inbreeding coefficient (*F*) with time (in generations) for a pair of alleles at a neutral locus in four populations of various sizes. Note the rapid increases in *F* for populations of small effective size (N_e). (After Parkin 1979)

local environmental conditions. Some species are predominantly selfing, others exhibit mixed mating systems, whereas some are highly outcrossing (reviewed in Schemske and Lande 1985; Barrett and Eckert 1990).

Inbreeding in any form (selfing, sib-mating, etc.) increases homozygosity in a population above levels expected under random mating. Wright (1933, 1977) defined the coefficient of inbreeding (*F*) as a measure of this increase in homozygosity, resulting from the correlation of uniting gametes. With inbreeding, genotype frequencies in a population for a single locus are

$$(p^2 + pqF) + (2pq - 2pqF) + (q^2 + pqF) = 1$$

Thus Hardy–Weinberg frequencies obtain when *F* = 0, and all individuals are homozygous when *F* = 1. Given genotype and allele frequencies, *F* is estimated as

$$F = 1 - \frac{H}{2pq}$$

where *H* is the proportion of heterozygotes. The *F* referred to here is a within-population measure. The mean of *F* over all populations is termed F_{is}. The related measure F_{st} is the amount of species-level deviation from Hardy–Weinberg expectations associated with between-population differences in allele frequencies. Detailed treatments of *F* statistics and the theory of inbreeding in general are available in Wright (1922, 1977), Fisher (1949), Cockerham (1969, 1973), Malécot (1975), and Weir and Cockerham (1984).

If *s* is the proportion of self-fertilized progeny, then *s* and *F* are related by

$$F' = \frac{s(1+F)}{2}$$

where F and F' are the inbreeding coefficients of the parents and progeny before selection, respectively. In the absence of inbreeding depression, an equilibrium value of F is reached when

$$F = \frac{s}{2-s}$$

Measurement of F and s in plant populations has been greatly facilitated by the use of gel electrophoresis and the development of statistical procedures for the analysis of electrophoretic data (reviewed in Brown and Weir 1983; Ritland 1983; Brown 1989c). F is calculated from allele and genotype frequencies, whereas s is estimated from the segregation of marker loci in open-pollinated progeny arrays from maternal parents whose genotypes are either assayed directly or inferred from progeny genotypes. The most commonly employed model to estimate selfing and outcrossing ($t = 1 - s$) rates is the mixed-mating model (Fyfe and Bailey 1951; Clegg 1980). The model assumes that matings are of two types: self-fertilizations and random matings. In most natural plant populations, however, outcrossing between related individuals will commonly occur. The single-locus selfing rate, s, estimates the homozygosity in progeny that is due to the combined effects of self-fertilization and all forms of biparental inbreeding. Thus s gives a measure of what the selfing rate would be if self-fertilization were the only form of consanguineous mating practiced by a population.

Procedures for estimating selfing rates have been extended to make use of the greater information provided by multilocus data (Ritland and Jain 1981; Shaw et al. 1981). Multilocus models give a more accurate estimate of the true selfing rate because outcrossing events are more reliably detected, even if they are consanguineous. Hence the multilocus estimate of s will often be lower than estimates based on single loci. In self-compatible populations, the difference between the multilocus estimate and the mean of the single-locus estimates of s can be used to assess the degree of biparental inbreeding (Ritland 1984; Brown 1989c).

In plants that cannot self, owing to male sterility or self-incompatibility, single-locus estimates of s can be used to assess the degree of biparental inbreeding. For instance, male-sterile individuals in seven populations of gynodioecious *Bidens* spp. in Hawaii had average "selfing" rates of 15% (range 0–25% [Sun and Ganders 1988]). This is equivalent to saying that the average degree of relatedness between male-steriles and their mates was about that of first cousins. The rate of apparent selfing in self-incompatible *Helianthus annuus* varied from 9% to 46% among populations and years (Ellstrand et al. 1978) and was positively correlated with plant density. Not all studies of outcrossing levels in self-incompatible or sexually dimorphic species have revealed significant values of s (e.g., Ennos 1985a; Barrett and Shore 1987).

Measures of Inbreeding Depression

The major effect on fitness of self-fertilization in plants is *inbreeding depression*, which can be defined as the relative reduction in fitness of selfed offspring compared with outcrossed offspring, or

$$1 - \frac{\text{fitness of selfed offspring}}{\text{fitness of outcrossed offspring}}$$

Inbreeding depression can be measured in plants, as originally undertaken by Darwin (1876), in experiments that compare the relative fitness of progeny obtained from artificial self- and cross-pollinations. A growing body of empirical data (reviewed below) from experiments that have employed this approach is now available for both cultivated and wild plant populations.

The extent of inbreeding depression can also be assessed in natural populations without resorting to experimental studies. One approach recently developed by Ritland (1989, 1990) involves the use of electrophoretic data. If the inbreeding coefficient and the selfing rate are measured in a given generation, and the former is remeasured when progeny reach adulthood, then inbreeding depression is expressed as

$$1 - \frac{(1-s)F'}{(F'-F'') + (1 - s)F''}$$

where F' and F'' are the inbreeding coefficients of the progeny generation before and after selection, respectively. This method would be difficult to use for long-lived plants, especially those with overlapping generations, but could be profitably employed in studies of annual species.

An additional method by Ritland (1989, 1990) assumes that the population under study maintains an equilibrium F value in each adult generation. Thus if inbreeding depression is present, selfing will increase F values measured in juveniles (seeds, seedlings) over those values for adults. Selection against selfed progeny then returns the population to equilibrium. In this case, the inbreeding coefficient of the parental generation and its selfing rate are sufficient to estimate inbreeding depression as

$$1 - \frac{2F(1-s)}{s(1-F)}$$

In the absence of electrophoretic information, comparison of the relative performance of hand-selfed and -outcrossed seed with that of open-pollinated seed has been suggested as a way to estimate the selfing rate of populations (Charlesworth 1988). By this approach, the selfing rate is

$$s = \frac{\text{outcrossed performance} - \text{open-pollinated performance}}{\text{outcrossed performance} - \text{selfed performance}}$$

Thus inbreeding depression and the selfing rate can be estimated with electrophoretic data if experimental studies are impractical; alternatively, both can be estimated from field data when electrophoretic techniques are not available.

Inbreeding Depression and Mating Patterns
Theoretical Considerations

Fisher (1941) first pointed out that a selfing individual has an automatic advantage over an outcrossing individual owing to the greater number of genes that it contributes to its own seeds. In a population of outcrossers, each individual, on average, passes one haploid copy of its genome through each of its own seeds and one such copy each time it fertilizes the seeds of another plant. A variant that can self-fertilize, without losing the ability to donate pollen to others, will pass on two haploid copies of its genome in each selfed seed as well as one copy through pollen donation. This automatic advantage will result in the spread of the variant to fixation if inbreeding depression is <0.5. If inbreeding depression is >0.5, complete outcrossing is favored (Lloyd 1979; Lande and Schemske 1985). Since this model gives a quantitative prediction of the amount of inbreeding depression necessary to favor selfing or outcrossing, it has stimulated much empirical work on the relationships between the selfing rate and inbreeding depression in natural populations (see below). The prediction that at equilibrium plant populations should practice either complete selfing or complete outcrossing conflicts with the frequent observation of mixed mating patterns in nature (Schemske and Lande 1985; Aide 1986; Waller 1986; Barrett and Eckert 1990). This discrepancy has motivated theoretical work that attempts to find conditions under which mixed mating systems are stable and how changes in the rate of inbreeding affect the level of inbreeding depression (Campbell 1986; Holsinger 1986, 1988a, 1988b; Uyenoyama 1986; Charlesworth and Charlesworth 1987, 1990).

The most commonly assumed genetic basis for inbreeding depression is the presence in a gene pool of lethal or highly deleterious recessive alleles. In an outcrossing population, these are maintained by the balance between mutation and selection. When an individual self-fertilizes (or mates with a relative), these alleles are often made homozygous, resulting in inbreeding depression. If lethal or highly deleterious recessive alleles are the primary cause of inbreeding depression, even relatively low rates of selfing (e.g., 10–20%) should rapidly purge a population of most of its genetic load (Figure 1.6A; Lande and Schemske 1985; Charlesworth and Charlesworth 1987). Thus, according to this model, populations with a long history of inbreeding would be expected to show little inbreeding depression. Controlled inbreeding to reduce genetic load has been proposed as a conservation strategy for mammalian zoo populations, and a test of this approach in Speke's gazelle did indeed lead to a significant reduction in inbreeding depression in a few generations (Templeton and Read 1984).

If inbreeding depression is modeled as a quantitative trait (Lande and Schemske 1985) or, similarly, as being due to many partially recessive genes, each of small deleterious effect (Charlesworth and Charlesworth 1987), then selfing also reduces genetic load. However, modeled in this way, selfing reduces inbreeding depression much more slowly than under the recessive lethal model, and even under complete selfing, populations may still exhibit significant inbreeding depression (Figure 1.6B).

Finally, if overdominance is the basis of inbreeding depression, selfing can actually increase the severity of inbreeding depression. The reason is that selfing increases homozygosity, whereas random mating always restores Hardy–Weinberg genotype frequencies. Thus the difference between the average heterozygosity of selfed and outcrossed progeny increases with the selfing rate (Charlesworth and Charlesworth 1987, 1990). In the case of balanced overdominance, where each homozygote is equally fit, inbreeding depression will always increase with increased selfing (Figure 1.6C). With asymmetrical overdominance, inbreeding depression increases with the selfing rate until selfing reaches a level too high to maintain the polymorphism. Above this level, one of the alleles is lost and inbreeding depression falls to zero (Figure 1.6D).

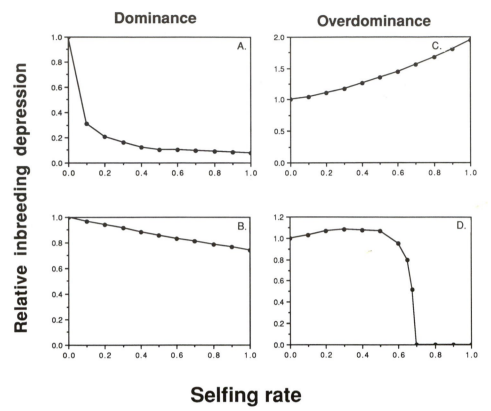

Figure 1.6. Relationships between inbreeding depression and the selfing rate under models assuming different genetic bases for inbreeding depression (redrawn from Charlesworth and Charlesworth 1987). Inbreeding depression values are expressed relative to those of an outcrossing population. The assumed bases of inbreeding depression are: (A) lethal or highly deleterious recessive alleles; (B) many genes, each of small deleterious effect; (C) symmetrical overdominance; (D) asymmetrical overdominance. The shapes of the curves change somewhat with parameter values, but these are representative. See Charlesworth and Charlesworth (1987) for model details.

Empirical Evidence

Prior to the review by Charlesworth and Charlesworth (1987), summaries of work on inbreeding depression in plants were made by Darwin (1876), Crumpacker (1967), and Wright (1977). Many earlier studies were conducted on cultivated species and involved only measurements of yield components. Despite these limitations, the results in general indicated that inbreeding depression usually occurs in outcrossing plants, is less severe in species that are partially self-fertilizing, and may be absent altogether in species that are highly selfing (Wright 1977). During the past decade, more sophisticated work that has involved joint estimates of both the selfing rate and inbreeding depression have provided more extensive data in which to assess the relationships between inbreeding depression and mating systems in natural populations of plants.

Figure 1.7 summarizes recent work on inbreeding depression in wild plant species,

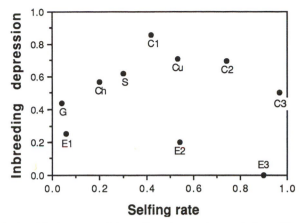

Figure 1.7. Relationship between inbreeding depression and the selfing rate in several species of herbaceous angiosperms. Species are: G = *Gilia achilleifolia* (Schoen 1983); Ch = *Chamaecrista fasciculata* (Fenster 1991a, 1991b); S = *Sabatia angularis* (Dudash 1990, unpubl. data); Cu = *Cucurbita foetidissima* (Kohn 1988, unpubl. data); C = *Clarkia tembloriensis* (Holtsford 1989); E = *Eichhornia paniculata* (P. Toppings and S. C. H. Barrett unpubl. data). For *Clarka* and *Eichhornia,* data from three populations are plotted. Inbreeding depression for *Clarkia* and *Eichhornia* was measured in the lath house and glasshouse, respectively; all others were measured in the field.

where an attempt has been made to estimate mating system parameters. As can be seen, values of inbreeding depression can be quite high, often exceeding 0.5. The relationship between inbreeding depression and rates of selfing varies considerably. In several cases (e.g., *Clarkia tembloriensis, Cucurbita foetidissima, Sabatia angularis*), species with moderate to high selfing rates exhibit strong inbreeding depression, indicating that the common perception that these species will suffer little from inbreeding effects (e.g., Templeton and Read 1984) may be largely unfounded. The occurrence of inbreeding depression in species that frequently self-fertilize supports the idea that recessive lethals are not the sole cause of inbreeding depression in nature.

In two studies that examine the relationship between selfing rates and inbreeding depression within species, evidence was obtained indicating the importance of mating patterns in influencing levels of inbreeding depression. In *Eichhornia paniculata*, levels of inbreeding depression for most life-history characters were weak. However, differences in flower production of outcrossed and selfed progeny were largest in tristylous populations with little selfing, smaller in populations with mixed mating systems, and absent altogether from a highly selfing population from Jamaica (Figure 1.8). The relatively low levels of inbreeding depression observed in this study may have occurred because comparisons were conducted under glasshouse conditions (see below) or because natural populations of *E. paniculata* are often small and experience frequent bottlenecks (Figure 1.1; Barrett et al. 1989); these effects reduce genetic load (Lande and Schemske 1985). Studies by Holtsford (1989) on inbreeding depression in three populations of *Clarkia tembloriensis* with different selfing rates revealed strong inbreeding depression for survivorship and fertility in each population (Figure 1.7). The highest levels were found in the most outcrossing population ($t = 0.58$; inbreeding

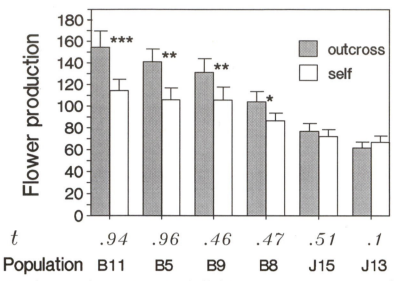

Figure 1.8.   Flower production ($\bar{x} \pm 2$ *SE*) of selfed and outcrossed offspring in populations of *Eichhornia paniculata* with different outcrossing rates (*t*). Four populations from Brazil (B) and two populations from Jamaica (J) were compared under uniform glasshouse conditions. (From Toppings 1989)

depression $= 0.77$) and the lowest levels were evident in the most selfing population ($t = 0.03$; inbreeding depression $= 0.51$). Comparisons were conducted in a lath house, indicating that high levels of inbreeding depression can be detected in some selfing populations, even under artificial conditions.

Levels of inbreeding depression can depend on the environment in which it is measured. In *Sabatia angularis,* for example, values of inbreeding depression increased when comparisons were conducted in the glasshouse, common garden, and field, respectively (Figure 1.9). Similarly, higher levels of inbreeding depression in *Clarkia tembloriensis* were found in garden plots as compared with lath-house conditions (Holtsford 1989). Overall, measures of inbreeding depression in the field tend to be higher than those in glasshouse experiments (Charlesworth and Charlesworth 1987), suggesting that conditions in the field, where plants are exposed to pests, diseases, drought, and competitive interactions, are likely to be more stressful. Kohn (1988) found only 1% survival after one year for seeds of *Cucurbita foetidissima* planted in the field, but outcrossed seeds were three times as likely to establish as selfed seeds. Such low rates of establishment are likely to be common in plants with high fecundity, yet few investigators would tolerate such severe mortality in controlled glasshouse studies, even under stress conditions.

Another common finding of inbreeding depression studies in herbaceous plants is that inbreeding effects are often manifest most strongly late in the plants' life cycle. Studies in which a multiplicative fitness function (e.g., seed set $\times$ germination $\times$ survival $\times$ flower or fruit production) was employed to compare selfed and outcrossed progeny (e.g., Toppings 1989; Dudash 1990; Fenster 1991b) found the largest relative differences in flower or fruit production. Schoen (1983) and Kohn (1988) found small

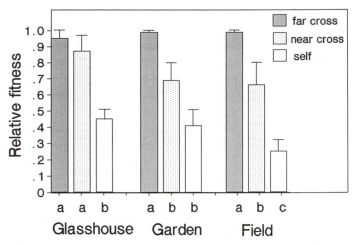

Figure 1.9. Relative fitness of offspring of the herbaceous biennial *Sabatia angularis* resulting from hand self-pollinations, cross-pollinations within a subpopulation (near cross), and cross-pollinations between subpopulations (far cross) measured under glasshouse, garden, and field conditions. Seed set did not differ among pollination types. All seeds were germinated in the glasshouse and transplanted as seedlings into the appropriate environment. Relative fitness ($\pm$ 2 SE) was estimated from the product of germination, survival, and seed set (see Dudash [1990] for details).

differences between selfed and outcrossed seeds in their ability to germinate in the field, but strong effects on the probability of postestablishment survival. These findings support the idea that in these groups, inbreeding depression does not result solely from a few genes with major effects. Rather than being highly dysfunctional, most selfed offspring that germinate appear to be somewhat less vigorous than outcrossed offspring. This small effect, however, becomes compounded during growth and reproduction and can ultimately result in large reductions in fitness. On the other hand, differences in seed set following selfing and outcrossing suggest that inbreeding depression is commonly expressed during embryo development in certain plants, particularly woody groups such as the gymnosperms (Charlesworth and Charlesworth 1987). Thus inbreeding depression may result from the combined effects of a few lethal recessive genes, and perhaps many other genes of small effect. As yet there are no genetic models of inbreeding depression that take into account a mixture of causes.

Outbreeding Depression

Outbreeding depression usually refers to the reduction in fitness that occurs following intraspecific hybridization between individuals from spatially separated genetic sources (demes, subpopulations, populations). Outbreeding depression may occur when populations are adapted to local environmental conditions and hybrids are less fit in either location, or when combinations of genes are *intrinsically coadapted*; that is, they are selected to perform well in the internal genetic environment defined by other interacting genes in the population (Templeton 1986). When genotypes from different populations are crossed, the coadapted complexes of genes are broken apart, particularly in later generations, and most new hybrid combinations perform poorly (Templeton et al. 1986).

The genetic and mating structure of plant populations is particularly favorable for the development of locally adapted races as well as intrinsic coadaptation. Subpopulation structure, small effective population size, limited gene dispersal, and inbreeding favor restricted recombination and facilitate the evolution of locally adapted gene complexes (Levin 1978). A considerable literature beginning with the genecological studies of Turesson (1922) and Clausen et al. (1940) documents examples of local adaptation by the use of transplant studies, and, in some cases, genetic differentiation has been demonstrated over distances of a few meters in response to small-scale changes in soil conditions (Snaydon 1962; Snaydon and Davies 1976), local competitors (Turkington and Harper 1979), and pathogens (Parker 1985). These considerations would suggest that many plant species would be likely candidates to exhibit outbreeding depression when crosses between plants from different locations are made. However, the limited data currently available provide no clear answer to this question.

Price and Waser (1979) suggested that because of limited gene dispersal and local environmental heterogeneity there would be an optimal outcrossing distance for many plant species. Crosses between near neighbors would suffer from inbreeding depression because the plants would likely be related, whereas crosses from distant plants would suffer from outbreeding depression through the disruption of favorable gene complexes. Thus crosses at some intermediate distance would result in the fittest offspring. Experimental field studies of *Delphinium nelsonii* and *Ipomopsis aggregata* have confirmed some of these predictions. Measurements of pollen and seed dispersal were consistent with the idea that neighborhood areas for both species were small, since dispersal distances were mostly less than a meter. Local adaptation would thus be favored because of restricted gene flow and environmental heterogeneity. Evidence for local adaptation was obtained for both species by using reciprocal transplants of seeds in the case of *D. nelsonii* (Waser and Price 1985), or by planting seeds near and distant from their maternal parent in the case of *I. aggregata* (Waser and Price 1989). In both species, crosses between plants from intermediate distances (3–10 m) set more seed than crosses between nearby (1 m) or more distant (100–1000 m) plants (Waser and Price 1983, 1989). Furthermore, seeds produced by crosses between plants separated by 10 m performed better than seeds from crosses between plants occurring in close proximity to each other, or from more distant pairs of plants (Waser et al. 1987; Waser and Price 1989). In *D. nelsonii,* pollen tube growth rates are apparently faster for pollen grains from plants at the optimal distance (10 m) than for pollen from 1 or 100 m. This raises the possibility that pollen from the optimal distance may preferentially fertilize ovules when in competition with pollen from other distances.

Most other experimental studies in which investigators have conduced crosses between plants at different spatial scales have failed to detect the occurrence of an optimal outcrossing distance or evidence for outbreeding depression (see Table 9 in Sobrevilla 1988; Newport 1989). In *Chamaecrista fasciculata,* seed set and the progeny performance of field transplants increased with interparent distance and did not decline even at distances of 1 km, well beyond the scale of local differentiation since neighborhood areas were estimated to have a radius of 3.1 m (Fenster 1991a, 1991b). Studies of pollen tube growth in crosses involving maternal and paternal parents of *C. fasciculata* from different spatial scales also failed to detect any location effects (Fenster and Sork 1988). Similarly, in *Sabatia angularis*, the total relative fitness of progeny transplanted back into the field was higher for between-subpopulation crosses than within-subpopulation crosses (Dudash 1990). In *Phlox drummondi*, seed set increased and levels of abortion decreased with interparent distances up to 200 m (Levin 1984), whereas in *Espeletia*, distance showed no clear effect on patterns of seed set (Sobrevilla 1988).

The complex relationships between proximity and fitness in plants are exemplified by studies of *Mimulus guttatus* by Ritland and Ganders (1987a, 1987b). Different fitness components showed varying relationships with distance, depending on the local population genetic structure and degree of relatedness of colonies. The variable responses they observed indicate some of the difficulties in predicting whether outbreeding depression is likely to occur in particular populations or species. Many factors—including the magnitude and scale of local selection pressures, the extent of genetic drift, and the mating system—will influence genetic differentiation; thus the lack of consistent results is not particularly surprising. Since all species mentioned above are animal pollinated and primarily outcrossing, it appears, however, that the mating system alone does not reliably predict whether outbreeding depression will occur.

Few data are available bearing on outbreeding depression in predominantly selfing species (but see Svensson 1988). In a study of *Impatiens pallida*, D. W. Schemske (pers. commun.) made crosses using pollen donors from within the same transect (0–18 m distance) and donors from a second transect (>32 m distance). He then planted these two types of seeds, and selfed seeds from cleistogamous flowers, back into the shaded habitat of maternal plants, as well as into a treefall gap some 500 m away. In the maternal environment, there were no fitness differences among the three seed types. In the treefall, however, the relative fitness of seeds from the most distant crosses was 60% higher than for either of the other seed types. Thus neither inbreeding nor outbreeding depression was evident when seeds were planted in their maternal environment, but wide crosses appeared to provide an advantage for exploitation of a novel environment.

The degree of genetic divergence among populations of some selfing species can complicate attempts to predict the outcome of interpopulation crosses. This problem was encountered in a study of the tropical disjunct *Eichhornia paradoxa* (Cole and Barrett 1989). Fewer than a dozen isolated collections of this rare species have been made, ranging from Guatemala to southern Brazil. Two morphologically differentiated populations from northeastern Brazil located some 500 km apart were artificially hybridized, resulting in F_1 progenies with a high degree of heterosis for vegetative traits. The hybrids were highly sterile, however, producing no viable pollen or seeds. Electrophoretic analysis of the two populations revealed that they were nearly uniform, but fixed for alternative alleles at 14 of 27 loci screened. Hence, the genetic identity of populations was 0.657, a value normally associated with congeneric species. Clearly, the two populations are distinct biological species, raising the difficult issue of taxonomic delimitation in highly selfing groups and the problem of how to sample variation of this type in rare plants.

IMPLICATIONS FOR CONSERVATION

Despite strong genetic foundations to conservation practices employed for domesticated plants and their wild relatives (e.g., Frankel and Bennett 1970; Frankel and Hawkes 1975), the role of population genetics in rare plant conservation has only recently been appreciated. Most management practices have been directed toward habitat preservation, particularly understanding the demographic features of endangered species, such as the size of local populations and their growth rates. However, if the long-term survival of rare plant populations is to be considered, strategies that maintain genetic variation must be attempted. Fortunately, since the demography and genetics of populations are not independent properties, management practices that aim to maintain large population sizes will tend to conserve genetic diversity. Loss of genetic variation through stochastic forces and the deleterious effects of inbreeding in small populations are the major threats that compromise attempts to maintain the long-term viability of populations of rare plant species.

Where habitats are in immediate danger of destruction, the collection from the wild of plant material (e.g., seeds, clonal fragments) and its maintenance in botanical gardens become necessary. Here the input from genetics becomes even more important both at the sampling stage (Brown and Briggs, Chapter 7) and where attempts are made at controlled breeding. If manipulation of genetic resources is considered important to enhance the chances of survival of an endangered species, this should be undertaken only after some understanding of the breeding biology and population genetic structure has been obtained. The availability at botanical gardens of glasshouses and other facilities, where studies of plant reproductive biology can be conducted, should enable basic information to be obtained on the breeding systems of rare plants. Data on population genetic structure require more sophisticated approaches, although the use of electrophoretic methods and various molecular techniques is now routinely employed in many studies in systematics and population biology. Collaboration with university research departments, or the installation in botanical gardens, of laboratory facilities equipped with biochemical and molecular tools, would greatly assist the monitoring of genetic variation in rare plant species.

The plant conservation biologist faces a series of decisions when developing a plan for the preservation of a species *ex situ,* or for its reestablishment into the wild. It is of importance to determine to what extent local populations are genetically differentiated, whether such differences have adaptive value, and whether the mixing of gene pools from different populations will increase or decrease successful establishment and long-term survival. There is abundant evidence that most plant populations are genetically differentiated from one another (Antonovics 1976). However, population biologists are far from being able to make general predictions on whether such differences will be of adaptive value under new environmental conditions, or how novel genetic combinations are likely to fare in nature. Because of our overall ignorance of the population biology and genetics of rare plants, it seems likely that conservation biology will be a "learn as we go" endeavor. We therefore advocate the use of experimental approaches to speed up the learning process. We believe that conservation plans that use experimental approaches are more likely to succeed and will increase our overall knowledge of population biology to the benefit of future conservation undertakings.

In addition, work of this type is more likely to increase potential funding sources for conservation over that which would be obtained from descriptive studies and the simple monitoring of rare plant populations.

For instance, if a few small populations of a rare plant species remain, and the goal is to reestablish new wild populations in sites that are conserved, experimental approaches are required. One might establish pure samples from extant populations, as well as composite mixtures with greater genetic variation, even at the expense of breaking up coadapted gene complexes. Despite considerable theoretical work on the relationship between environmental heterogeneity and genetic variability, few experiments with plants have investigated whether colonization and persistence are related to genetic variability (Martins and Jain 1979; Bell 1985). We have little knowledge of whether genotypes found in local environments are finely adapted or whether they represent chance combinations of genes resulting from founder effects and genetic drift. Well-planned experimental reintroductions could increase the likelihood of success, particularly if multiple seed stocks are used. Such experiments would require only that hybrids be produced and that different seed sources be involved. Experiments of this type could encourage greater interchange of views between conservation biologists and experimental population biologists, and may potentially attract funds not normally directed at either discipline.

A recent conservation program in Illinois illustrates how experimental approaches may be required to enable reintroduction of rare plants into the wild (De Mauro 1989 and pers. commun.). The last two remaining populations of *Hymenoxys acaulis* var. *glabra* were recently extirpated from the state as a result of habitat destruction. Although material was collected from the last remaining population of 30 plants, no seeds could be synthesized because of strong self-incompatibility and the fact that all plants were apparently of the same mating group. However, controlled crosses between Illinois plants and material from the only known population in Ohio resulted in abundant viable seed, owing to the presence of different mating groups in the Ohio population. The hybrid seed has been used in experimental plantings in Illinois in an effort to reestablish populations in the state. Although, in this case, interpopulation hybridization was undertaken out of necessity in order to restore fertility to the Illinois material, carefully controlled studies, similar to those conducted by Templeton and associates in Missouri Ozark populations of collared lizards (Templeton 1986), would be likely to provide valuable information on the extent of local adaptation in rare plants, as well as furthering conservation efforts.

Genetic manipulations of rare plants will obviously lead to the disruption of locally adapted gene combinations. This may be of concern when local races of a species are morphologically or ecologically distinct and are recognized taxonomically. In some cases, the attempt to preserve population differentiation may conflict with practices aimed at species preservation. Such cases require individual judgment, but we believe that, in the long-term, species conservation is, in general, of greater importance. Attempts to preserve population distinctiveness should be undertaken only when they do not endanger species conservation. Should reestablishment practices be successful at a number of sites, population differentiation will almost inevitably follow.

It should be clear from the evidence presented above that knowledge of a species' selfing rate or inbreeding coefficient does not necessarily allow us to predict with accuracy the level of inbreeding depression that it is likely to suffer. Thus inbreeding depression might occur even in species with a history of inbreeding. In addition, the level of inbreeding depression will usually depend on the environment in which it is assessed. Thus although little or no inbreeding depression may be evident when a plant is grown under the benign conditions of a glasshouse or botanical garden, its effects may be more severe when seeds are reintroduced back into the wild. Seeds produced under uncompetitive conditions may have lower "ecological value" than their numbers might suggest. Small samples of rare plants maintained in botanical gardens may be inbred and of inferior genetic quality, complicating attempts at reintroduction. Thus conservation of wild plants in botanical gardens may lead to unconscious "domestication," particularly if many sexual generations under benign conditions are allowed to occur. Occasionally challenging seed stocks to natural or seminatural conditions could aid in assessing the durability of species to the rigors of nature, even at the expense of losing some genetic diversity. A major challenge of *ex situ* conservation will be to ensure that sexually propagated samples of rare plants do not become museum specimens incapable of surviving under natural conditions.

ACKNOWLEDGMENTS

We thank Deborah Charlesworth, Michele Dudash, Marcy De Mauro, Charles Fenster, Tim Holtsford, Brian Husband, Doug Schemske, and Peter Toppings for providing unpublished data and manuscripts. Brian Husband and Bill Cole helped with the figures. The manuscript was prepared while S.C.H.B. was a recipient of an E.W.R. Steacie Memorial Fellowship funded by the Natural Sciences and Engineering Research Council of Canada, and J.R.K. was supported by an NSERC of Canada Postdoctoral Fellowship.

2

Ecological Implications of Genetic Variation in Plant Populations

LAURA FOSTER HUENNEKE

In recent years, the protection of genetic diversity within species has become a priority for conservation efforts. The long-term objective is to maintain the evolutionary viability of the taxon, maximizing its chances for persistence in the face of changing environments. Genetic considerations may also have short-term importance in our strategies for managing rare plants. When taxa have become critically endangered, and direct manipulations of field populations or the establishment of *ex situ* populations are being planned, what genetic factors should be considered? Here I review evidence concerning the relevance of genetic variation to the ecology and persistence of plant populations, with the goal of developing guidelines for the genetic management of rare plants.

Three lines of evidence are presented here. First, I explore how the pattern and extent of genetic variation within a taxon are related to ecological range and performance. Second, I discuss how genetic variation within a population is related to that population's chances of persistence and ecological success. Third, the genetic and ecological problems facing a number of rare plants are reviewed, to determine how often genetic concerns are crucial to short-term conservation goals. I close with some conclusions about what "types" of rare species might require genetic management, and what guidelines should be followed in sampling, propagation, and genetic manipulations.

GENETIC VARIATION AS A DETERMINANT OF GEOGRAPHICAL AND ECOLOGICAL RANGE

Geographical variation is presumably tied to variation in the physical environment, and thus to variable ecological conditions, for those taxa with widespread distribution. Plants with a wide geographical or ecological range often display considerable phenotypic variation; this variation may be determined by phenotypic plasticity, by genetic variation across the range, or by some combination of these (Jain 1979). Traditionally the contrast has been viewed as one between species with broadly adapted "general-purpose genotypes" (Baker 1974) and species with many distinct, locally adapted genotypes. To what extent does genetic variation constrain the total geographical or ecological range of a taxon?

Geographical Variation: Ecotypes

Distinctive variations in growth form have long been noted for wide-ranging taxa. Local "ecotypes" whose form or physiology appeared to vary in an adaptive way with environmental conditions, and that represent genetic differences among populations, have been described in numerous studies (Turesson 1922; Clausen et al. 1940, 1948; Van Dijk 1989). Distinct ecotypes presumably represent discontinuous genetic variation, correlated with specific habitats. Quinn (1978) has argued, though, that ecotypes are not ecological or evolutionary units. According to his view, each population is a unique realization of the genotype–environment interaction (e.g., Schwaegerle and Bazzaz 1987). In those species in which genetic variation is more clinal, genetic control of physiology may still be an important determinant of geographical and ecological range.

Whether labeled as ecotypes or not, genetically determined differences in ecological performance clearly exist among populations of many plants. Such differences are often responses to unusual edaphic conditions. A classic example is the clear genetic determination of copper tolerance in grasses growing on mine-spoil dumps (McNeilly and Bradshaw 1968). Metal tolerance is ubiquitous in mine populations, but almost completely lacking in adjacent populations growing on normal soils; despite gene flow across the abrupt environmental boundary, differentiation is maintained by selection. Such differentiation may arise quite rapidly; in the Rothamsted Park Grass experiment, where plots of lawn grasses received different fertilizer treatments, genetic differences could be detected between populations in limed plots and those receiving no lime, only six years after a single application of lime (Snaydon and Davies 1982).

Attempts to correlate a species' ecological breadth with its overall genetic variability have met with mixed success. Babbel and Selander (1974) compared edaphic specialists with more widespread congeners in two genera of composites. In both cases, the more widespread species did possess higher total variation (both allelic diversity and heterozygosity), but the widespread species possessed no more variation among populations than did the restricted species. Levin et al. (1979) compared several species of *Oenothera* and found no strong correlations between range (or extent of edaphic variation) and genetic variation in a taxon. Mashburn et al. (1978) surveyed two widespread *Typha* species and found no detectable electrophoretic variation at all, despite their substantial ecological amplitude. Hamrick et al. (Chapter 5 and previous reviews) have found only weak correlation between geographical range and among-population differentiation. It is unclear whether this result is a relative one (less differentiation among populations relative to the increased total diversity in a widespread species), or reflects ecological and genetic characters that in themselves contribute to a broad geographical range (e.g., weediness).

On the other hand, some taxa display substantial genetic differences correlated with specific ecological conditions. Allard et al. (1978) found electrophoretically distinct genotypes in *Avena barbata,* each restricted to either xeric or mesic California grasslands. Recognizable soil "races" were detected by Heywood and Levin (1985) in the widespread *Gaillardia pulchella.* Rozema et al. (1978) found genetic differentiation according to soil salinities in a *Festuca*. The implication is that genetic variation over a wide geographical (or habitat) range is often substantial and may be of adaptive value.

Transplant Experiments

Transplant or immigration experiments may reveal the selective advantage of genetic differences among populations. Transplant experiments of tundra perennials (Chapin and Chapin 1981; McGraw and Antonovics 1983) showed significant genotype–environment interactions, with residents invariably performing better than nonresidents in growth and survivorship. Schmidt and Levin (1985) performed reciprocal transplants among annual *Phlox* populations; nonresident survivorship and fecundities were each only 70% those of resident genotypes. Bradshaw (1984) reviewed a number of transplant experiments; in general, the fitness of transplants, moved to an environment similar to that of their origin, was about half that of residents. If plants were moved to noticeably different environments, realized fitness values were 0.1 or less those of residents. Bradshaw (1984) concluded that genetic heterogeneity among populations is frequently related in an adaptive manner to variation in abiotic environmental factors.

In at least some cases, then, genetic variability appears to be associated with geographical and ecological range. Here one confronts the question of causation: Are widely distributed plants genetically variable because of their distributions, or are they widely distributed because of their variability? The answer has great relevance to conservation efforts. If wide distribution permits variability to develop, then preservation of a portion of the taxon's genetic diversity might later be the basis for "reconstitution" of that diversity. However, if adaptive genetic variation is essential in determining the plant's persistence in a range of environments, then conservation efforts aimed at maintaining the species' full potential should preserve a wider range of its variation. Low fitness of transplants and existence of diversified local genotypes suggest that preserving a broad range of the species' variability might be prudent for maximizing persistence.

CONTRIBUTIONS OF GENETIC VARIATION TO POPULATION VIABILITY

Genetic variation is the basis for potential evolutionary change in a taxon. Genetic characteristics also influence the physiological and demographic performance of a population. In this section, I first discuss the interaction of genetic variation and phenotypic plasticity in plant response to environment. Next I review evidence that genetic differentiation allows efficient occupation of differing microhabitats within a population. Third, genetic variability is related to a population's ability to survive stochastic threats (such as disturbance or pathogens). Fourth, genetic variation is considered with regard to demographic performance of a population or of individuals. Finally, I discuss evidence that genetic variation of a founding population influences the likelihood of success of that colonizing group; such data are of relevance to *ex situ* conservation or reintroduction attempts.

Plasticity

Studies of genetic variation in plants are often confounded by phenotypic plasticity. Plasticity allows a single genotype to succeed in a range of environments and may conceal the true extent of genetic differentiation. Of course, plasticity itself has a genetic basis and varies among individuals and among populations; in a recent review, Schlichting (1986) concluded that there was no necessary trade-off between the plasticity of individuals and the genetic variation of a population or taxon. One comparative study of several congeners (Primack 1980) found that species with restricted ranges can show as great phenotypic variation as those with more widespread ones.

Bazzaz and co-workers have explored the relative importance of plasticity and genetic variation by assembling populations of known genotypic constitution, by growing these populations along resource gradients (e.g., soil moisture gradients), and by measuring the total width of the gradient used by the population. Results have been mixed: Zangerl and Bazzaz (1984) found a significant contribution of genotype diversity to total population performance in an inbreeding weed. However, when Garbutt et al. (1985) grew artificial populations of five different genotypes on environmental gradients, they observed that between-genotype differences actually contributed very little (relative to within-genotype plasticity) to niche breadth. In another species, Garbutt and Bazzaz (1987) found that some genotypes had very plastic offspring, whereas other genetic lines in the same species produced a range of specialists. It remains unclear whether phenotypic plasticity can fully compensate for low genetic variability in a heterogeneous environment. However, where a species lacks plasticity or where environmental variation is extensive, genetic variation may well contribute substantially to population performance.

Microhabitat Differentiation

Within a single plant population, the physical and biotic environment may be quite heterogeneous. To what extent does genetic variation increase a population's ability to succeed in the full range of available microsites? The presence of genetic differences within a population is not sufficient evidence for concluding that differences represent adaptive responses to environmental patchiness; certainly, much spatial structure within populations is due to the extremely local nature of most gene flow (Levin and Kerster 1974; and Schaal 1975).

However, evidence is accumulating that in natural populations much genetic variation is related to natural selection arising from small-scale environmental heterogeneity, as plants sample the environment in a finely grained way (Menges, Chapter 3). A number of cases have documented the restriction of particular genotypes to specific microhabitats within a site. Examples include *Spartina* clones specialized for marsh, swale, or dune microhabitats (Silander 1979, 1985b); distinct genotypes of an arid-zone bunchgrass restricted to roadsides, salt flats, or valley floor (Shumaker and Babbel 1980); and annual grasses with predictable genetic differences between those in relatively xeric microsites and those in moister sites, even on a single hillside (Allard et al. 1978; Hamrick and Holden 1979; Nevo et al. 1986). In small populations of an aquatic *Veronica*, Linhart (1974) demonstrated genetic differences in growth form and reproductive patterns between individuals found in the center of the pond and those found at the margins.

The biological environment may vary within a site as well; genetic differentiation according to neighboring species has been documented recently for several plants. The best-known examples come from white clover (*Trifolium repens*) growing in diverse pasture communities. "Biotypes" of *T. repens* taken from the vicinity of a particular grass species perform best when transplanted to the vicinity of that same grass species; the biotypes are differentiated for competitive performance against a spectrum of potential neighbors (Turkington and Harper 1979). This variability is superimposed on clover's tremendous differentiation of other traits of selective value (growth rate, growth form, production of distasteful chemicals), to the extent that variation among clones within a population is as great in magnitude as variation between populations (Burdon 1980a). Variation in the physical environment and in neighborhood composition, together with selection for balanced competitive ability among species pairs (Aarssen and Turkington 1985), maintains high genotypic diversity, even in this clonal species.

The *Trifolium* example highlights the importance of genetic variation in clonal species, particularly in patchy environments. Many clonal plants appear to maintain selectively advantageous variability within populations (e.g., bracken fern [Wolf et al. 1988]; a mixture of competing perennial grasses [Kelley and Clay 1987]; general review by Ellstrand and Roose 1987). Schmid (1985) speculated that for clonal species with a *phalanx* (central clump) growth form, plasticity is critical in coping with heterogeneity. However, species with a *guerrilla* habit (ramets moving fair distances and portions of genetic individuals intermingling) may show local or genetic differentiation, since at least some ramets reach favorable patches. (In nonclonal plants, the prolific production and dispersal of genetically variable seed crops effects a similar exploration of heterogeneous environments.)

Even in a homogeneous environment, selective pressures may favor maintenance of variation. For example, Ennos (1985b) found that genetic variation in root growth in *T. repens* resulted in different genotypes exploiting different soil depths or volumes. Because of this spatial separation, a mixture of genotypes showed superior performance relative to stands of pure genotypes.

To what extent can specific alleles or genotypes be correlated with environment? Brown (1979) presented a comprehensive (at that time) review of correlations between particular electrophoretic genotypes and ecological or physiological parameters (germination rate, survivorship, flood tolerance, etc.). Little of this variation has been explained in mechanistic terms, however. A study by Verkleij et al. (1980) focused on polymorphisms for seed amylases in a weedy species with a dormant seed bank. The amylases were extremely variable and linked to variation in germination physiology and dormancy (the critical ecological behaviors of arable weeds).

Transplants within a site demonstrate that microhabitat differentiation influences plant performance in the same way as does between-site heterogeneity. Silander (1985b) transplanted *Spartina* genotypes among microhabitats in coastal marshes; resident genotypes had higher survivorship and growth rates than did genotypes from other microhabitats within the marsh (e.g., dune vs. swale genotypes). Even transplants within a seemingly homogeneous population cause similar effects; Waser and

Price (1985) found that seeds planted close to their maternal plant had lower mortality and more rapid growth than seeds planted only 50 m from the parent. Selection against "nonresidents" averaged 34%. Such small-scale genetic differentiation allows more efficient exploitation of a heterogeneous environment than would be possible for a genetically uniform population.

Genetic Variation and Population Response to Stochasticity

Stochastic events constitute major threats to the viability of small populations. As Menges (Chapter 3) points out, these influences may be environmental (e.g., the occurrence of devastating storms), demographic (e.g., chance differences among individuals in survivorship or fecundity), or genetic (e.g., loss of alleles due to drift). For managers, the challenge is to minimize the chances of extinction of a given population in the short term due to these effects. What role does a population's genetic structure play in determining vulnerability to stochastic events? Relevant examples come from three kinds of studies: the genetic structure of populations in variable environments, the impact of pathogens and herbivores on populations, and the role of variation in population and individual fitness.

Variation in a Fluctuating Environment

General reviews of variation have concluded that great heterozygosity characterizes species living in variable environments (e.g., Nevo 1978; Beardmore 1983; Loveless and Hamrick 1984; Brown and Burdon 1987). Where generations overlap (either as perennials or as a soil seed bank), temporal changes in the environment may also be countered by genetic differences among generations. Burdon et al. (1983) observed successive seedling cohorts of an annual plant becoming established at different times in a single growing season. The cause and intensity of mortality varied among cohorts, and the resulting adult plants differed genetically. The existence of variation in the initial pool contributed to the survival of at least some seedlings from each cohort despite great differences in microclimate at the time of establishment.

Even within a single cohort, environmental conditions may act in different ways at different points in the life cycle. Clegg and Allard (1973) and Clegg et al. (1978a, 1978b) documented differences in selective pressure and direction at various life-history stages. Thus viability selection (causing differential survivorship at the seedling stage) favored different genotypes than did fecundity or fertility selection.

Pathogens and Herbivores

Genetic uniformity clearly renders animal populations more susceptible to disease epidemics (O'Brien and Evermann 1988). The same appears true of plants, although there are few documented examples from natural populations. Futuyma (1983) reviewed agricultural examples, suggesting that the widespread cultivation of uniform preferred genotypes has resulted in regional crop failures due to attack by pathogens and parasites. Because these "natural enemies" are themselves variable genetically, with short generation times and high reproductive rates, genetic diversity within a crop is important as an evolutionary response to rapid shifts in the pathogen.

Significant variation within and among natural plant populations for genetic resistance to particular pathogens has been documented (Burdon 1980b; Parker 1985; reviewed in Burdon 1985). Disease outbreaks cause heavy mortality on occasion in field populations (e.g., see Silvertown [1985] and the enormous literature on epidemics of forest trees). Variation in resistance to particular pathogens is presumed to be an important selective force for increased genetic variation (Bremermann 1980); the existence of a range of genotypes in a population may result in the survival of (at least) a few individuals after pathogen attack.

Herbivory may present the same sort of stochastic challenge. Some evidence exists for differentiation among individuals for vulnerability to grazing. For example, several of the variable traits in *Trifolium repens* result in differential levels of herbivory by sheep or by slugs, or differential tolerance of grazing (Burdon 1980a; Dirzo and Harper 1982). Scott and Whalley (1984) demonstrated differentiation for adaptive traits in a perennial grass exposed to varying intensities of grazing. Burdon and Marshall (1981) suggested that outcrossing (and thus genetically variable) weeds have proved more difficult subjects for biological control than have selfing or apomictic weeds. Clegg and Brown (1983) cite the success against *Opuntia* in Australia, where the plant's genetic uniformity (perhaps arising from clonal propagation and a small range of introduced material) made biological control effective. Again, the presence of genes for resistance or tolerance within a population (even at low frequencies) may determine the chance of persistence in the face of insect outbreaks or other herbivory episodes.

Of course, some stochastic occurrences act on a population's members without regard to their genetic constitution. For example, major disturbances (e.g., fire or land-use conversions) may be devastating to all genotypes in the population. However, if disturbances are more moderate, genetic variability may result in the survival of at least a few individuals.

Demographic Stochasticity

Plant survival and fecundity are determined largely by the size and physiological status of the individual plant, which are presumed to be a function of local conditions and resource availability. Gottlieb (1977a) clearly demonstrated for one population that differences in size (and in fitness components) were influenced far less by genotype than by environmental patchiness. Nevertheless, more recent quantitative studies suggest that growth rate and other physiological characters influencing plant size and fitness do sometimes have substantial heritabilities (general review by Sarukhan et al. 1984), despite the masking effects of environmental heterogeneity. For example, Antlfinger et al. (1985) found that growth rate and size (and thus survival and fecundity) had significant heritable components in a violet.

Genetic Variation and Fitness: Inbreeding

Genetic considerations are critical for obligately outbreeding species. Well-developed incompatibility systems result in lowered fecundity of matings with close relatives. However, self-compatible plants are frequently cited as examples of taxa where inbreeding would produce no deleterious effects, and where genetic considerations need not be given high priority. How widespread is inbreeding depression in different sorts of plants? Is heterozygosity advantageous to an individual? In populations that may be either outbreeding or selfing, what are the ecological implications of the two breeding systems?

Inbreeding Depression, Selection Against Homozygotes, and Heterosis

Inbreeding depression appears to be due primarily to the expression of deleterious recessive alleles in the homozygous condition. In their review, Charlesworth and Charlesworth (1987) concluded that gymnosperms frequently exhibit inbreeding depression despite high frequencies of self-pollination and the lack of effective self-incompatibility systems; selfed progeny appear to have low viability. The same review suggested that in angiosperms even partially selfing species frequently experience inbreeding depression (see also Barrett and Kohn, Chapter 1). Selection against selfed progeny may happen early in embryonic development. Comparisons of genotypes in embryos, juveniles, and adults show differential survival and apparent strong selection against (selfed) homozygotes (e.g., Burdon et al. 1983 [*Echium plantagineum*]; Yazdani et al. 1985 [*Pinus sylvestris*]). Some studies have reported differentially high survivorship and/or fecundity of heterozygotes (e.g., Clegg and Allard 1973; Schaal and Levin 1976). (Unfortunately, comparisons of heterozygosity levels between juveniles and adults of long-lived perennials may be difficult to interpret; if homozygotes are more common among juveniles than among adults, selection may be constant and acting against those homozygotes, or selective pressures may have changed in the environment such that homozygotes are actually being favored increasingly through time.)

It has been suggested that heterozygosity itself confers a fitness advantage, independent of the avoidance of inbreeding depression. Levin (1983) and Ledig (1986b) have reviewed this topic and have suggested that heterosis occurs frequently when the parents are not too distantly related. Ledig (1986b) concluded that heterozygosity is frequently associated with higher growth rates and survivorship, particularly for forest trees.

Heterozygosity may not enable superior performance in a given stable environment, but may provide the ability to cope with fluctuating environments. Mitton and Grant (1984) reviewed studies suggesting that heterozygosity could play such a buffering role. Linhart and Mitton (1985) found that heterozygosity in *Pinus ponderosa* was not correlated with female cone production or with growth, but was correlated with low variance in these performance measures. Nevo et al. (1986) found that the most heterozygous genotypes of wild barley occurred in microsites exposed to the greatest microclimatic fluctuation.

Implications of Selfing Versus Outcrossing

There are numerous comparative studies of progeny from selfed and outbred matings within a population. For example, some plants produce both *cleistogamous* (closed, self-pollinating) and *chasmogamous* (open, receptive to outcrossed pollen) flowers. Outcrossed offspring are presumed to be more variable genetically than the progeny of self-fertilizations. Schoen (1982) found pronounced decreases in survivorship for selfed progeny in a normally outcrossed species (likely an example of inbreeding depression). In those taxa where selfing is frequent, however, results vary. Mitchell-Olds and Waller (1985) documented higher seed survival and seedling growth rates in chasmogamous *Impatiens* relative to those in cleistogamous progeny. In contrast, Clay and Antonovics (1985a, 1985b) found no differences in mean survivorship or reproduction between progeny from chasmogamous flowers and those from cleistogamous flowers. Antlfinger (1986) reviewed several cases where cleistogamous offspring were observed to be at no fitness disadvantage relative to chasmogamous ones.

Other relevant comparisons consider the relative fitness of sexually produced off-spring (considered as a group or population) versus asexually produced, nonvariable progeny (e.g., Antonovics 1984). Case and Taper (1986) used a modeling approach to demonstrate that intense resource competition would favor sexual (variable) off-spring; the competitive advantage gained by the occasional rare genotype would give the sexual group competitive dominance over the asexual (genetically identical) group, despite the latter's reproductive advantage. Between-species comparisons suggest the same result: Variation may confer competitive advantage in field situations. For ex-ample, Russell and Levin (1988) compared closely related *Oenothera* species that had different breeding systems, and found that the obligate outcrosser was competitively superior to facultatively selfing taxa and to permanent translocation heterozygotes, whose progeny are genetically identical to the parent. Similarly, Yazdi-Samadi et al. (1978) grew two *Avena* species together and found that a polymorphic population of either species had the competitive advantage over a monomorphic population of the other.

Founder Events

In a review of founding events in plants, Clegg and Brown (1983) concluded that most such events are associated with declines in total variation relative to that in the parent population. This decline in variation could be attributed to sampling (absence of rare alleles in the founding group) and to population bottlenecks after colonization (when heterozygosity may be lost). Decreased genetic variation did not appear to reduce population viability or colonization success. However, as Clegg and Brown (1983) pointed out, most plant colonizations are by invasive weedy species, many of which are selfing or apomictic; the breeding system itself leads to lower variation and de-creased vulnerability to genetic problems. They concluded that weeds succeed despite low levels of variation in founding populations, but that most outcrossers would have greater colonization success if the founding group included a range of genetic diversity.

Carson (1987) confirmed that founder events in colonizing species, characterized by low variation overall, rarely lead to genetic divergence in the new group. However, in taxa where the founding group is variable (or is even a single outcrossed propagule), genetic rearrangements and shifts are common in the new environment, and differen-tiation or proliferation of new taxa is likely. Barrett and Kohn (Chapter 1) discuss a founder event in *Echinochloa*, where Australian populations (based on founders from California) show bottleneck effects in both allozyme and quantitative traits.

One implication of Carson's argument is that rapid changes could be expected within a population removed for *ex situ* propagation, particularly if the environment differs from the native site and if genetic recombination is rapid. Most *ex situ* popu-lations are maintained in environments that differ substantially from the original site, of course; a handful of botanic gardens and arboreta can scarcely duplicate a wide range of natural environments. To slow the rate of *ex situ* differentiation, managers might focus on careful limitation of the rates of sexual recombination.

A classic example of a founding event was reported by Schwaegerle and Schaal (1979). An artificial population of a pitcher plant (*Sarracenia*) was known to have originated with a single transplanted individual. After several generations, the large population had lower levels of heterozygosity and polymorphic loci on average than did natural populations. However, several of this species' small, isolated natural populations had equally low levels of genetic variation.

In sum, then, genetic variation within a population can influence the ability of the population's members to exploit a patchy environment, to survive stochastic events, to maintain high levels of reproductive performance, or to adjust to novel or fluctuating environments.

IMPLICATIONS OF GENETIC VARIATION IN RARE TAXA: CASE STUDIES

The above findings suggest that genetic variation influences the ecological performance of a population's members. How frequently do these effects occur in rare and threatened populations? Selected rare plants are discussed below, with attention to the role that genetic variation may play in population viability.

Eucalyptus Species in Australia

In Australia, the genus *Eucalyptus* has proliferated into many species, some of them narrow endemics. Several eucalypts have been well studied genetically (e.g., Moran and Hopper 1983, 1987; Sampson et al. 1988). The data allow comparison of genetic structures of widespread, regional, and localized (narrowly distributed) taxa. Localized species have little genetic diversity; the small, isolated populations frequently have low within-population diversity, but at least some among-population variation. This between-population diversity is chiefly the occurrence of "common local" alleles—those found in only one or two populations, but common in those. In contrast, widespread species possess few localized alleles, although overall allelic diversity is greater. Interestingly, regional species showed the lowest overall diversity. Some of these appear to have been long isolated in small, naturally disjunct populations (i.e., not fragmented as a result of human activity); genetic distances among populations are proportional to geographical distance. Sampson et al. (1988) described one such species with a mixed breeding system (a facultative selfer); genetic data suggested selection against homozygous, selfed progeny, and low observed fecundity.

Astragalus

Diverse species of *Astragalus* in the western United States provide additional comparisons of species with differing distributional patterns. Karron (1987b, and Chapter 6; Karron et al. 1988) has contrasted pollination biology and genetic structure in two geographically restricted species with those in two widespread species sympatric with them. One of the localized species, which does occur on variable soil types, contains moderate levels of genetic variation. However, the other restricted species (whose distribution apparently resembles that of the regional eucalypts just described) contains very little genetic diversity. There is little differentiation among populations for any of the four taxa. All are pollinated by generalist bees, but at least one of the restricted species is visited far less frequently than its sympatric widespread congener; both restricted species increase seed set upon hand pollination, indicating limitation by pollinator services. Karron suggests that management of pollinator populations would be an important part of any management plan for rare plants such as these.

Annual Grasses of California Vernal Pools

Griggs and Jain (1983) published a genetic and ecological study of the rare *Orcuttia* species, annual grasses of isolated vernal pools in California's Central Valley. Many populations have been extirpated by agriculture or development, and few sites remain for any species. These facultatively outcrossing grasses showed moderate genetic diversity within populations, and some differentiation among populations. Transects established in one site for demographic monitoring were "obliterated by grazing" in 1977, and the plants were "totally consumed by grasshoppers" in 1978. These stochastic factors, and others such as unpredictable drying of the pools, are likely far more important influences on short-term population performance than are inbreeding considerations. Genetic variation within the dormant seed bank, on the other hand, may be critical in providing potential recruits each year despite such catastrophic events.

The Large-Flowered Fiddleneck

Amsinckia grandiflora is extant in a single population in central coastal California, although other historical sites are known. The species is *heterostylous*—that is, with two distinct flower forms (pin and thrum). Because pin pollen produces viable seed only on thrum stigmas, and vice versa, fecundity may be low despite considerable pollen movement among individuals. The small size of the population has led to observed fluctuations in the relative abundance of pin and thrum flowers from year to year (a form of demographic stochasticity). When the ratio of pin to thrum forms is too skewed, most pollination is ineffective, and seed set may be quite low (U.S. Fish and Wildlife Service 1987). Low effective population size (in the genetic sense) may be limiting reproductive output for the population.

The Mauna Kea Silversword

Argyroxiphium sandwicense is a giant rosette plant endemic to the higher slopes of Mauna Kea in Hawaii. Total population size, density, and areal extent on the slopes have all declined severely in recent decades. Most of the slow-growing, long-lived individuals are monocarpic; current small population size results in very few plants flowering simultaneously in any one year. Experiments by Powell (1985, pers. commun.) demonstrate that the silversword is nearly self-incompatible; observed seed set and seedling recruitment are very low. Both demographic stochasticity and ineffectual pollinator services are important determinants of this small population's behavior.

Outplanting by the state of Hawaii increased population numbers substantially; these outplants have shown good survivorship and growth, and are flowering as they mature. Interestingly, most of the genetic material used for the outplanting apparently arose from a single maternal plant, one unusual in possessing a clumped or multiple-rosette growth form. This growth form is common among the outplanted progeny (84%, vs. about 25% in the original population) and is resulting in polycarpic behavior by these multiple-stemmed individuals (E. Powell pers. commun.). Clearly, the selection of genetic material for propagation and reintroduction has resulted in a shift in population characteristics.

TYPES OF RARITY AND OF RARE PLANTS

Not all rare plants are rare for the same reasons or in the same way. Stebbins (1980) wrote that endemism is usually the result of adaptation to specific environmental features that themselves are quite restricted; however, our understanding of endemism now encompasses a greater diversity of ecological situations and causation (Kruckeberg and Rabinowitz 1985). Rabinowitz (1981) clarified the concept of rarity by pointing out that "rare" may mean narrow geographical range, narrow habitat specialization, or low local population densities—or any combination of these. Some of these categories tend to receive more attention as endangered taxa than do others; my impression is that most of the federally listed threatened or endangered taxa are quite restricted geographically, but may or may not be rare according to the two other criteria.

We can speculate on the ecological and genetic implications of the various types of rarity, although the literature does not as yet permit firm conclusions. Those species restricted to patches of specific habitat have likely survived in isolated populations for a long while, and may be relatively tolerant of inbreeding and of increasing distance between populations. "Sparse" species—those plants never found anywhere in great numbers—likewise possess specific traits that enable them to tolerate low local densities (features of dispersal, establishment, and competitive ability, for example, in the sparse prairie grasses studied by Rabinowitz).

More vulnerable, perhaps, are species whose distribution patterns have been altered historically. Such taxa include those, formerly found in large populations, whose numbers have been severely reduced (e.g., the silversword). The vulnerable might also include those with wide historical distributions now restricted to small patches (true of many prairie and California grassland plants). Some of today's rare taxa are endemics that previously ranged more widely (e.g., in the mesic conditions of the American Southwest during Pleistocene times) and are now "trapped" in relict pockets of suitable habitat. Predictions of substantial climatic change in the coming decades suggest that many more plants will soon encounter similar restrictions.

Some rare plants are definitely habitat specialists; for example, *Howellia aquatilis,* an aquatic annual of regional distribution but narrow habitat requirements, showed zero detectable genetic variation within or among populations (Lesica et al. 1988). In other cases, historical range shifts and isolation, rather than (or in addition to) ecological specialization, seem to account for low genetic diversity within and among populations; the California fan palm is such an example (McClenaghan and Beauchamp 1986). Narrow endemics, with small population sizes and sometimes limited reproductive potential, might seem best suited to stable or refugial environments; however, at least some rare plants are found in unstable or successional sites (e.g., *Pedicularis furbishiae* [Menges and Gawler 1986] and *Plantago cordata* [Meagher et al. 1978], both plants of frequently disturbed streamsides).

Many threatened or endangered taxa have been proved to possess extremely low electrophoretic diversity (e.g., *Howellia*, as mentioned above; *Pedicularis furbishiae*; and *Purshia subintegra*, an Arizona endemic of calcareous soils, studied by Phillips et al. 1988). Sometimes extensive clonal behavior, coupled with low levels of observed seed set, suggest that self-incompatibility may be combining with low genetic diversity to constrain recruitment in rare plants. In other cases, even obligate outcrossers in low-density populations are able to maintain high levels of seed production. Clearly, it is difficult to draw conclusions about current population status from simple descriptions of distribution and breeding system (when these are known).

IMPLICATIONS FOR CONSERVATION OF RARE SPECIES

The foregoing discussion demonstrates that there are no simple rules of thumb regarding the importance of genetic variation in population viability. However, we can make some predictions about the relative vulnerability of various species to genetic problems, the relative importance of sampling plants from various portions of the range or the habitat, and the relative ease of various tactics for *ex situ* propagation or for reintroduction.

1. Taxa known to have suffered recent changes in distribution (number, sizes, or isolation of populations) may be experiencing patterns of gene flow radically different from those typical of the species' evolutionary history. It is difficult to estimate how many rare taxa fall into this category; in South Africa, Tansley (1988) has estimated that half of the threatened species of Proteaceae are not naturally rare, but have become restricted in numbers or distribution because of agricultural conversion of natural vegetation. The viability of these remnant populations is very much an open question.

High priority should be given to the investigation of breeding system and reproductive performance in these species. For example, obligate outcrossers recently fragmented into small, relatively homogeneous populations might experience reduced fecundity. Where there appear to be reproductive difficulties, electrophoretic or other data on the genetic structure within and between populations should be collected to analyze whether direct genetic management (of pollinators or of migration and/or seed dispersal) might be appropriate.

2. Geographical differentiation (the between-population component of genetic variation) is frequently related in an adaptive way to environmental differences across a species' range. Where environmental variation is marked or discontinuous, sampling for *ex situ* conservation efforts should strive to include material from all portions of the range or all distinct habitats. Because within-site environmental heterogeneity also exerts selective pressures on a population, sampling should include representatives from any distinctive microhabitats in a site (e.g., topographical features or soil heterogeneity). A conscious effort to include the full range of ecologically relevant variation may increase the chances of preserving the species' full ecological amplitude.

3. Within a microhabitat or a site, sampling should be strictly random (or perhaps systematic—e.g., on a grid). The goal is to create a sample representative of the entire

population. Haphazard or subjective sampling often focuses on especially robust or vigorous individuals, and may easily miss much of the genetic range present in the original population.

4. The design of transplant or reintroduction efforts should also incorporate consideration of adaptive variation. When material is transplanted to any site differing from the native location (whether another field site or a botanic garden), survivorship and fecundity will probably be low relative to performance in the native site. Careful matching of source population with transplant site is appropriate where feasible (where both are well characterized, with strong similarities).

5. If plants are to be maintained *ex situ* before transplant or reintroduction, care must be taken to maintain any genetic characteristics involved in the plant's specialization to a habitat. The troubling implication of rapid recombination rates, and of variability in sexually produced material, is that a few generations of selection in an *ex situ* conservation program may result in the accidental loss of the traits that confer success in the field. Work on metal tolerance suggests that constant strong selection is necessary to maintain tolerance at high frequencies in subsequent generations. We know very little about how rapidly adaptive traits might be lost in an artificial or a novel environment; we should remember this concern while maintaining populations in *ex situ* situations.

6. Conversely, when we know that introductions will encounter a heterogeneous or novel environment, greater variability in the founding or transplant population may increase the chances that at least some individuals will be successful. This consideration may enter into decisions about sampling seeds versus clonal material (for those species capable of clonal propagation). Seeds produced by sexual recombination contain a greater range of variation than does clonal material (McNeilly and Bradshaw 1968; Hume and Cavers 1981). As Barrett and Kohn (Chapter 1) suggest, deliberate crosses and recombination might be used to generate novel, variable material for introduction to a novel environment.

7. Last, one often encounters the generalization that self-compatible plants are rarely affected by inbreeding depression. Should this be true, concerns about small population sizes in transplant or propagation attempts would be minimized in many cases. However, inbreeding depression appears to be potentially detrimental to all but entirely selfing plants. The size and genetic constitution of a founding population, and the management of *ex situ* populations, should be carefully designed to prevent high levels of inbreeding.

3

The Application of Minimum Viable Population Theory to Plants

ERIC S. MENGES

GOALS OF POPULATION VIABILITY ANALYSIS

Conservation biology is a crisis-oriented science, yet faced with the difficult challenge of understanding and predicting the long-term dynamics of complex systems (Soulé 1985). At the population level, conservation biologists are largely concerned with population viability. The analysis of population viability holds the same challenge as conservation biology as a whole; that is, we need to understand the long-term viability (or survivability) of populations, but we must make immediate decisions on their protection and management. Therefore, as conservation biologists we must tolerate uncertainty and be prepared to make best guesses based on available data.

For endangered species, a major goal of population viability analysis is predicting extinction probability for the population of interest over a defined time period. From this extinction probability, one can derive as estimate of the minimum population size necessary to have an acceptably low extinction probability. This *minimum viable population* (MVP) is related to extinction probability because larger populations are better able than smaller ones to counter stochastic events that threaten populations with extinction. Myriad factors may have to be considered. In essence, we require detailed long-term autecological information (Burgman et al. 1988) to anticipate future threatening events. Obviously, this requirement conflicts with the crisis-management nature of conservation biology, and the definition of situations where short-cut analyses can be used remains a major practical need.

The impetus for the consideration of "minimum" thresholds in analyses of population viability lies in the general expectation that smaller populations are more likely than larger ones to go extinct for a variety of reasons (Pimm et al. 1988; Shaffer 1981). The theoretical roots for this conclusion were first explored in the equilibrium theory of island biogeography (MacArthur and Wilson 1967). Indeed, small population size predicts the risk of extinction among bird species (Diamond et al. 1987; Pimm et al. 1988).

However, even very large populations have finite extinction probabilities. Therefore, several different values for "minimum" viable population can be formulated, depending on "acceptable" probabilities of extinction. In addition, MVP and extinction probability values are likely to vary over different populations, in different environments, and given various assumptions for demographic and genetic parameters that cannot be measured precisely. As a result of this uncertainty, realistic population viability analyses will not generally conclude with a single MVP, and will instead analyze a series of scenarios given different assumptions. These general goals of population viability analysis are the same for any organism. However, MVP analyses for plants will necessarily differ from those used for animal populations, simply because of the differences between plants and animals. To date, the conceptual underpinnings of population viability analysis have been heavily influenced by experience with particular animal species. Therefore, a goal of this chapter is to examine critically the extent to which minimum viable population theory and its applications need to be modified when considering plant species.

COMPONENTS OF POPULATION VIABILITY ANALYSIS

The minimum viable population concept was first explained in detail by Shaffer (1981), who discussed various sources of uncertainty that may threaten populations with extinction. Systematic threats are often very dangerous to populations and imply obvious management responses: correction and control of these threats. Once the systematic imbalances are addressed, the analysis of stochastic factors becomes relevant. The focus of most scientists studying population viability has been the sources and effects of stochasticity.

Shaffer (1981) described four types of stochasticity. One type, *environmental stochasticity*, refers to variation over time in the population's operational environment (Mason and Langenheim 1957). Variation in the physical environment and variation in predators, parasites, diseases, and competitors are all part of environmental stochasticity. Modeling studies have concluded that environmental stochasticity is extremely important, affecting populations of all sizes (e.g., Sykes 1969; Cohen 1979; Leigh 1981; Goodman 1987b; Menges 1990). Greater environmental stochasticity is expected to increase extinction risk (Tuljapurkar and Orzack 1980). Populations with a given set of mean demographic parameters (survivorship, growth, and fecundity rates) will be subject to increasing extinction risk as the parameters vary more widely (this being a reflection of environmental stochasticity) (Figure 3.1). Variation in population size itself increases the risk of extinction among British land birds (Pimm et al. 1988), and variance in population growth rate through time is more crucial to population dynamics than density-dependent population growth (Goodman 1987b). Clearly, it is difficult to understand the effects of all sources of environmental stochasticity for any organism. However, by defining the environment operationally, one can analyze the effects of stochasticity by observing variation in demographic parameters within the same population over time.

Shaffer (1981) also described *natural catastrophes* as potential threats to the survival of populations. Periodic or random events such as fire, floods, disease outbreaks, drought, windstorms, earthquakes, and ice storms can have extremely important influences on the survival of populations and metapopulations (e.g., P.S. White 1979; Menges and Loucks 1984; Pickett and White 1985). In a systems sense, these natural catastrophes can be considered to be an extreme form of environmental stochasticity. The effects of natural catastrophes on populations can be difficult to predict because it

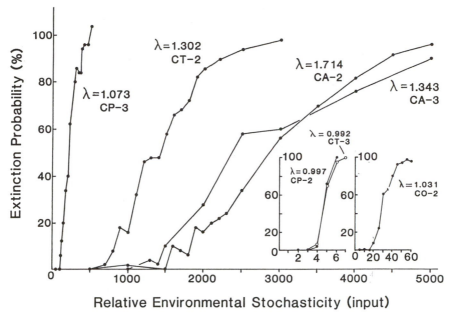

Figure 3.1. Extinction possibilities increase with the level of environmental stochasticity for four species of *Calochortus* studied by Fiedler (1987) and modeled by Menges (1991). CP = *C. pulchellus*; CT = *C. tiburonensis*; CA = *C. albus*; CO = *C. obispoensis*; 2 = 1982–3; 3 = 1983–4. Species with the lowest deterministic growth rate (lambda) are most sensitive to environmental stochasticity. Note that the two insets at the lower right have different *x* axes.

is difficult to sample enough of an entire disturbance regime to make specific predictions. Theory predicts that the exact effects of environmental disturbance will depend on the degree of autocorrelation and the speed of population response (Tuljapurkar 1985), but also suggests that natural catastrophes may be the greatest threat for many populations, necessitating populations of thousands or millions of individuals to ensure survival (Shaffer 1987). Future analyses of population viability will likely include scenarios that both include and delete the effects of natural catastrophes.

A third type of stochasticity, termed *demographic stochasticity,* involves a sampling problem of very small populations. Chance events affect the survival and reproductive success of individuals, even if overall demographic parameters remain constant. Individuals either survive or not; either successfully reproduce or not. As a result, in very small populations, demographic stochasticity may cause variation in population size independent of variation in the environment. The effects of demographic stochasticity are usually important only in populations of 50 or less (Pollard 1966; Keiding 1975). For example, modeling exercises used by Menges (1991b) on data of Pinero et al. (1984) compare environmental and demographic stochasticity (Figure 3.2). Thus demographic stochasticity does not generally act alone in causing extinction (Gilpin and Soulé 1986; Shaffer 1987).

Finally, *genetic stochasticity* needs to be considered in analyzing population viability. Small populations may be genetically depauperate as a result of changes in gene frequencies, owing to founder effects or inbreeding. If a population suffers from in-

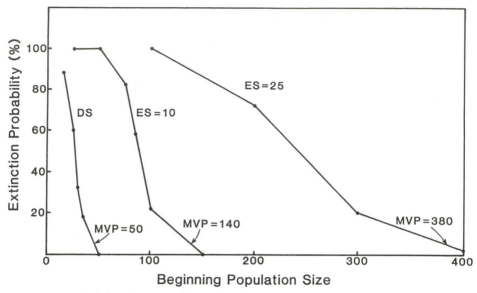

Figure 3.2. Modeling demographic versus environmental stochasticity effects on extinction and minimum viable population sizes (MVP). MVP is defined as less than 5%. Original data for *Astrocaryum mexicanum* at site BB. (From Pinero et al. 1984; modeling from Menges 1991b)

breeding depression, then its short-term viability may be compromised. The effects of inbreeding on populations have been used to recommend a general effective MVP of 50 individuals. In addition, genetic stochasticity may be implicated in the loss of long-term evolutionary flexibility (for discussions on plants, see Antonovics 1984; Harris et al. 1984; Ledig 1986b). The analysis of this loss provides the basis for an effective MVP of 500. Early analyses of genetic stochasticity resulting in both "magic numbers" (Franklin 1980) have been widely used in management considerations (e.g., Lehmkuhl 1984); such dependence has been criticized by several authors (e.g., Soulé and Simberloff 1986; Lande 1988b). The assumptions underlying the "magic numbers" do not likely hold for all organisms, and, as will be discussed later, some are highly unlikely for plants. Therefore, studies on specific organisms will be needed to understand the effects of genetic stochasticity on minimum viable population.

Finally, MVPs to buffer environmental stochasticity and natural catastrophes in wild populations are on the order of 10^3–10^6 (Shaffer 1987; and others). Attaining this population size will likely make genetic considerations secondary, given that little loss in heterogeneity is evident for effective population sizes less than 500 (Barrett and Kohn, Chapter 1).

INTEGRATIVE MODELS

The integration of the various types of stochasticity and their effects on population viability remains a challenge for conservation biology. In general, most of the early attempts have involved conceptual or heuristic models. For example, Gilpin and Soulé

(1986) describe vortices of species extinction resulting from the interaction of various types of stochasticity with population characteristics such as morphology, physiology, behavior, spatial distribution, age and sex structure, and demographic parameters. Although their analysis clearly shows that population extinction can be a complex process and that many of the types of stochasticity have interactive and positive influences on one another, their model cannot predict the viability of particular populations. Instead, it provides a framework by which future studies can be structured. Unfortunately, future studies are unlikely to consider all the factors that Gilpin and Soulé (1986) have suggested as potentially important.

More mechanistic models of population dynamics and extinction for specific plant species are being formulated, although none have considered multiple stochastic factors. The effects of environmental stochasticity have been modeled for Furbish's lousewort (*Pedicularis furbishiae*), using a stochastic modeling approach (Menges 1990). Population projections, based on traditional deterministic matrix-projection techniques (Leslie 1945), are modified to be structured by life-history stage, rather than age (Lefkovitch 1965; Caswell 1982a, 1982b). For example, transitions among six life-history stages between 1984 and 1985 show that changes among many stages are possible (Table 3.1). Different transition matrices were formed from other years, for individual populations, and for plants growing in specified environments (Menges 1990). Data are based on monitoring of individual plants as well as on more detailed observations and experiments on seed production and seed germination. To produce an empirically based environmental stochasticity, matrices representing different time periods were randomly alternated. Many populations of Furbish's lousewort are not viable, and conditions favoring persistence of individual populations are ephemeral (Menges 1990).

Population viability analysis, including numerical estimates of MVPs, do require substantial empirical data and understanding of links among environmental variability, demography, and genetics. In some cases, this analysis may be essential to design conservation strategies. However, most species will be protected by habitat preservation and conservation strategies based on available autecological information. Population viability analyses of a few species may help to indicate what sorts of auteco-

Table 3.1.   Transition Matrix, 1984–1985, for All Populations of *Pedicularis furbishiae*

	From seedling	From juvenile	From vegetative	From reproductive		
				Small	Medium	Large
To seedling	—	—	—	2.45	7.48	29.93
				(1.64)	(5.02)	(20.08)
To juvenile	0.39	0.47	0.14	0.09	0.04	—
	(0.03)	(0.02)	(0.02)	(0.01)	(0.02)	
To vegetative	0.01	0.21	0.24	0.24	0.16	0.01
	(0.00)	(0.02)	(0.02)	(0.04)	(0.02)	(0.00)
To small reproductive	—	0.11	0.45		0.26	0.00
		(0.02)	(0.03)	(0.04)	(0.03)	(0.00)
To medium reproductive	—	0.00	0.11	0.21	0.42	0.28
		(0.01)	(0.02)	(0.04)	(0.03)	(0.00)
To large reproductive	—	—	—	0.01	0.10	0.61
				(0.01)	(0.03)	(0.00)

Note: Values in parentheses are standard errors over 15 populations. Further information can be found in the source.
Source: Menges (1990).

logical data are most relevant to similar species. To date, MVP studies on plants have not considered all the possible types of stochasticity that may affect populations. However, a few well-studied animals have been analyzed for several types of stochasticity in pioneer studies of population viability. In the following section, I summarize some of the main points to be gained from these analyses.

VIABILITY ANALYSES FOR ANIMALS: EXAMPLES AND TRENDS

Grizzly Bear

Perhaps the best-studied organism for population viability is the grizzly bear (*Ursus arctos* L.). Population viability analyses for grizzly bears consider environmental and demographic stochasticity using age- and sex-structured simulation models based on data sufficient to estimate levels of environmental stochasticity (Shaffer 1978; Shaffer and Samson 1985). The simulations predict that grizzly bear populations of 50–90 would be viable (defined as a 95% probability of survival for 100 years). The average extinction time for a population of 50 bears is 114 years, reflecting a skewed distribution with a few populations surviving a long time (see also Goodman 1987b). The current Yellowstone National Park grizzly bear population has a 64% chance of extinction in the next 100 years (Shaffer and Samson 1985). An independent stochastic simulation by Knight and Eberhardt (1985) finds a 47% chance of extinction over the same time period.

Genetic effects on grizzly bears are just now being considered (Samson et al. 1985; Allendorf and Servheen 1986). Augmentation of gene flow into isolated populations of grizzly bears is now being recommended, as are integrative models incorporating both the genetic and demographic consequences of small populations. Even small populations of grizzly bears require vast areas of land, implying that the largest U.S. parks will not support viable populations (Newmark 1985).

Spotted Owls

Spotted owls of the Pacific Northwest require areas of old-growth forest for survival. Demographic models attempt to predict minimum (old-growth) habitat requirements for survival of the spotted owls (Lande 1988a). These models include classic demographic models of deterministic population growth and the sensitivity of rate of increase to changes in demographic parameters. Lande (1988a) also considers modest temporal stochastic variation of certain population parameters (although little quantitative data are available to support specific levels of environmental stochasticity). Individual populations under these scenarios are fairly stable in size, owing in part to long life span, which buffers environmental stochasticity (Goodman 1984; Menges 1991b).

The viability of the spotted owl metapopulation (a group of interacting populations) is of more importance than the survival of particular populations. Not all potential spotted owl territories are utilized at any one time. Simple algebraic models can be used to describe the proportional occupancy of potential territories as a function of individual population demography, dispersal, and suitable habitat (Lande 1988a). However, proportional occupancy data, rather than actual dispersal data, are employed. Extinction of the spotted owl metapopulation is predicted if old-growth forests are reduced below 21% of the total area. Management plans by the U. S. Forest Service currently call for considerably less acreage in old-growth forests (7–16%).

Butterflies

An extensive and long-term study on the ecology of checkerspot butterflies (*Euphydryas* spp.) forms the basis for studies on the viability of individual populations and metapopulations (Ehrlich and Murphy 1987; Harrison et al. 1988). Because environmental stochasticity is extremely effective in causing extinctions of local populations, the key to survival of a species as a whole appears to lie in metapopulation dynamics. Unlike the spotted owl, checkerspot butterflies require more than one type of habitat patch for metapopulation survival. Thus extinction is common in large areas that lack topographic diversity. However, areas with both mesic north slopes and warmer south slopes allow larval development in most years, and subsequently population increases. Such areas have acted as reservoirs from which emigration has supplied butterflies to smaller, less topographically diverse, extinction-prone habitat islands (Harrison et al. 1988). Individual butterfly populations have gone extinct, but dispersal and colonization of different habitats in the same area allow long-term survival of the species. Ehrlich and Murphy (1987) have concluded that a minimum viable metapopulation approach to conservation may be needed for many species.

Genetic Studies

The viability of several animal species has been inferred from studies of genetic variation within species. Although these studies do not provide an integrative framework for understanding population viability, they suggest that small populations may be compromised by extremely narrow genetic diversity. For example, elephant seals exhibit no polymorphism in electrophoretic loci, based on blood sampling in five breeding colonies (Bonnell and Selander 1974). Negative effects of low genetic variation have been found for cheetahs (O'Brien et al. 1985) and lions (Wildt et al. 1987).

General Trends in Animal Studies of Population Viability

Most studies of population viability in animals have been unable to consider all possible threats to populations. This is inevitable, considering the difficulty in collecting sufficient data, and is also likely to be a problem in future viability studies on plant populations. Of particular concern is the difficulty in collecting detailed demographic data over long time periods to quantify levels of environmental stochasticity. Such data have been collected for only well-studied animals, such as the grizzly bear.

In studies of population viability in animals, the effect of land areas or habitat area seems paramount. This makes sense in light of the fact that many animals hold specific territories or need minimum areas to support individuals, families, or populations. Density-dependent population regulation may be important in many animal species. Analyses of population dynamics in grizzly bears have been extrapolated to nature-preserve design by considering the amount of land required to support the grizzly bears (Newmark 1985). In addition, many studies use information on the occupancy of patches of various sizes (incidence functions) to infer minimum island or habitat sizes required to support populations (Diamond 1975). Patterns of patch occupancy have also been used to infer extinction times for fragmented metapopulations—for example, in spruce grouse (Fritz 1979) and spotted owls (Lande 1988a). Several ecologists have also modeled extinction based on reserve size (Wright and Hubbell 1983) and territoriality (Lande 1987).

Studies of population viability of animals need to consider explicitly the *Allee effect*—whereby very small population numbers reduce the ability of animals to find mates and reproduce. The minimum density of populations and the resulting effect on extinction have been related to the Allee effect by several researchers (e.g., Phillip 1957; Dennis 1981). A review of population viability analysis identified the Allee effect as one of the major demographic influences on extinction (Lande 1988b).

PLANT TRAITS WITH IMPLICATIONS FOR MVP ANALYSES

There are no complete studies of the effects of various kinds of stochasticity on plant population viability. Nonetheless, it is clear that analyses of minimum viable population size for plants will almost certainly differ considerably from those for animals. Many of the trends in analyses of animal population viability will probably not be applicable to many plant species. For example, the key role of habitat area may be redefined to include microhabitat availability, and density-dependent effects may be unimportant for plants growing in mixed stands. The Allee effect will probably hold only when animal mutualists are involved, and many genetic effects will be discounted or be more complex. In addition, there are a number of plant traits that have specific implications for the way viability assessments will proceed. They are outlined in this section.

Small Neighborhood Sizes

Adult plants are sessile organisms, unlike most animals. Seeds, pollen, and vegetative dispersal units are the only motile stages in a higher plant's life cycle. Pollen dispersal and seed dispersal are often limited in spatial extent, with most seeds falling close to the parent and most pollen moving between closely spaced individuals, creating small neighborhood sizes (e.g., Meagher 1986; Golenberg 1987). Populations structured into small neighborhoods will be more inbred, with lower within-population genetic variation and more divergence among populations (Templeton, Chapter 12), than those in larger neighborhoods. Certain breeding systems tend to produce small neighborhood size, and breeding system does explain a portion of the distribution of genetic variation at several levels (Hamrick et al., Chapter 5). The viability of individual populations will be influenced by genetic structure and gene flow across the landscape.

Disturbance Ecology

The sessile nature of plants increases their vulnerability to disturbance events, including very fine-scale disturbance that may cause death or injury to individual plants. (However, underground parts of many plants can survive certain disturbances by resprouting, allowing populations to persist [e.g., Stohlgen and Rundel 1986].) Both large-scale catastrophes, such as fire, and small-scale catastrophes, such as small windthrows and locally spreading diseases, will affect plant populations. Analyses of plant population viability will need to explore disturbance processes at a variety of scales both within and between plant populations.

Microsite Specialization

Plants experience the environmental mosaic in a very fine-scale way, being unable to average environmental variation over large areas. Not surprisingly, environmental variation affects demographic parameters, such as rates of mortality, growth, and reproductive output among individuals (e.g., Klemow and Raynal 1981; Sarukhan et al. 1984; Schmidt and Levin 1985) and among populations (e.g., Werner and Caswell 1977; Pinero et al. 1984). Individual differences in demographic parameters (within the same time period) increase both equilibrium population size and population persistence (Holsinger and Roughgarden 1985). Thus microsite variation can act to buffer populations from environmental stochasticity. Microsites are especially important for seeds, seedlings, and small plants. The concept of a safe site or regeneration niche for seed germination and seedling establishment has proved very useful in plant ecology (Grubb 1977).

Microsite variation implies that assessments of population viability should ideally be based on environmental states and variation at the scale of individual plants. Unfortunately, this effort may make precise assessment of population viability quite time-consuming in fine-grained environments and may be an unaffordable luxury in most circumstances. However, although the requirement for fine-grained environmental information seems daunting in analyzing plant, as compared with animal viability, it may also be an advantage. The sessile nature of plants allows more precise characterization of local operational environments than is possible for animals. Such characterization of how local environments affect population dynamics and thus population viability will directly suggest novel management strategies for increasing the viability of plant populations. For example, environmental changes caused by prescribed fire alter demographic parameters of many plant species (e.g., Abrahamson 1984; Knapp and Hurlbert 1986).

Not all species will be able to utilize multiple microsite types and be buffered from environmental stochasticity. Such niche breadth will require either genetic variation or phenotypic plasticity. Narrowing of microsite-type occupancy (through selection) may contribute to an overall decline in species genetic variation. This may be a factor in low population viability of rare species that are edaphic specialists. Edaphic specialists may have small population sizes, and thus reductions in genetic variation due to inbreeding and genetic drift further reduce population viability.

Microsite specialization has many important practical ramifications for conservation biology. It is difficult to apply biogeographical concepts for nature-reserve design to the distribution of plant species that grow in specific microsites. For plants, the number and distribution of microsites may be more important considerations than total reserve area. This may be why many small reserves will support as many plant species, and even as many endangered species, as fewer large reserves of the same total size (Jarvinen 1982).

Phenotypic Plasticity

Perhaps as a consequence of being restricted to microsites, plants have evolved re-markable levels of phenotypic plasticity (Schlichting 1986), which may buffer popu-lations from environmental variation. In heterogeneous environments, plasticity could also compensate for low genetic variation, but it is unclear how often this is important (Huenneke, Chapter 2). Plasticity is enhanced by modular architecture, which allows indeterminate growth by reiteration of modules (J. White 1979). In predicting effects on population dynamics, plasticity can be represented by a negative covariance struc-ture among elements of a projection matrix, which mathematically buffers environ-mental stochasticity and increases population viability (see Caswell 1983). For exam-ple, reproductive output could increase with light levels, whereas survival is higher under shadier conditions. This plasticity could allow occupancy of and expansion in a spatially heterogeneous site and also buffer the effects of variation in light over time, due to gap formation and closure.

Clonal Growth: Strategies and Plasticities

Clonal growth in plants in another aspect of an indeterminant growth strategy. There are a range of mechanisms for clonal growth in plants, including rhizomes, stolens, root shoots, and offshoots. Clonal organisms exhibit characteristics and dynamics at two levels: that of the genetic individual, or *genet,* and that of the clonally produced module, or *ramet* (Harper 1977) (Table 3.2).

Clonal growth in plants complicates classic demographic analyses. First of all, expensive and time-consuming genetic procedures such as electrophoresis are required to define genetic individuals. In addition, it is difficult to define physiological individ-uals in clonal plants because physiological integration varies among clonal plant spe-cies. Physiological connections vary in duration, and the degree to which young or disadvantaged ramets are provided with photosynthate, nutrients, or other resources by other ramets varies among and within plant species across environments (Ashmun and Pitelka 1984; Salzman and Parker 1985).

Table 3.2. Comparison of Characteristics of Genets with Ramets

Characteristic	Genets	Ramets
Single-cell stage	Zygote	Meristematic cell or none
Resting and dispersing stage	Seeds, sexual spores	Meristem (buds, asexual spores)
Limitation to size	Possibly none (reiteration of ramets)	Usually limited
Reproduction	Sexual	Clonal growth
Alternating generations	Yes	No
Potential for exponential increase in number	Yes	Yes
Competition between individuals	Yes	Yes
Genetic variation between individuals	High (recombination, mutation)	Low or none (somatic mutation, recombination)

Source: Modified from Hamilton et al. (1987).

Despite the fact that individual ramets in clonal plants are not genetic individuals and may or may not be physiologically independent, demographic projections can be made at the level of the individual ramet. Population viability assessments based on demographic components will probably consider numbers of ramets and minimum viable populations of ramets. In contrast, genetic analyses of clonal plants will probably be based on the minimum number of genetic individuals or genets. Thus demographic and genetic approaches to minimum viable population size may necessarily be different from each other for some clonal organisms.

At face value, this difference between the two approaches might be disturbing. However, the actual effects of clonal growth on population viability parallel this methodological problem. For example, jack-in-the-pulpit (*Arisaema triphyllum*) produces offspring both sexually and clonally (Bierzychudek 1982). Its sensitivity to environmental stochasticity was buffered as compared with nonclonal species (Menges 1991b). Similarly, clonal reproduction in Yucca (*Yucca glauca*) is negatively associated with fruit set, buffering oscillations in population dynamics that would result from poor pollinator service (Kingsolver 1986). In these cases, clonal growth serves to buffer ramet populations from environmental stochasticity that otherwise would threaten local extinction.

Variation in allocation to clonal reproduction is but one aspect of potential plasticity in clonal plants. For example, different parts of the same genet may vary in growth form in such a way as to exhibit divergent "foraging" in environments with different resource levels (Slade and Hutchings 1987). Such foraging may be enhanced by physiological exchange among ramets in different microenvironments (Salzman and Parker 1985). In clonal plants, the number of ramets, especially physiologically independent "adult" ramets, may be the "minimum" to consider in understanding short-term population viability. In contrast, for assessing long-term evolutionary potential, understanding the distribution of genetic variation among genes is probably important.

Spatial Aggregation and Metapopulations

Motile animals that move through the landscape often achieve regular spatial dispersion, using behavioral adaptations such as territoriality to enforce these distributions. As described earlier, the spatial regularity of some animals allows viability analyses to translate animal numbers to landscape areas required to support minimum viable populations.

Plants, however, do not always attain a regular distribution. Plants tend to be distributed in aggregated fashion over the landscape because of limited dispersal and the sessile nature of the adults. Competitive effects are predictable largely as the effects of close neighbors (Mack and Harper 1977; Waller 1981). As competition causes mortality, extremely aggregated patterns of seedlings and small plants may become more random or occasionally regular (Pielou 1960; Leps and Kindlman 1987).

The overall lack of regular spatial pattern and the extremely large variation in naturally occurring densities complicate the analysis of spatial requirements for viable plant populations. Analyses of MVP in plants will more often consider numbers of individuals of various ages or sizes than the amount of area available in the landscape.

Because plant species are patterned at several scales, metapopulation analyses of plants will very often be valuable. An example is a study of the heart-leaved plantain (*Plantago cordata*) in two watersheds in Illinois. Populations in an agriculturally altered watershed suffered local extinction without recolonization, owing to disturbance-mediated habitat loss. The metapopulation in an unaltered watershed, although dynamic, was not declining (Bowles and Apfelbaum 1989).

Biologists considering plant population viability will need to obtain statistics for both within-population and between-population processes. Between-population rates such as immigration, emigration, and gene flow will define the boundaries of interacting populations (metapopulations) or indicate where species may form single functioning populations even when spatially fragmented. Genetic links among metapopulations will act favorably for population viability because the contribution of even one or a few migrants per generation can reverse the loss of genetic variation in small populations, although at a cost to total variation in the metapopulation (Lacy 1987; Lande and Barrowdough 1987). Newly subdivided populations with altered gene flow may be at particularly high risk of genetic rearrangements (Templeton, Chapter 12) that could decrease fitness.

Demographic movement of propagules among populations may allow persistence of metapopulations even when individual plant populations are not viable (Harrison et al. 1988; Menges 1990). Metapopulation structure also has the potential to feed back to influence individual population survival. For example, more isolated populations have lower survival probabilities than partially connected populations (Fahrig and Merriam 1985).

For a species, fates of local populations may not be sufficient to predict overall viability. Subdivided metapopulations may be buffered from the negative effects of environmental variation (Gilpin 1987; Goodman 1987a; Quinn and Hastings 1987) if catastrophes do not simultaneously destroy linked populations. In the endemic Furbish's lousewort (*Pedicularis furbishiae*), catastrophic mortality of whole populations by riverine disturbance is common (Menges and Gawler 1986) and is more important than within-population processes in affecting overall population dynamics (Menges 1990). Therefore, conservation efforts will need to consider encouraging conditions that favor a positive balance between establishment of new populations and local extirpation (Menges 1990). These probably would involve maintenance of appropriate hydrologic regimes and land uses in the watershed.

Similarly, metapopulation dynamics of checkerspot butterflies have been analyzed by Harrison et al. (1988). Here, some dispersal data, together with analysis of habitat suitability, occupancy, and distribution, established the spatial extent of interacting populations (metapopulation) and distance effects on patch occupancy. In Santa Clara County, California, butterfly occupancy of individual patches required threshold levels of habitat quality and shorter distances from a large reservoir population.

Lande's (1987, 1988a) metapopulation analysis for spotted owls, previously described, may provide a useful approach for demographically based metapopulation models for plants. In his approach, dispersal data are not directly estimated or used. Instead, the percent of suitable patches occupied, together with classic demographic statistics, allows an estimate of the minimum number of patches required to sustain a metapopulation. This treatment acknowledges that all suitable patches will not be occupied at any one time, even under a regional equilibrium. Because many plant species require patchy specialized habitats or now occupy patches of human-fragmented habitat, metapopulation analyses will be required to complete a picture of overall species viability.

Resource Generality and Niche Definition

Many animals have specific requirements for food or nesting sites that simplify analyses of population viability. For example, analyses of spotted owl population viability are very strongly tied to the number and distribution of old-growth forest remnants, owing to nesting and foraging requirements. Analyses of population viability in animals will often be simplified by reference to limiting resources such as specific host plants or nest sites.

In contrast, niche dimensions for plant species may be harder to quantify, may overlap, and may involve temporal partitioning, disturbance gradients, or other life-history differences. Where plants differ in components of physically or chemically based niche, such niche dimensions may be difficult to quantify. For example, analyses of soil nutrient requirements may involve extensive and expensive sampling. In addition, quantifying the niche of the plant on a disturbance–succession gradient may require extensive landscape sampling. Many plants are active only at certain times of the year or only in certain years. This partitioning of a temporal gradient may require long-term studies.

Because plants use more general resources than animals (light, water, space, nutrients) and because many of the niche dimensions important for plant species are difficult or time-consuming to quantify, understanding resource requirements and their effects on extinction may be challenging. Such analyses certainly will depend very heavily on previous studies. In the final analysis, it may be difficult to relate population viability of plants to specific resources at all.

The Importance of Size

Plant growth is largely indeterminant because of the availability of meristematic tissue in roots, rhizomes, and stems. Because of this indeterminant growth, many plant individuals do not age in the sense that animals do. Most viability analyses of animals have used age as a useful covariate of demographic parameters and of fitness in general. Classic demographic analyses are based on age-structured populations (Leslie 1945), and these analyses can provide very useful measures of population dynamics and fitness components. For plants, age is not generally such a useful measure and often cannot be quantified nondestructively or at all. Therefore, classic age-structured demographic analyses will not be useful for many plant species.

For plants, size is often the superior predictor of the demographic parameters of individuals (e.g., Werner 1975; Gross 1981; Lacey 1986), and size structures of tree populations have long been used in predicting future dynamics (e.g., Goff and West 1975; Enright 1982). Size advantages often begin as early as larger seeds (Stanton 1985). Fortunately for plant ecologists, size-based demographic models have now become available (Lefkovitch 1965). Such models can provide integrative measures of population growth and fitness components and be used as a basis for modeling population viability (Caswell 1982a, 1982b). Analyses of the demography of plants and their population viability have considered (Menges 1990) and will consider size-based or stage-based classification systems.

Cryptic Phases in Life History

Both animals and plants may progress through stages in the life cycle of individuals that are difficult to observe or understand. In plants, many life-history stages do not allow easy estimation of mechanistic demographic data such as rates of mortality, growth, and reproduction. For example, many perennial plants (e.g., orchids) may exist for several years as dormant underground individuals (Gilbert and Lee 1980). This sort of dormancy is difficult to distinguish from mortality during short-term studies.

Seed dormancy will require special attention in population viability studies. Dormant seeds and variation in properties of seeds in a single population may buffer populations from environmental stochasticity and thus augment population viability (Cohen 1966).

Mechanistic understanding of fecundity in plants may be difficult in several species. Included in this list would be species with very small seeds whose fecundity is difficult to quantify. Understanding what phases of fruit production are limiting helps in designing mitigation strategies and in engineering species recovery, but such studies are not feasible for every species. Finally, certain plant species have complex life cycles, with free-living stages that are difficult to study. These would include lycopod gametophytes, fern gametophytes, and early stages in establishment of lichens.

Overall, these difficulties would imply that our best initial attempts to assess population viability will be based on the biology of convenient plants, most likely angiosperms and conifers. It also implies that population viability assessments of certain species will require experts in those groups and that, even so, the mechanisms of population viability for some species will remain opaque.

Mutualisms Specific to Plants

Just as host plants may be important to understanding viability of animals, mutualists of many plant species may be prime determinants of their population viability and distribution. Many plant–pollinator and plant–disperser relationships are highly specific. Loss of pollinators or dispersers or decreases in their population sizes may have important effects on plant viability. In some cases, long-term viability may already be nil as important mutualists have become extinct (Janzen and Martin 1982). Particular mutualists that are important to many species are known as keystone mutualists, and their extinction would also bring down many other species (Terbourgh 1986). To understand population viability where mutualists are involved will first require knowledge of the natural history of both mutualists. In many cases, analyses of MVP as affected by mutualism are likely to be complicated and time-consuming.

Diversity in Breeding Systems and Life History

Plant species have a staggering array of breeding systems and life histories. Breeding systems of plants will affect population dynamics and the formation and maintenance of genetic diversity. Both will affect extinction probability (Karron, Chapter 6). Any generalizations about plant population viability extrapolated from one species to another will need to carefully consider similarity in breeding systems. Plasticity in life history will also affect population persistence in stochastic environments. In facultative biennials (semelparous species that vary in life span or time before reproduction, depending on growth rate and resulting size [Werner 1975]), delay of flowering can decrease the rate of extinction in variable environments (Klinkhamer and deJong 1983). Therefore, facultative (vs. obligate) biennials have higher MVPs in a more variable environment. In general, plasticity in life-history components and breeding systems could buffer plants from environmental stochasticity.

Potential for Selfing

Many plant species are capable of producing self-seed and may even have specific selfing mechanisms. For example, amphicarpy ensures some seed production even under difficult conditions (Cheplick 1987). In species capable of selfing, fecundity will not be limited by pollinators bringing pollen from other individuals. The ability to self and the occurrence of high levels of selfed seed will increase the demographic components of population viability by ensuring or increasing seed set. It will also buffer environmental stochasticity due to variation in pollinator service. However, selfing populations will have lower heterozygosity and may have lower population viability in the long run. Like clonal growth, selfing may be a two-edged sword with tradeoffs for population viability measured over short versus long time periods.

Effects of Inbreeding on Genetic Stochasticity

Low levels of within-population genetic variation in small isolated populations are predicted by theory and are worrisome to conversation biologists. In many, but not all, plant species, small and isolated populations and species contain low levels of within-population variation (e.g., Ledig and Conkle 1983; Loveless and Hamrick 1984; Karron 1987a; Waller et al. 1988). A key question is whether such narrowness in genetic variation has negative effects on population viability. The analysis of genetically based MVP rests on the link among small effective population size, low genetic variation within populations, and negative effects of low genetic variation on demographic parameters such as survivorship and fecundity. Such effects have been demonstrated in animals, but few studies have been conducted in plants.

Many plant species have a history of inbreeding due to specific adaptations for selfing, persistent small neighborhood sizes (Ehrlich and Raven 1969; Levin 1981), and strong local differentiation (e.g., Linhart 1974; Turkington and Harper 1979). Thus many deleterious recessive genes have been exposed to selection and subsequently lost from individual populations. Most genetic variation will exist between, not within, such populations. This effect will be further pronounced when populations have undergone lengthy bottlenecks (Barrett and Kohn, Chapter 1; Templeton, Chapter 12). Therefore, one might expect little subsequent loss in population viability in small populations of a previously inbreeding species. In a sense, plant species may often be adapted to further inbreeding. Adapting Speke's gazelle (*Gazella spekei*) to high levels of inbreeding (rather than avoiding it) has been successfully used in a breeding program for this very rare animal (Templeton and Read 1983). However, inbreeding depression actually does occur in plant populations with a variety of breeding systems and histories (Barrett and Kohn, Chapter 1; Huenneke, Chapter 2), and it may well be that inbreeding depression is a possibility even for inbred plant species. Because of genetic constraints, the relationship of breeding systems to inbreeding depression is not as simple as once thought (Holsinger 1988b).

Although long-term, evolutionary response may be hampered in populations with little genetic variation, abundant plasticity has allowed many purely asexual taxa (e.g., dandelions) and largely asexual species (e.g., bracken fern) to be quite successful for fairly long periods of time. Therefore, short-term analyses of population viability will probably more often strongly consider demographic than genetic components. The emphasis on demography is also being made in reviews of viability assessments for animal species (Lande 1988b).

IMPLICATIONS FOR CONSERVATION

A whole series of conservation activities (including management of *in situ* metapopulations and populations, management of *ex situ* populations, reintroduction, and nature-reserve design) would benefit from assessment of minimum viable populations (MVP), using population viability analyses, for individual species. This chapter has outlined a series of plant traits (or traits well expressed in plants) that will tend either to buffer plants from stochastic factors (traits such as adaptability to disturbance, microsite variation, phenotypic plasticity, indeterminate growth, clonal growth, resource generality, metapopulation structure, seed dormancy, ability to set selfed seed, tolerance to inbreeding) or to enhance their vulnerability (traits such as small neighborhood size, sessile growth habit, edaphic specialization, population isolation, specific mutualisms, inbreeding depression).

Unfortunately, viability assessments are likely to be complex, and our current understanding of the interaction of demographic and genetic effects is fragmentary. As more complete analyses are made, it may be increasingly possible to use a few key indicators of population health to make quick, preliminary estimates of MVP. Until such research is accomplished, we are left with informed speculation. Based on the current state of MVP theory and the special traits of plants, the implications for plant conservation include the following:

1. Environmental stochasticity and natural catastrophes will be the primary threats to most *in situ* plant populations that are not already threatened by systematic trends. These types of variation imply that MVPs may be on the order of 10^3–10^6 individuals (Shaffer 1987). Full MVP analyses will be needed most in species subject to environmental stochasticity and natural catastrophes. The assessment of the role of natural catastrophes on plant population dynamics and extinction will require a view toward long time frames and ecosystem context.

2. Genetic stochasticity will be less important in most viability analyses than those involving environmental variation. That is, MVPs necessary to buffer environmental stochasticity in most *in situ* populations will be sufficient to protect the genetic integrity of plant populations. Although ample theory exists regarding the relationship of population size to inbreeding, the actual effects of inbreeding on plant population vi-

ability components are difficult to demonstrate, and generalizations are difficult to make.

3. Demographic stochasticity can be largely ignored. Environmental and genetic stochasticity predicates population sizes that will forestall the negative effects of demographic stochasticity in natural populations. However, if these factors are minimized in very small *ex situ* populations, demographic stochasticity must be considered.

4. For *ex situ* plantings, populations of 10^3–10^6 will usually be unattainable. However the reduction of environmental variation that should be possible in botanical garden settings will reduce necessary MVPs. No analyses of MVPs to counter environmental stochasticity and natural catastrophes in garden populations have been made, but it seems reasonable to expect 10^2–10^4 plants to counter uncontrollable weather, herbivores, and pathogens. Genetic considerations may be relatively more important in *ex situ* populations of plants than in larger wild populations. Clonal propagation will help preserve genotypes in the face of environmental variation.

5. For *ex situ* seed storage, where environmental variation is negligible, population size considerations will focus on sampling genetic variation and attaining MVPs based on genetics. Care should be taken to sample variation from different genetic neighborhoods and microsites.

6. Concepts involving minimum island size and Allee effects, central to animal population viability, will not be important in plant assessments.

7. Because plants are sessile, microsite requirements will need to be understood to assess population viability and help engineer recovery. The patterning of catastrophic events and disturbance regimes in general will need to be considered as well.

8. Phenotypic plasticity in growth and life history allows plants to buffer a great deal of environmental variation; this will be an advantage in conservation efforts.

9. Specific viability studies of plants will explicitly consider important traits of plants, such as subtle differences in resource use, size-dependent demography, crucial life-history phases, specificity of plant–pollinator and plant–disperser relations, ability to produce selfed seed, and clonal growth. In clonal species, MVP analyses will proceed at both ramet and genet levels.

10. Metapopulation analyses, perhaps based in part on proportional occupancy of suitable microsites, will be important in understanding plant population viability. Metapopulation considerations involving both demography and genetics will be included in MVP analyses.

ACKNOWLEDGMENTS

In preparing this chapter, I have benefited from discussions with Laura Huenneke, Bruce Pavlic, Marlin Bowles, Marcella DeMauro, Douglas Schemske, Mark Deyrup, Kevina Vulinec, Donald Falk, and Kent Holsinger. Christie Prinz, Patricia Bache, and Marcia Moretto helped prepare this manuscript. I also thank the other authors of this book, moderators Leslie Gottleib and Kent Holsinger, and Donald Falk for contributing to the conference discussion that improved this paper.

4

Conservation of Rare Trees in Tropical Rain Forests: A Genetic Perspective

K. S. BAWA and P. S. ASHTON

Conservation of tropical biota is one of the most pressing environmental issues of our time. Whereas many species of organisms are endangered with extinction in the temperate zone, species are becoming extinct every day in the tropics. For example, Myers (1988) has identified ten hotspot areas around the world that have unusually large proportions of endemic plant species and where the rates of deforestation are also very high. Myers estimates that forested areas in these hotspot regions, which contain 34,000 endemic species, will be reduced to 10% of the original area during the next 20 years. Using species–area relationships, Myers calculates that this 10% area will ultimately contain only 17,000 of the existing 34,000 endemic species. The rate of extinction of plant species in these ten hotspot areas alone, which account for only 3.5% of the existing tropical forests, according to Myers, will be two species per day. Myers's estimates of extinction might seem excessive, but others have come up with similar figures. Simberloff (1986), again, on the basis of species–area relationships, estimates that the number of plant species at equilibrium in the area projected to remain forested in the Neotropics by the end of the century would be 14,000 fewer than the present number of 92,000 species. Under higher rates of deforestation, the number of species lost would be 60,000.

Thus the protection of species that face imminent extinction is a matter of high priority. However, conservation and management of biodiversity even in protected areas in the tropics are challenging and cumbersome tasks for a wide variety of reasons.

1. The patterns of distribution and abundance of species in most tropical areas have not been well documented. For instance, in the British Isles every plant species has been mapped on a 10 × 10 km grid in a computerized system, whereas tropical areas have not been mapped even on a 100 × 100 km grid (Prance 1984). Thus the habitat and geographical ranges of species are poorly known. Approximately 13% of tree species in Australian rain forests, for example, are known from only one site and another 10% from two sites (Tracey 1981). The number of species per unit area is so large that new species continue to be discovered even from the most intensively studied areas.

2. Wherever detailed studies have been conducted in tropical rain forests, an extraordinarily large proportion of species have been found to be rare in the sense that they exist in very low population densities. For example, Poore (1968) showed that in a Malaysian rain forest sample of 22.3 hectares (ha) where all trees over 28 cm in diameter at breast height (dbh) were marked, 38% of the species were represented only once. In a 1-ha plot in Brazil, the total number of individuals equal to or greater than 10 cm dbh was only twice the total number of species (Gentry 1988). In a data set from 665 plots, each of 0.2 ha in size and spread over a range of the lowland forests of Malaysia, only 1014 of the 4005 species occurred in ten or more of the plots (Ashton unpubl. data).

3. Almost all tropical plants engage in complex mutualistic interactions to complete several phases of their life-history cycle, starting from association with mycorrhizae for initial establishment to pollination and seed dispersal by a wide variety of animals. The mutualists also interact with other species in the community, adding further structural complexity to the interactions. Thus conservation of tropical plants must take into account a variety of other animal (and plant) species with which they interact directly or indirectly.

4. Information on population genetic structure, which is central to the design of appropriate conservation strategies, is poorly developed or not available at all for most tropical forest plant species.

Overall, then, some unique biological features of tropical forest plants as well as the threat of imminent extinctions make the problem of conserving such biota more complex than that of other organisms.

This chapter is concerned with rarity and the consequences of rarity on the reproductive output and the genetic structure of populations. The focus is tree species in tropical lowland forests. We first examine different types of rarity, then review the genetic consequences of various patterns of abundance, and finally explore the *in situ* and *ex situ* conservation strategies for conservation and management of rare populations of tropical trees.

RARITY IN TROPICAL PLANTS

There is no evidence that tropical forest ecosystems differ *qualitatively* from other ecosystems. Rather, their interest lies in their manifestation of the extremes of some ecosystem phenomena. What interests us here is their extraordinary richness of species, which has an obvious corollary that because space is finite, species populations of trees are *on average* dispersed in lower population densities and have fewer total individuals than tree species in the temperate zone forests. In that sense, the species are rarer, and, except for habitat generalists, individual populations tend to be smaller. In principle, then, tropical forests will be among the most difficult ecosystems in which to conserve biodiversity, because proneness to extinction tends to be related to population size (MacArthur and Wilson 1967). Populations with low density will also require larger areas for their safe conservation.

There are, however, many kinds of rarity, and the gross generalizations outlined above do not apply equally to all types. Kruckeberg and Rabinowitz (1985) have pointed out that rarity is influenced by three aspects of spatial distribution: whether species have wide or narrow geographical ranges; whether they have wide or narrow ecological ranges; and whether the populations always exist in low effective density throughout their geographical range, or in concentrations of high density with scattered individuals in between. Schoener (1987) called these last two alternatives *suffusive* and *diffusive rarity*, respectively. To this should be added populations that, for reasons intrinsic to the biology of the plant, are usually evenly dispersed, or aggregated into clumps that accumulate and disperse over time.

In an analysis of 160 rare species occurring in the British Isles, Rabinowitz and her colleagues found that 59% were habitat specialists; only 14% had a narrow geographical range; and a mere 7% universally had small populations. We have no comparable data from tropical forests. Nevertheless, our current knowledge of rain forest plants allows us to make preliminary generalizations. We distinguish four types of rarity for tropical forest trees.

1. The first category includes species that are uniformly rare. Such species typically occur in low population densities with an average of one reproductively mature individual or less per hectare. The examples include pioneer species of such genera as *Bombax, Ceiba, Pajanelia,* and *Oroxylum,* and emergent species of *Durio* and *Coelostegia* in the mature phase in the Far East. In Magnoliaceae, an archaic family with many rare species, all lowland rain forest trees seem to occur universally as scattered individuals, although many are geographically rather widespread. Species with low population densities usually have mechanisms that allow long-distance dispersal of pollen and seed. For example, species of *Bombax, Ceiba, Oroxylum,* and *Durio* are pollinated by bats and have their seed dispersed by wind (by mammals in the case of *Durio*). Pollination in the Magnoliaceae is by beetles,and the seeds are dispersed by birds.

2. The second category contains species that are common in some places but rare in between. This condition corresponds to Schoener's (1987) diffusive rarity. Areas in which a species is common may serve as "sources" for propagules that establish in "sink" habitats where the species is not abundant and where in the long run reproductive output is low and local mortality exceeds the recruitment of new individuals (Pulliam 1988). Many tree species in tropical rain forests at a given site may fit this category.

3. In the third category are the species that are local endemics. Good examples in Asia occur in the Hamamelidaceae, the great majority of whose genera contain one or a few species and have more or less restricted geographical ranges. Are the species of the archaic families, whose survival in tropical forests is so celebrated among naturalists, generally rare, and do they generally betray their antiquity with relict or disjunct distributions, as in the Hamamelidaceae? It cannot be said that this is the case, any more than it is among clearly derived, modern families such as Dipterocarpaceae and Myrtaceae. Nevertheless, although examples like the monotypic hamamelid *Maingaya malayana,* known from a handful of trees on Pengang Island off the coast of Malaya and now possibly extinct in the wild, do exist, more often these plants are actually quite abundant where they occur. Examples include *Neostrearia,* a tree of the south Queensland rain forests.

Species that are rare on account of narrow geographical range vary greatly in representation among life forms. Epiphytes, ground herbs, and to a lesser extent shrubs, which tend to be concentrated in the relatively fragmented landscape of the mountain forests, are very much richer in narrow endemics than trees. These life forms are the principal cause of the uniquely rich flora of the northern Andes (Gentry and Dodson 1987).

Differences in the representation of rare species and the various forms of rarity in a flora can be explained by several factors. Shorter life cycles allow more rapid evolution. Genetic systems favoring rapid differentiation among small populations—including self-compatibility, retention of interspecific compatibility, and sometimes polyploidy—appear to be more frequent among epiphytes, ground herbs, and shrubs than among trees in rain forests. The orchids, begonias (C. C. Townsend pers. commun.), and Melastomaceae (Renner 1986) serve as striking examples of such genetic systems. Gentry and Dodson (1987) have suggested that many species in such groups may be ephemeral, arising through chance hybridization of incompletely genetically isolated forms that leads to distinctive new forms in tiny numbers that are easily eliminated before they have a chance to multiply and spread. If this is so, then the concept of "species" in such instances becomes stretched almost to absurdity.

4. In the fourth category are species that are clumped even when overall population density is very low. A special form of clumping occurs on soils exceptionally low in nutrients even by the standards of the humid tropics (Ashton in prep.). Here, small differences in nutrient status may have a major influence on the ecological distribution of species, such that populations become fragmented into soil-determined ecological islands of various sizes. Although overall population density may be low, local semipermanent aggregations of high density can occur. Generally, individuals do occasionally become successfully established in the suboptimal habitats that separate the islands, and may even breed and thus act as stepping stones for gene exchange. The size of the clumps may be very large, and, in this respect, the population structure may closely resemble species in the second category. On the other extreme, the clumps may be very small, consisting of few individuals. In the case of habitat specialists, the number of individuals will be determined by the size of the habitat and the size of mature plants.

GENETIC CONSEQUENCES OF RARITY

The genetic consequences depend on the type of rarity. We now consider the effects of the four types of rarity outlined above.

In the first category of species, which are uniformly rare almost throughout their range to the extent that the average density of trees may be one adult tree per hectare or less, there is evidence that many such species are outbred with a high rate of outcrossing (Bawa and O'Malley 1987; O'Malley and Bawa 1987; O'Malley et al. 1988). Presumably, such species have efficient mechanisms for long-distance pollen and seed dispersal. Gene exchange occurs over large distances, a property that increases neighborhood size and neighborhood area (Barrett and Kohn, Chapter 1). Rarity in such cases is not expected to result in depletion of genetic variation within populations. In fact, populations of such species may exhibit a considerable amount of genetic variation. Because of extensive gene flow, populations are also not expected to show much genetic divergence. Evidence for considerable genetic polymorphism within populations of large canopy trees exists for several species from the Old as well as the New

World tropics (Gan et al. 1977; Bawa and O'Malley 1987; Buckley et al. 1988; Hamrick and Loveless 1989). Many of these species—for example, *Pithecellobium pedicellare* in the Mimosoideae (O'Malley and Bawa 1987) and *Bertholletia excelsa* (Brazil nut) in the Lecythidaceae (Buckley et al. 1988)—are known to occur in low densities, although it is not known whether they are uniformly rare throughout their range. Evidence for little genetic divergence in this category of species is scant because genetic differentiation has been studied in only one such species, the Brazil nut. Two populations separated by a distance of 150 km show little genetic divergence (Buckley et al. 1988). On a smaller spatial scale of 2–3 km, Hamrick and Loveless (1989) have noted substantial gene flow among populations of several tropical trees. However, the number of species sampled by Hamrick and Loveless that fall into the category of rarity considered here is not known.

Localized, minor, temporal fluctuations in dispersion and population density of uniformly rare species should have little or no effect on reproductive output of individual trees because pollen can be dispersed over long distances.

Obviously, conservation of uniformly rare species requires large areas (see below) because individuals scattered over an extensive area are held together by gene flow. Reduction in the size of the area could gradually lead to inbreeding among the limited number of individuals in a given reserve. However, inbreeding would not result in substantial loss of genetic variation unless it continues for several generations. Fragmentation and reduced area of the reserve could lead also to a decrease in reproductive output. Lowered reproductive output might result from inbreeding as well as the lack of enough pollen movement among widely spaced plants. Decrease in the size of the area is likely to adversely influence the population size of pollinators and seed dispersal agents; hence, seed and fruit set as well as regeneration may decline, owing to indirect effects of mutualists.

In the second category, in which species have low population densities at some places but are common elsewhere, we are concerned with only that part of the range of distribution where population densities are low. Even there, as mentioned earlier, population densities can vary widely in time as well as in space, depending on immigration, regeneration, and mortality. Thus genetic consequences will depend largely on the prevailing population density. Reduction in population size could lead to a decrease in the level of heterozygosity, and if the small population size were to persist over several generations, gene fixation would occur via inbreeding and random genetic drift. Outcrossing rates seem particularly sensitive to changes in population density. It has been shown in several species that with a decrease in density there is an appreciable decline in the rate of outcrossing (Knowles et al. 1987; Vaquero et al. 1989; Murawski et al. 1990). At present, evidence for fluctuation in the population densities of rare plant species over time is scarce, although Primack (pers. commun.) has inferential evidence that rare species in tropical forests are likely to experience more fluctuations than abundant ones.

Reduced population density in this category of species may result also in a decrease in seed and fruit set, particularly if pollen is dispersed over limited distances. The problem could be particularly acute in self-incompatible and dioecious species in which the nearest conspecific flowering tree may not belong to the appropriate compatibility type or sex, respectively. It is interesting that dioecious species are disproportionately poorly represented among species with low population densities in a Central American forest (Hubbell and Foster 1986a). However, the effect of density on the reproductive output of trees that often exhibit low density has not been empirically determined.

Thus the combined effects of small effective population size and reduced outcrossing rate may lead to the depletion of genetic variation. Genetic depauperation coupled with demographic stochasticity in small populations could lead to further population decline and eventual extinction. The transitory small populations therefore may not persist for long periods unless there is continual immigration from areas where the species is more abundant.

Hamrick et al. (Chapter 5) show that rare tropical trees have less genetic variation than the more common species. However, it is not known if the rare species fit the category considered here. Ideally, one would like to compare the level of genetic variation and inbreeding in populations of different densities to determine the effect of diversity on genetic parameters.

The boundaries of a protected area may contain a large number of rare species that exist as transitory populations. In the absence of immigration from "source" areas outside the reserve, a large number of such species may be doomed to extinction, even in the absence of disturbance, as a result of inbreeding and stochastic genetic and demographic processes. Artificial enrichment with the aim of enlarging population size might be one management intervention to sustain the species in the protected area.

In the third category, that of local endemics, population size may be small or large (Kruckeberg and Rabinowitz 1985). When populations are large, they should harbor considerable genetic variation unless they have gone through a bottleneck of small population size and inbreeding. There are no data on the population genetic structure of locally endemic tropical tree species, but the data from other tree species indicate that no general statements can be made about the level of genetic diversity. For example, in Torrey pine (*Pinus torreyana*), which is locally endemic in California, there is a low level of genetic variation within populations, but the populations are highly differentiated from one another (Ledig and Conkle 1983). On the other hand, populations of some other conifers that are also locally endemic to California are quite variable (Ledig, in Kruckeberg and Rabinowitz 1985). Similarly, several species of *Eucalyptus* that are localized in their distribution have been found to contain moderate to high levels of genetic variability (Moran and Hooper 1987). Thus in tree species that are generally widely outcrossed, local endemics may not necessarily show lower levels of genetic variability.

Management intervention for *in situ* conservation may be necessary only if population sizes fall below a certain critical point. If the intervention calls for enrichment to increase population size, seeds must be obtained from the same population.

Finally, there are species that are clumped in distribution with very few individuals dispersed between the clumps. In this case, the clumps are intrinsic to the biology of the species rather than to specialized physical habitat requirements. Clumps therefore move, disperse, aggregate, and coalesce over time. As mentioned in the previous section, the clumps themselves may consist of a very limited number of individuals. Many such species may be habitat specialists. This category of species in a qualitative sense has the same dispersion pattern as the second category, except that even in habitats where the species is abundant, population sizes are very small.

Small population size in clumps, if it persists over several generations, may lead to loss of genetic variation. However, because of the presence of occasional individuals between clumps, gene exchange among clumps may occur, albeit much less frequently than within clumps. Such gene movement may prevent loss of allelic diversity.

Thus the genetic consequences of this type of rarity depend on the size of the clumps, the time over which small population size in the clumps persists, and the amount of gene flow among clumps. Unfortunately, not for a single species do we have information for such parameters. In the absence of data about dispersion and demography of such species, it is difficult to suggest how they might be managed to maintain genetic diversity in protected areas.

CONSERVATION OF RARE TROPICAL PLANTS

Conservation of rare tropical forest plants, particularly trees, may require large protected areas of hundreds of square kilometers. The requirement of large conservation areas by no means is a new or novel conclusion. It is the justification for such a conclusion that is relevant to this discussion. Three principal arguments can be advanced to justify the conclusion.

1. Many rare species in tropical forests at a given site probably represent occasional emigrants from population centers occurring elsewhere (Hubbell and Foster 1983). Such individuals, as mentioned earlier, in many cases may not constitute viable populations in the long run. Populations in "sink" areas are expected to undergo extinction without continual immigration from "source" areas (Pulliam 1988). The nonequilibrium view of community structure also predicts random extinctions at a local scale (Hubbell and Foster 1986a, 1986b). Populations in sink areas may be particularly prone to extinction. Large conservation areas may be required to ensure the inclusion of population centers or the "source" populations.

2. Most tropical forest trees engage in mutualistic interactions with fungi (mycorrhizae), pollen, and seed vectors for their survival. In those cases where a large number of individuals occur over a small area, the minimum limit on areas of pristine forest for conservation may depend on the area required for conservation of nectarivorous bats, birds, and primates that serve as seed dispersal agents. Start and Marshall (1976) have shown that the nectarivorous bat *Eonycteris spelaea* forages up to 60 km from its communal roost and feeds on a succession of species in inland forests throughout the year. It also feeds on the nectar of frequently or continuously flowering species of the mangrove habitat. The moths that pollinate a number of plant species in a tropical dry deciduous forest utilize, during the dry season, plant resources in the moist forests many kilometers away (Janzen 1987). In topographically diverse areas, frugivorous birds may move back and forth between upland and lowland areas, depending on the availability of fruit at a given site (Terborgh 1986). These examples emphasize the importance of maintaining diverse habitats and communities for the conservation of tropical biodiversity. Fortunately, however, wide-ranging pollinators such as bats, birds, and large bees universally feed on a diversity of plant species, and their needs, as demonstrated by a long history of plant domestication, can be accommodated in cultivated vegetation. Thus it is also possible to retain substantial amounts of tropical biodiversity if a pristine area is carefully integrated into a landscape containing man-

aged natural forests and agroforestry systems. Nevertheless, in such cases, too, the conservation unit remains a large area.

Small areas may also suffer from another type of cascade effect. It is well known that conservation of top predators requires large areas. The elimination of top carnivores from a habitat may result in an increase in the populations of seed eaters, with attendant effects on the composition of plant species that are the targets of these seed predators (Terborgh 1988).

3. Population genetic studies of some canopy species show that these trees are widely outcrossed (O'Malley and Bawa 1987; O'Malley et al. 1988). If one were to accept that the effective population size for long-term evolutionary potential for outcrossing species is 500 (Lande and Barrowclough 1987, and references therein) and that the ratio of effective population size to census size in trees is 1:4, then 2000 reproductively mature individuals may constitute the minimum size for survival. For canopy species that occur at densities of one adult or less per hectare, a large area would be needed to maintain minimum viable populations. However, the above generalizations are complicated by at least two factors.

First, there is evidence that outbreeding is not universal among tropical forest plants. It appears that many herbaceous taxa and also understory shrubs may be highly self-compatible (Kress and Beach 1990), and many tree species with bisexual flowers seem to have at least limited self-compatibility (Bawa et al. 1985; Murawski et al. 1990). Also, some herbaceous and gap-phase species spread vegetatively through adventitious embryogeny, often without pseudogamy. To date, the degree to which this apoximis is facultative is little known, and its influence on the population genetics of tropical forest plants quite unknown. It can reasonably be supposed, though, that apomixis can allow species to survive both in very low population densities and in many small isolated population segregates, without the loss of fecundity through failed cross-pollination. Heterosis will at least be maintained by apomixis, but increasing genetic uniformity may reduce the proportion of successful crosses and thereby ultimately hasten the demise of such species (Ashton 1984).

Second, in several tree species, populations seem to have survived reductions in size. For instance, extremely small isolated populations of the wind-pollinated bunya pine (*Araucaria cunninghamii*) have almost certainly survived in moist gorges in the desserts to the east of the Great Dividing Range of Queensland since the beginning of the Holocene. Minute populations of mountain summit plant species are known to all of us, some of them morphologically distinct; but we cannot guess how old they are. Indeed, the rate of turnover of species in ecological or true islands of different size is unknown among plants. The remarkable predictability with which habitat specialists occur in once continuous but now isolated habitat fragments in tropical forests does suggest that chance extinction takes a long time in the equable climates of the equatorial tropics. Such species comprise, for instance, a whole tree flora of specialized soils in the coastal hills of northwest Borneo. The hills have become increasingly fragmented by river systems since the Pliocene, yet geomorphological dating of the landscape suggests that populations have survived and remained remarkably constant, on fragments as small as 50 km^2 (5000 ha) for at least 500,000 years (Ashton in prep.).

Conservation of rare tropical trees *ex situ* is also difficult for a number of reasons. First, only a limited number of large woody plants can be cultivated in a botanical garden or an arboretum. Second, the complex sexual systems and pollination mechanisms found in many species may not always ensure seed set under *ex situ* conditions. Third, the seeds of most species are fleshy, and techniques for their cryogenic preservation are not well developed (Eberhart et al., Chapter 9).

CONCLUSIONS

The tropical habitats are being altered at an unprecedented scale. At the current rates of deforestation, mass extinctions of species seem imminent. Furthermore, a considerable amount of genetic diversity within species that might survive is likely to be lost. The problem of conserving or managing biodiversity in the tropics is compounded by the presence of a large number of rare plant species, a wide variety of mutualistic interactions between plants and other organisms, and the difficulties in *ex situ* conservation.

It is apparent that tropical forests are extraordinarily rich in the number of rare species. However, there are many different types of rarity. Detailed inventories at local and regional levels are needed, first, to determine the exact number of rare species and, second, to assess the frequencies of various types of rare species, because the genetic consequences of rarity may vary with the type of rarity.

Equally important is the need to determine the effects of deforestation and fragmentation of habitats on the fate of rare species. Rare species in fragmented habitats may be reduced to such low numbers that they may not constitute viable populations. In such populations, genetic drift and inbreeding may prevail and reduce the level of genetic variation. Inbreeding may result in inbreeding depression, with deleterious effects on reproductive output. Thus a combination of demographic and genetic factors may hasten the extinction of rare species in small, isolated fragments. We need comparative data on levels of genetic variation, inbreeding, and inbreeding depression for populations of rare species in habitats of different sizes.

Extensive deforestation and fragmentation of habitats can also influence the abundance of many common species and the genetic variation contained in such species. Spatial organization of genetic variation in most plant species follows a hierarchical order. Widespread species may exhibit ecotypic differentiation on regional or geographical scales. Much of the geographical variation may be lost if protected areas are restricted to a part of the species range. Furthermore, within geographical regions, even at local scale, seemingly large continuous populations may be genetically differentiated (Lande and Barroclough 1987). The boundaries of a nature reserve may not coincide with the fluctuating boundaries of a metapopulation, determined by changing selection regions and patterns of gene flow. A sufficient knowledge base to examine the impact of deforestation and habitat fragmentation on the genetic architecture of tropical tree populations in general does not exist.

The problem of the reserve size and number is further compounded by the host of mutualistic interactions between tropical forest plants and other organisms. These interactions involve several critical stages in the life cycle of plants, from seedling establishment (mycorrhizae) to pollination and seed dispersal. Criteria for determining the size of conservation units must take into account not only the dispersion and density of trees, but also the minimum area required to maintain diversity and viable populations of mutualists, particularly pollinators and seed dispersers.

Ex situ conservation of tropical forest plants is hampered by the very large number of taxa that require protection, the large area needed for the cultivation of woody plants, and the lack of adequate methods for long-term storage of seeds of many species. Moreover, once in cultivation, continual propagation by seed would be limited in many species because pollinators may not be abundant or even present at the new site. On the positive side, the long life cycles of woody species[1] should ensure survival of the original material for many years.

Overall, conservation of genetic diversity in tropical plants, rare as well as common, is a challenging and an urgent task. It requires knowledge about the distribution and abundance of species, key mutualistic interactions among organisms, amounts and patterns of genetic diversity in natural populations, and progress in technology of long-term *in situ* storage of propagules. Such knowledge is beginning to accumulate for tropical plant species. A key factor in further progress in the management of biodiversity, including intraspecific genetic diversity, will be the rate at which this knowledge expands in the future.

ACKNOWLEDGMENTS

Our research on genetics of tropical forest trees has been supported by grants from the National Science Foundation; Conservation, Food, and Health Foundation; and AID's Program in Science and Technology Cooperation.

II

DISTRIBUTION AND SAMPLING
OF GENETIC VARIATION

5

Correlations Between Species Traits and Allozyme Diversity: Implications for Conservation Biology

J. L. HAMRICK, M. J. W. GODT, D. A. MURAWSKI,
and M. D. LOVELESS

One goal of conservation biology is to preserve the evolutionary potential of species by maintaining natural levels of genetic diversity. Knowledge of the levels and distribution of genetic variation thus becomes a prerequisite for the establishment of effective and efficient conservation practices (Beardmore 1983; Frankel 1983; Allendorf and Leary 1986). Without such knowledge, the efforts of conservation biologists, no matter how well meaning, may not be effective (Asins and Carbonell 1987). For conservation purposes, information can best be obtained by empirically determining the level and distribution of genetic variation for morphometric and physiological traits, allozyme loci, and DNA sequences. However, since few endangered species have been studied, and since the resources for such studies are frequently limited, we must often rely on previously accumulated knowledge to make educated guesses pertaining to the genetic diversity of these species.

Studies of allozyme variation in plants offer several advantages over other measures of genetic variation (Hamrick 1989; Schaal et al., Chapter 8). Starch gel electrophoresis, the technique most frequently used to study allozyme variation, is relatively inexpensive and can be applied to most plant species. Sampling for electrophoresis is generally nondestructive, since relatively little material is needed and most enzyme loci are expressed at several life-cycle stages. Thus even endangered species can be analyzed without harm. In addition, most isozyme loci have discrete Mendelian inheritance. As a result, genetic interpretations of natural populations can be made without concern for environmental effects on the expressed phenotype, in contrast to quantitatively inherited morphological and physiological traits. Since most isozyme loci are codominant, allele and genotype frequencies can be calculated directly. In contrast, most single-gene morphological traits have dominant—recessive expression; breeding studies are necessary for the estimation of allele and genotype frequencies for these traits. Finally, since the same isozyme loci can be analyzed in every population or across related species, estimates of the levels and distribution of genetic variation can be directly compared. Thus allozyme loci have great utility as markers to describe patterns of genetic diversity. It should be noted, however, that allozyme diversity may not be well correlated with other measures of genetic diversity (e.g., quantitative traits) that may be of equal or greater importance in conservation. Yet, in the absence of this more pertinent information, generalizations derived from the allozyme literature can be used to develop strategies for the conservation of genetic diversity. Allozyme diversity can also be used as a yardstick to measure the effectiveness of *in situ* and *ex situ* conservation programs.

In this chapter, we summarize a recent review of the plant isozyme literature (Hamrick and Godt 1989) with the aim of suggesting generalizations useful to conservation biology. As a test of the predictions arising from the review, we compare the levels and distribution of allozyme variation of 32 species belonging to a tropical forest community.

DISTRIBUTION OF GENETIC VARIATION

The plant allozyme literature has been reviewed several times (Brown 1979; Hamrick et al. 1979; Gottlieb 1981; Loveless and Hamrick 1984). The consensus is that widespread, long-lived, wind-pollinated, outcrossing species maintain more genetic variation within populations than do species with other combinations of traits. Species with this particular suite of traits also have a lower proportion of their genetic diversity among populations.

A drawback of earlier reviews was that within-population variation was reviewed in one paper and among-population variation was reviewed in another (e.g., Hamrick et al. 1979; Loveless and Hamrick 1984). Brown (1979) attempted to summarize both aspects of genetic diversity, but the number of species he reviewed was limited. None of the earlier reviews considered genetic diversity at the species level. Hamrick and Godt (1989) analyzed the plant allozyme literature to determine if earlier generalizations were valid, given the larger current data base. Their review included 653 taxa-studies that summarized allozyme variation for 449 species representing 165 genera. Genetic diversity was estimated at the species level, for the first time, and within populations of species. Partitioning of genetic diversity within and among populations was also considered. The larger data base of this review increased the accuracy of comparisons among groups of species with different attributes. As a result, generalizations generated by this review should be more reliable.

Procedures

Hamrick and Godt (1989) classified each species for eight traits:

1. Taxonomic status: gymnosperms, dicotyledons, monocotyledons
2. Regional distribution: boreal–temperate, temperate, temperate–tropical, tropical
3. Geographical range: endemic, narrow, regional, widespread
4. Life form: annual, short-lived perennial (herbaceous or woody), long-lived perennial (herbaceous or woody)

5. Mode of reproduction: sexual, sexual and asexual
6. Breeding system: selfed, mixed mating (animal- or wind-pollinated), out-crossed (animal- or wind-pollinated)
7. Seed dispersal mechanism: gravity, gravity and animal-attached, explosive, wind, animal-ingested, animal-attached
8. Successional status: early, mid, late

Classification of each species was based on information gleaned from the original papers or by consulting regional floras.

Plant allozyme variation is usually estimated at the population level. However, for species with considerable population differentiation, population means may not accurately reflect total genetic diversity. For example, consider two populations for which ten loci were studied. If five loci were polymorphic in population 1 and two different loci were polymorphic in population 2, the proportion of loci polymorphic (P) for the two populations would be 0.5 and 0.2, respectively. The mean within-population value would be 0.35. In contrast, the P value for the species would be 0.7. Because similar discrepancies arise for other genetic parameters, Hamrick and Godt (1989) estimated levels of genetic diversity at both the population and the species level.

Four genetic parameters were calculated at the species and within-population levels: percent polymorphic loci (P), mean number of alleles per locus (A), effective number of alleles per locus (A_e), and a genetic diversity index (H_e), which is equivalent to the proportion of loci heterozygous per individual under Hardy–Weinberg expectations. Variation among populations was estimated with Nei's (1973) genetic diversity statistics. Total genetic diversity (H_T) and mean diversity within populations (H_S) were calculated for each polymorphic locus. The proportion of the total diversity residing among populations ($[H_T - H_S]/H_T$) was expressed as G_{ST} values. Statistical procedures are given in Hamrick and Godt (1989).

Genetic Diversity at the Species Level

Plant species are, on average, polymorphic at 50% of their allozyme loci. Mean genetic diversity for the 473 entries reviewed was 0.149. Taxonomic status, geographical range, life form, breeding system, and seed dispersal mechanism had significant effects on the levels of genetic diversity (H_e) maintained by species (Table 5.1). Based on a multiple regression analysis, the eight traits accounted for 24% of the variation in genetic diversity at the species level. Geographical range accounted for the largest proportion of the explained variation (32%, or 8% of the total). Life form accounted for an additional 25% (6% of the total), and breeding system and seed dispersal mechanism each added approximately 17% (4% of the total).

Endemic species had less than half the genetic diversity ($\bar{H}_e = 0.096$) of widespread species ($\bar{H}_e = 0.202$). Narrowly and regionally distributed species had intermediate values. Differences in genetic diversity between endemic and widespread species were largely due to differences in the proportion of polymorphic loci ($\bar{P} = 40\%$ and 59%, respectively). Widespread species also had more alleles at polymorphic loci ($\bar{A} = 3.19$) than did endemic species ($\bar{A} = 3.00$).

Table 5.1. Associations Between the Characteristics of Species and Genetic Diversity at the Species Level

Characteristic	Level of genetic diversity	
	Low	High
Taxonomic status	Dicots	Monocots and gymnosperms
Life form	Short-lived perennials and annuals	Long-lived perennials
Geographical range	Endemic species	Widespread species
Regional distribution	No significant differences	
Breeding system	Selfing species; mixed-mating, animal-pollinated species	Outcrossing species
Seed dispersal	Explosively dispersed seed	Animal-attached seed
Mode of reproduction	No significant differences	
Successional status	No significant differences	

Source: After Hamrick and Godt (1989).

Genetic Diversity Within Populations

Within populations, plants are polymorphic, on average, at 34% of their loci. Mean genetic diversity within the 468 plant taxa reviewed was 0.113. Plant populations are thus more genetically heterogeneous than either invertebrates ($\bar{H}_e = 0.100$) or vertebrates ($\bar{H}_e = 0.054$) (Nevo et al. 1984). Cumulatively, the eight traits accounted for 28% of the variation in population-level genetic diversity (Table 5.2). Plant breeding system accounted for the largest proportion of this variation (33%, or 9% of the total). Geographical range accounted for an additional 27% (7.5% of the total) of the variation, whereas life form, taxonomic status, and seed dispersal mechanisms added smaller but significant proportions to the explained variation (13%, 12%, and 12%, respectively, or 3–4% of the total). Successional status and regional distribution accounted for a small proportion of the explained variation, while mode of reproduction did not exhibit significant differences among its categories.

Populations of selfing species had half the genetic diversity ($\bar{H}_e = 0.074$) of outcrossing, wind-pollinated species ($\bar{H}_e = 0.148$). Animal-pollinated species with mixed mating systems (i.e., partially selfed, partially outcrossed) had levels of genetic diversity that were similar to selfing species, whereas the genetic diversity of outcrossing

Table 5.2. Associations Between the Characteristics of Species and Genetic Diversity at the Population Level

Characteristic	Level of genetic diversity	
	Low	High
Taxonomic status	Dicots	Monocots and gymnosperms
Life form	Short-lived perennials and annuals	Long-lived woody perennials
Geographical range	Endemic species	Widespread species
Regional distribution	Boreal–temperate species	Temperate and tropical species
Breeding system	Selfing species	Wind-pollinated species
Seed dispersal	Explosively dispersed seed	Animal-attached and wind-dispersed seed
Mode of reproduction	No significant differences	
Successional status	Early successional species	Late successional species

Source: After Hamrick and Godt (1989).

animal-pollinated species was closest to the wind-pollinated outcrossing species. This suggests that the acquisition of self-incompatibility systems could have important effects on the genetic structure of animal-pollinated species. Endemic species had much less genetic diversity within populations ($\bar{H}_e = 0.063$) than did narrowly ($\bar{H}_e = 0.105$) or regionally distributed species ($\bar{H}_e = 0.118$). Populations of widespread species had the most genetic diversity ($\bar{H}_e = 0.159$).

Most of the differences in genetic diversity between selfing and outcrossing species were due to differences in the proportion of polymorphic loci. For selfing species, 20% of the loci were polymorphic per population, whereas for outcrossing, wind-pollinated species, nearly 50% of the loci were polymorphic within populations. Selfing and outcrossing species did not differ in the number of alleles per polymorphic locus or in the skewness of allele frequencies. Differences among species with different geographic ranges were also largely due to differences in the proportion of polymorphic loci within populations. Endemic species had significantly fewer polymorphic loci ($\bar{P} = 26\%$) than widespread species ($\bar{P} = 43\%$). Widespread species also had more alleles per polymorphic locus ($\bar{A} = 2.67$) than endemic species ($\bar{A} = 2.48$).

Genetic Diversity Among Populations

On average, the 406 plant taxa analyzed maintained 78% of their diversity at polymorphic loci within populations; 22% of the total diversity at polymorphic loci occurred among populations. The eight traits explained 47% of the heterogeneity in G_{ST} values among species. Breeding system and life form of the species were most highly associated with differences in G_{ST} values; together they accounted for 84% (39% of the total) of the variation explained by the model. Only geographical range and mode of reproduction had nonsignificant differences among their categories for G_{ST} (Table 5.3).

Selfing species had fivefold more diversity among populations ($\bar{G}_{ST} = 0.51$) than wind-pollinated, outcrossed species ($\bar{G}_{ST} = 0.10$). Animal-pollinated species generally had intermediate values ($\bar{G}_{ST} = 0.21$). Annual species also had nearly fourfold

Table 5.3. Associations Between the Characteristics of Species and the Genetic Diversity Among Populations

Characteristic	Proportion of genetic diversity among populations	
	Low	High
Taxonomic status	Gymnosperms	Angiosperms
Life form	Long-lived woody perennials	Annuals
Geographical range	No significant differences	
Regional distribution	Boreal–temperate species	Temperate and tropical species
Breeding system	Outcrossed wind-pollinated species	Selfing species
Seed dispersal	Gravity-dispersed and animal-attached seed	Gravity-dispersed seed
Mode of reproduction	No significant differences	
Successional status	Late successional species	Early and midsuccessional species

Source: After Hamrick and Godt (1989).

more variation among their populations ($\bar{G}_{ST} = 0.30$) than long-lived woody perennials ($\bar{G}_{ST} = 0.08$). Short-lived herbaceous species were intermediate ($\bar{G}_{ST} = 0.23$).

These results indicate that species with limited potential for gene flow have more differentiation among populations than do species with more potential for gene flow. Gene flow potential is correlated with pollination and seed dispersal mechanisms as well as with longevity and size. It is noteworthy that geographical range had no effect on the distribution of genetic diversity among populations. Endemic species partitioned genetic variation in much the same way as more widespread species.

The proportion of the total number of alleles occurring within each population may be of more interest to conservation biologists than genetic diversity per se. Hamrick and Godt (1989) did not calculate this parameter, but Hamrick (1983) showed that this parameter is closely associated with G_{ST} values. Hamrick (1983) estimated that selfing species maintain approximately 56% of their alleles within a single population, whereas outcrossed, wind-pollinated species maintained 74% of their alleles in each population. Differences between endemic (64%) and widespread species (70%) were less dramatic. Thus species with high G_{ST} values should maintain a lower proportion of their alleles within populations, while most of the alleles of species with low G_{ST} values will be shared among populations. This association is not unexpected since the presence or absence of alleles in populations is merely an extreme case of allele frequency differentiation.

Associations Among Genetic Parameters

Associations between species-level diversity, within-population diversity, and among-population diversity were examined by calculating correlations based on individual species values. Species-level diversity was highly correlated with population-level diversity ($r = 0.89$), but was negatively correlated with among-population diversity ($r = -0.11$). Population-level diversity was strongly negatively correlated with among-population diversity ($r = -0.45$).

These results suggested that species with high levels of genetic diversity are also likely to have populations with high levels of genetic diversity. This association is due to the influence of geographical range on variation at both levels. Within-population diversity is also a function of how total diversity is partitioned within and among populations. This is largely a function of the gene flow potential of species. Genetically vagile species have high amounts of diversity within populations and little heterogeneity among populations, while less genetically vagile species have relatively less variation within populations and greater differences among populations.

A TEST OF PREDICTIONS: ALLOZYME DIVERSITY IN A TROPICAL FOREST COMMUNITY

Although reviews can provide generalizations concerning the distribution of allozyme variation, they do not provide the detail needed to understand fully the interactions among species traits and genetic diversity. In such surveys, species from widely divergent taxonomic, historical, and ecological backgrounds are grouped on the basis of a series of traits. Such grouping may obscure interesting relationships between genetic diversity and the species' ecology. For instance, species from different communities may have adapted to local environmental conditions in rather different ways. Animal-mediated pollination may, for example, have different genetic implications for a plant population, depending on the animal communities serving as pollen vectors. Similarly, wind pollination might provide different predictions of genetic structure in tropical and temperate communities. This may explain why a relatively small proportion of the total variation among species in these broad surveys is explained by the traits examined. By examining ecological and genetic correlations among species with similar life forms from a single community or within a single taxonomic group, we may be able to dissect more finely the effects of life-history traits on patterns of genetic diversity.

With this in mind, we studied allozyme variation in trees of the tropical moist forest community of Barro Colorado Island (BCI), Republic of Panama. Tropical tree communities are especially useful for studies of genetic structure because they have a wide array of pollination syndromes and seed dispersal mechanisms. Barro Colorado Island is an ideal study site because its flora is well known (Croat 1978; Leigh et al. 1982b). In addition, the 50-hectare (ha) Forest Dynamics Plot (FDP) has been established on BCI in which every woody stem greater than 1 cm in diameter at 1.5 m has been identified and mapped.

Since 1983, we have studied the distribution of allozyme variation in 16 common species of trees and shrubs on BCI and the FDP. Species were selected to represent a wide array of pollination and seed dispersal mechanisms. A sampling design was established in which 50 or more individuals were sampled from four collection sites on the FDP and two additional sites on BCI, usually at least 1 km from the FDP and each other. This sampling scheme was completed for nine of the 16 species. The remaining species had varying subsets of this design.

An additional 16 species that are uncommon on BCI and the FDP were identified. Species having 5–25 individuals of any age on the FDP were sampled. The analyses of these species permitted a direct comparison between the levels of variation maintained within populations of common and uncommon species.

Standard starch gel electrophoresis was used to measure allozyme variation. Details of the collecting procedures and electrophoretic techniques are given in Hamrick and Loveless (1986). Genetic measures for each of the 16 common species are given in Loveless and Hamrick (1987) and Hamrick and Loveless (1989).

Levels of Variation

Estimates of the percent polymorphic loci and genetic diversity were calculated for the 16 uncommon species (Table 5.4). Two estimates of these parameters were also made for the 16 common species. The first was based on data from more than 250 individuals of each species sampled from the FDP. The second estimate was obtained by taking a random sample of ten individuals from the pooled FDP populations of each species. This second estimate was taken to eliminate the confounding effects of sample size on estimates of levels of variation in the uncommon and the common species.

The common tropical woody species maintain, on average, more genetic diversity than the average plant species, and also more than the average long-lived woody perennial species (Table 5.4). There was, however, considerable variation among the common species. Genetic diversity ranged from 0.340 for *Alseis blackiana* (an insect-

Table 5.4. Levels of Allozyme Diversity Within Populations
of Uncommon and Common Tropical Tree Species

Level	N	Mean no. populations	Mean no. loci	Proportion of polymorphic loci	Genetic diversity
All species[a]	468	12.7	16.5	34.2	0.113
		(1.3)	(0.4)	(1.2)	(0.005)
Long-lived woody perennial species[a]	115	9.3	17.0	50.0	0.149
		(1.4)	(0.9)	(2.5)	(0.009)
Common tropical trees and shrubs[b]	16	5.5	34.3	60.9	0.211
		(0.5)	(1.6)	(3.7)	(0.017)
Common tropical trees and shrubs[c]	16	1	34.3	56.7	0.185
			(1.6)	(3.6)	(0.015)
Uncommon tropical trees and shrubs[d]	16	1	25.8	42.1	0.142
			(1.8)	(6.1)	(0.018)
Uncommon species without strangler figs[d]	13	1	27.2	33.6	0.124
			(2.1)	(5.2)	(0.019)

Note: The data for the tropical species are compared with that for all higher plants, and long-lived woody species. Standard error in parentheses.

[a]From Hamrick and Godt (1989).

[b]From Hamrick and Loveless (1989).

[c]Ten individuals per species. From Hamrick and Murawski (unpubl. data).

[d]Mean number = 10 individuals per species. From Hamrick and Murawski (unpubl. data).

pollinated canopy tree with wind-dispersed seeds) to 0.075 for *Tachigali versicolor* (a bee-pollinated canopy tree with wind-dispersed seeds). *A. blackiana* is one of the most common trees on the FDP, whereas *T. versicolor* has few mature individuals, which suggested that genetic diversity might be related to adult population size. However, a rank correlation analysis on the 16 common species showed no relationship between genetic diversity and the density of flowering individuals on the FDP.

Significant differences ($p = 0.01$) existed between the estimates of genetic diversity for the unreduced samples of the common species and those of the uncommon species. Reducing the sample size of the common species to ten individuals reduced their mean genetic diversity (Table 5.4). However, mean genetic diversity for the 16 uncommon species was lower ($p = 0.10$) than that for the reduced sample of the common species (Table 5.4). The proportion of loci polymorphic for the uncommon species ranged from 14% to 84%, and genetic diversity values ranged from 0.026 to 0.257. Mean values for the uncommon species may be somewhat inflated because of the inclusion of three species of strangler figs (*Ficus* spp.). Recently, J. Thomson and C. Handley (pers. commun.) suggested that an individual strangler fig may be a genetic composite of several genetically distinct individuals. As a consequence, population sizes of strangler figs may be larger than previously estimated. When strangler figs were removed from the analyses, the mean proportion of polymorphic loci and the genetic diversity of the uncommon species decreased (Table 5.4). The difference between the mean diversity value of 13 remaining uncommon species and the mean value of the reduced sample of the common species was significant ($p = 0.01$).

Although the uncommon species maintain high amounts of allozyme variation on the FDP, their small effective population sizes apparently cannot maintain as much genetic diversity as the larger populations of the common species (Nei et al. 1975). It would be interesting to compare levels of variation found on BCI with levels of variation observed in communities where the uncommon species occur in higher densities or in areas where the common species on BCI are at lower densities.

Variation Among Collection Sites

Because the common species were collected from as many as six sites on BCI, it was possible to test generalizations concerning the influence of breeding system and seed dispersal mechanisms on the distribution of genetic diversity. Estimates of among-site differentiation were low ($\bar{G}_{ST} = 0.055$; range: 0.022–0.090), indicating that gene flow is substantial on BCI (Hamrick and Loveless 1989). A predicted rank of genetic divergence was developed, based on the pollination mechanisms and seed dispersal characteristics of each species. Wind-pollinated species with gravity-dispersed seeds were expected to have the least potential for gene flow. Species pollinated by vertebrates or large insects and having animal-ingested seeds were predicted to have the least divergence among sample sites. A positive and statistically significant relationship ($r_s = 0.60$; $p<0.01$) occurred between the observed and predicted ranks (Hamrick and Loveless 1989). Thus knowledge of the reproductive biology of these species allowed reasonably accurate predictions of the distribution of genetic diversity.

IMPLICATIONS FOR CONSERVATION BIOLOGY

One of the aims of any conservation program should be the preservation of genetic diversity. However, detailed studies of the level and distribution of genetic variation are not available for most plant species. Thus it is difficult to provide the minimum information needed—that is, a baseline to monitor losses in genetic diversity over time. In plants, there is also considerable variation among species for ecological traits that impact the distribution of genetic variation (e.g., population size, degree of isolation, pollination mechanism, seed dispersal mechanism). As a result, a genetically effective management strategy for one species may not be effective for another. Generalizations made from reviews can provide rough predictions of the genetic structure of unstudied plant species, and can suggest basic management approaches for plant species with different suites of characteristics.

Such reviews provide valuable information to conservation biologists in at least two ways. First, they identify characteristics (or suites of characteristics) associated with significant differences in the distribution of genetic variation. The conservation biologist can then focus on these characteristics when studying the natural history of species. For example, reviews of the plant allozyme literature have consistently shown that the breeding system significantly influences genetic diversity within and among populations (Brown 1979; Hamrick et al. 1979; Loveless and Hamrick 1984; Hamrick and Godt 1989). Thus an initial step in the design of an effective management strategy for a rare or an endangered plant species should be the determination of its breeding system. Our results also suggest that a study of a widespread congener could be used to predict the distribution of genetic variation among populations of an endemic species, if the breeding systems of the two species are similar.

Second, generalizations concerning the distribution of genetic variation can be used to design strategies for the preservation of genetic diversity. At the species level, taxa with limited geographical ranges have considerably less variation than widespread species. Endemic species have a lower proportion of polymorphic loci, often with fewer alleles per polymorphic locus and less even allele frequencies within populations. This conclusion supports and extends the results of Karron (1987a) and Karron et al. (1988) on *Astragalus* species and other restricted and widespread congeners. Endemic species might be expected to consist of smaller, more ecologically limited populations that may have experienced population bottlenecks. Thus the recent evolutionary history of a species can play a critical role in determining its contemporary genetic composition. Whenever possible, conservation biologists should document historical changes in population size and distribution as a means of better predicting amounts and patterns of genetic diversity.

Genetic diversity within populations is a function of the amount of genetic diversity present in the species and the partitioning of this variation among populations. At the within-population level, the two factors having the greatest influence on genetic diversity are geographical range and genetic mobility. Knowledge of natural species and population-level genetic diversity is essential if the conservation biologist wishes to monitor the effects of *in situ* or *ex situ* conservation policies on the maintenance of genetic diversity. Any loss of allozyme diversity during sampling or storage, or as a result of human-induced bottlenecks in natural populations, can be demonstrated only if original levels of diversity are known.

Knowledge of the distribution of genetic diversity among populations is critical in the formulation of sampling or management strategies. For species with high G_{ST} values, more populations will need to be preserved or sampled to ensure that allelic and genotypic diversity is retained. For example, a species with low genetic mobility and a G_{ST} value of 0.60 would need six populations to retain 95% of the genetic diversity in the species. In contrast, the same goal could be met by preserving two populations of an outcrossing species with 20% of its variation among populations.

Thus generalizations based on reviews of the allozyme literature can be used to develop basic strategies for the conservation of genetic diversity. Conservation biologists must be aware, however, that guidelines based on these generalizations are subject to error, since less than 50% of the species-to-species heterogeneity in genetic diversity has been explained by our analyses. For instance, knowledge of the reproductive biology of a species allows fairly accurate rank ordering the extent of population differentiation, but it does not allow accurate predictions of the absolute amount of population differentiation. Therefore, since much of the species-to-species variation in genetic diversity is due to the specific ecological and evolutionary history of a species, any management plan developed without knowledge of the species' natural history will be subject to error (e.g., see Avise and Nelson 1989). Furthermore, an important consideration is whether patterns of allozyme variation can be used to predict patterns of genetic variation in other traits. The results of the few studies that have compared morphometric and allozyme variation have not consistently demonstrated positive associations among different types of traits (Hamrick 1989). Price et al. (1984) compared estimates of population differentiation based on allozyme polymorphism and morphometric traits in the self-pollinating species *Avena barbata, Hordeum vulgare,* and *H. jubatum,* and an outcrossing species, *Clarkia williamsonii.* Rank correlations between morphometric and allozyme distance measures either approached significance or were significant for the selfing species. The rank correlation for *C. williamsonii* was not significant. These results suggest that enzyme loci provide more information about the distribution of genes controlling morphometric traits in selfing plants than in outcrossing species. This conclusion has a firm foundation in theory, since the breeding system employed by a species should influence how allozyme variation is linked with other traits. Compared with outcrossers, selfing species should experience slower decay of linkage disequilibrium, slower increases in the variance of coancestry, and slower increases in direct gene expression, owing to increased homozygosity; furthermore, all the above-mentioned factors should increase the potential for associations among allozyme and phenotypic variation (Brown and Burdon 1987).

Conservation biologists should therefore be cautious in applying the generalizations developed in this chapter. Furthermore, decisions should not be based solely on allozyme data; whenever possible, studies of geographical variation in morphological or physiological traits should be used to further refine management policies. Additional information can only improve plans for the long-term conservation of genetic diversity.

A final consideration should be just what is to be preserved and how many individuals are needed to achieve this goal. If the goal is to preserve overall genetic diversity or common alleles at allozyme loci, relatively few individuals (e.g., ten) or populations (e.g., five) will need to be preserved (Holsinger and Gottlieb 1989; Brown and Briggs, Chapter 7). If the goal is to preserve allele and genotype frequencies in their natural frequencies, however, larger population samples (e.g., 50) may be necessary. We feel that the latter goal should be adopted for all but the most extreme situations. With small sample size (e.g., ten), natural allele and genotype frequencies will not be retained. If the goal is to reintroduce the species into natural habitats, changes in the genotypic composition of the species may prevent its successful reintroduction. Although recombination and natural selection could regenerate the original genetic composition if the alleles are available, these processes may take several generations and may lead to poorly adapted populations in the meantime (Templeton 1986). Reductions in population size due to the scarcity of adapted genotypes could increase the chances of demographic extinction. Annual and selfing species may be especially susceptible to extinction during this period (Gilpin and Soulé 1986). It is our expectation that conservation biologists will have the resources for only a limited number of reintroductions and that failures to recolonize will seriously deplete the seed reserves of many species. We suggest, therefore, that to preserve the evolutionary potential of species, conservation biologists should make every attempt to preserve the natural genetic composition of populations.

CONCLUSIONS

Management guidelines for the preservation of genetic variation can be developed from generalizations based on reviews of the plant allozyme literature. In addition to describing the distribution of variation within and among populations, allozyme markers can be used to estimate quantitatively the population parameters that influence the maintenance and distribution of genetic diversity. Allozyme markers can also be used to monitor the loss of variation from populations under *in situ* or *ex situ* management. Thus to devise consistently successful programs of gene conservation in naturally occurring plant species, detailed studies of the population genetic structure as well as the evolutionary factors that initiate and maintain this structure are essential.

6

Patterns of Genetic Variation and Breeding Systems in Rare Plant Species

JEFFREY D. KARRON

Rare plants have long fascinated evolutionary biologists, who developed a wealth of theories concerning the causes and consequences of rarity (Darwin 1859; Fernald 1926; Stebbins 1942, 1980; Cain 1944; Drury 1980). Only recently, however, have the genetics and population biology of these taxa been documented in detail. Often the most informative studies involve comparisons between closely related rare and common congeners (Mehrhoff 1983; Kruckeberg and Rabinowitz 1985; Karron 1989). Such species pairs are likely to have similar ecological features and phylogenetic histories. Thus they may approximate "natural history experiments," with the common species serving as a control.

This approach has been used to study the three classes of rarity (Rabinowitz 1981):

1. Sparse species, which typically occur at low local density, have been the subject of research by Rabinowitz and her colleagues (e.g., Rabinowitz et al. 1984).
2. Habitat specialists, such as edaphically restricted taxa, have been studied by several workers, including Kruckeberg (1984) and Fiedler (1985).
3. Geographical restriction, the focus of my review, includes the majority of threatened and endangered species in the United States. Many of these taxa have very small ranges with only one to five populations. Often these species comprise fewer than 20,000 individuals.

This chapter compares the biology of geographically restricted and widespread congeners. Although these groups are often referred to as a dichotomy, they are actually extremes of a continuously distributed variable: geographical range. Most of the examples cited here concern highly restricted or very widespread species. However, it is important to remember that many taxa are not readily assigned to one of the two groups.

My review begins with a comparison of patterns of genetic variation in restricted and widespread congeners. This is followed by a description of breeding systems and levels of inbreeding depression in these taxa. The chapter concludes with a discussion of the implications of these findings for the conservation of rare plants.

In addition to surveying the literature, I will draw on the results of my own research with two locally endemic and three geographically widespread species of *Astragalus* (Fabaceae). This genus is well suited to such comparisons, since more than 50 of the

360 species in North America are either listed or under review for listing as threatened or endangered taxa (U.S. Fish and Wildlife Service 1985). For example, *A. linifolius* is limited to three populations in a 60-km² area near Grand Junction, Colorado. I have also studied *A. osterhouti*, which is known from only four populations in a 25-km² area near Kremmling, Colorado. Many other *Astragalus* taxa, however, are widely distributed across several states (Barneby 1964). One such species is *A. pectinatus*, which ranges from Saskatchewan, Canada, to southern Colorado. Barneby (1964) considers *A. pectinatus* to be one of the closest relatives of rare *A. linifolius* and *A. osterhouti*. Because the rare species are not sympatric with *A. pectinatus*, I have also studied two additional widespread taxa with ranges that include the distribution of the restricted species: *Astragalus linifolius* grows with widespread *A. lonchocarpus*, and *A. osterhouti* is associated with widespread *A. pattersoni*. These additional comparisons are valuable because co-occurring congeners are likely to be exposed to similar ecological constraints and selection pressures.

LEVELS OF GENETIC VARIATION

Two factors that could account for reduced levels of genetic variation in rare species are (1) changes in allelic frequencies due to chance events (genetic drift and the founder effect) (Wright 1931; Nei et al. 1975; Franklin 1980; Carson 1983; Levin 1984; Barrett and Kohn, Chapter 1), and (2) strong, directional natural selection toward genetic uniformity in a limited number of environments (Van Valen 1965; Babbel and Selander 1974). In addition, high levels of inbreeding followed by selection against individuals homozygous for rare alleles may further reduce genetic variation. Thus we may expect geographically restricted species to exhibit lower levels of polymorphism than are found in their widespread congeners.

To what degree is this predicted pattern evident in nature? Table 6.1 shows data on the extent of electrophoretically detectable genetic variation in 11 groups, each containing both locally endemic species and geographically widespread congeners. Since the data are presented as paired comparisons, it was necessary to exclude several thorough surveys that included only rare taxa (e.g., Griggs and Jain 1983; Schwartz 1985; Waller et al. 1987; Lesica et al. 1988).

In general, the restricted species exhibit lower levels of genetic variation than do the widespread taxa (Figures 6.1 and 6.2). However, the extent of genetic polymorphism varies considerably among the rare species. In several cases, geographical range is a poor predictor of levels of genetic diversity.

Some of this heterogeneity in levels of genetic variation may be due to historical factors. For example, the present range of a species may not correspond to its past distribution (Levin et al. 1979; Harper 1981; Kruckeberg and Rabinowitz 1985). Whereas some restricted species were once widespread and have more recently declined in range, others are of relatively recent origin and have never occupied a large range (Stebbins and Major 1965). In addition, the extent of genetic variation is likely to have been influenced by the severity of past genetic bottlenecks (Bonnell and Selander 1974; Franklin 1980; Frankel and Soulé 1981; Critchfield 1984; McClenaghan and Beauchamp 1986).

An example of a taxon with very low levels of polymorphism is *Pinus torreyana*, which is known from only two populations along the coast of southern California. No electrophoretically detectable variation is present in either population (Ledig and Conkle 1983). However, the populations are homozygous for different alleles at two of 59 loci. Ledig (1986b) has suggested that these low levels of polymorphism resulted from severe population bottlenecks.

In contrast to rare *Pinus torreyana,* most widespread species of *Pinus* have very high levels of genetic variation (Hamrick et al. 1979; Hamrick 1983). A notable exception is *P. resinosa,* which is remarkably uniform with respect to both allozymes and morphology (Fowler and Morris 1977; Allendorf et al. 1982; Millar and Libby, Chapter 10). Although this species is presently distributed throughout much of the northeastern United States, it is thought to have undergone a severe bottleneck during the Pleistocene glaciations (Davis 1976; Critchfield 1984).

Populations of widespread *Lisianthius skinneri* also lacked electrophoretically detectable enzyme polymorphism. However, Sytsma and Schaal (1985a) believe that the small populations they sampled were recently founded and do not necessarily reflect genetic diversity in this taxon.

A second rare species with very little genetic polymorphism is *Limnanthes macounii,* which is restricted to wet meadows and vernal pools on southeastern Vancouver Island and adjacent Trial Island in British Columbia (Kesseli and Jain 1984). In addition to low levels of genetic variation within populations, there was no evidence of genetic differentiation among the eight stations surveyed. The northernmost species in the genus—*L. macounii*—is also the most disjunct. Because alleles at three loci are not present in any other member of the genus, this taxon appears to have had a long history of isolation from its congeners.

In contrast to *Pinus torreyana* and *Limnanthes macounii,* rare *Astragalus linifolius* and *A. osterhouti* exhibit moderate levels of genetic variation (Karron et al. 1988). Polymorphism in these restricted *Astragalus* spp. is comparable to that in widespread *A. pattersoni,* but is lower than the levels recorded for widespread *A. pectinatus* (Figures 6.3 and 6.4). Populations of the rare species were not strongly differentiated. However, a surprising finding was that an enzyme that was polymorphic in the smallest population of *A. osterhouti* (population "C" has fewer than 400 individuals) did not vary in two larger populations.

Although most of the 23 rare species surveyed exhibit low to moderate levels of polymorphism, *Layia discoidea* is a rather remarkable exception. This serpentine species is restricted to 155 km² of the inner South Coast Ranges of California. Despite its narrow endemism, *L. discoidea* is polymorphic at 19 of the 21 gene loci surveyed (Gottlieb et al. 1985). This species is quite distinct morphologically from its widespread relative, *L. glandulosa.* However, electrophoretic data indicate that the two species are quite similar genetically. This finding supports the conclusion that *L. discoidea* is recently derived from *L. glandulosa* (Clausen et al. 1947; Gottlieb et al. 1985).

BREEDING SYSTEMS AND INBREEDING DEPRESSION

Certain aspects of the biology of rare plants may favor the evolution of self-fertility. In many locally endemic taxa, population size fluctuates dramatically (e.g., Vandermeer 1982; Griggs and Jain 1983; Menges et al. 1985; Huenneke, Chapter 2). If few

Table 6.1. A Comparison of Levels of Genetic Polymorphism for Geographically Restricted and Widespread Congeners

No. in Fig. 6.1	Taxon	Geographical distribution[a]	N loci	Percentage of loci polymorphic[b]	x̄ alleles per polymorphic locus	Source
1.	*Astragalus linofolius*	R	12	33.3	2.3	Karron et al. 1988
	Astragalus osterhouti	R	12	16.7	2.5	Karron et al. 1988
	Astragalus pattersoni	W	12	25.0	2.4	Karron et al. 1988
	Astragalus pectinatus	W	12	33.3	3.0	Karron et al. 1988
2.	*Capsicum cardenasii*	R	24	25.0	2.2	McLeod et al. 1983
	C. annuum var. aviculare	W	26	38.5	2.3	McLeod et al. 1983
	C. baccatum var. baccatum	W	25	24.0	2.3	McLeod et al. 1983
	Capsicum eximium	W	23	43.5	2.2	McLeod et al. 1983
3.	*Clarkia franciscana*	R	13	7.7	2.0	Gottlieb 1973a
	Clarkia lingulata	R	8	62.5	2.8	Gottlieb 1974
	Clarkia amoena ssp. huntiana	W	11	27.3	2.7	Gottlieb 1973a
	Clarkia biloba	W	8	62.5	3.4	Gottlieb 1974
4.	*Eucalyptus caesia*	R	18	38.8	2.1	Moran and Hopper 1983
	Eucalyptus cloeziana	W	20	30.0	ᶜ	Moran and Bell 1983
	Eucalyptus delegatensis	W	20	85.0	ᶜ	Moran and Bell 1983
	Eucalyptus grandis	W	20	65.0	ᶜ	Moran and Bell 1983
	Eucalyptus saligna	W	20	65.0	ᶜ	Moran and Bell 1983
5.	*Gaura demareei*	R	18	27.8	3.0	Gottlieb and Pilz 1976
	Gaura longiflora	W	18	33.3	3.2	Gottlieb and Pilz 1976
6.	*Layia discoidea*	R	21	90.5	2.7	Gottlieb et al. 1985
	Layia jonesii	R	17	64.7	3.0	Warwick and Gottlieb 1985
	Layia munzii	R	17	52.9	3.2	Warwick and Gottlieb 1985
	Layia leucoppa	R	17	58.8	3.2	Warwick and Gottlieb 1985
	Layia glandulosa	W	21	81.0	3.6	Gottlieb et al. 1985
	Layia platyglossa	W	17	82.3	3.7	Warwick and Gottlieb 1985

7.	*Limnanthes bakeri*	R	19	36.8	2.3	Kesseli and Jain 1984
	Limnanthes macounii	R	18	11.1	2.0	Kesseli and Jain 1984
	Limnanthes vinculans	R	17	41.2	3.0	Kesseli and Jain 1984
	Limnanthes douglassi "inland"	W	19	94.7	4.2	Kesseli and Jain 1984
8.	*Lisianthius aurantiacus*	R	12	33.3	2.0	Sytsma and Schaal 1985a
	Lisianthius habuensis	R	12	8.3	2.0	Sytsma and Schaal 1985a
	Lisianthius jefensis	R	12	16.7	2.0	Sytsma and Schaal 1985a
	Lisianthius peduncularis	R	12	33.3	2.0	Sytsma and Schaal 1985a
	Lisianthius skinneri	W	12	0	—	Sytsma and Schaal 1985a
9.	*Oenothera organensis*	R	15	6.7	2.0	Levin et al. 1979
	Oenothera biennis I	W	20	30.0	2.3	Levy and Levin 1975
	Oenothera grandis	W	18	33.3	2.8	Ellstrand and Levin 1980
	Oenothera hookeri	W	20	0	—	Levy and Levin 1975
	Oenothera parviflora I	W	20	40.0	2.5	Levy and Levin 1975
	Oenothera strigosa	W	20	25.0	2.2	Levy and Levin 1975
10.	*Pinus balfouriana*	R	23	57.6	2.1	Hiebert and Hamrick unpubl., cited in Hamrick et al. 1981
	Pinus longaeva	R	14	78.6	2.6	Hiebert and Hamrick 1983
	Pinus radiata	R	22	45.4	2.2	Moran et al. 1980
	Pinus torreyana	R	59	3.4	2.0	Ledig and Conkle 1983
	Pinus contorta	W	25	92.0	3.2	Yeh and Layton 1979
	Pinus ponderosa	W	20	60.0	2.3	O'Malley et al. 1979
	Pinus resinosa	W	27	11.1	2.0	Allendorf et al. 1982
	Pinus rigida	W	21	76.2	3.1	Guries and Ledig 1982
11.	*Stephanomeria malheurensis*	R	25	12.0	2.3	Gottlieb 1973b, 1979
	S. exigua spp. *coronaria*	W	25	60.0	3.0	Gottlieb 1973b, 1979

[a]R = restricted, W = widespread

[b]For a locus to be considered polymorphic, the frequency of the most common allele < 0.99.

[c]Not reported by authors.

Source: Modified from Karron (1987a).

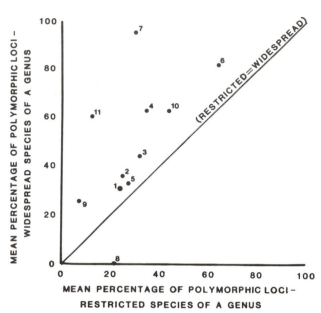

Figure 6.1. A comparison of the mean percentage of loci exhibiting polymorphism for the restricted and widespread species of each genus. If the mean values are identical for the restricted and widespread species, they will fall on the diagonal line. The percentage of loci exhibiting polymorphism is significantly higher for the widespread species (Wilcoxon's signed-rank test; $p < 0.05$). The numbers next to the data points refer to the genera listed in Table 6.1. (From Karron 1987a)

mates are available for cross-fertilization (Baker 1955, 1967; Jain 1976) or if pollinator service is unreliable (Tepedino 1979; Karron 1987b), natural selection may increase the frequency of self-pollinating individuals. Also, such species may exhibit lower levels of inbreeding depression due to the chance loss of deleterious alleles (Allard et al. 1968). Because inbreeding depression is often thought to constrain the evolution of selfing, a reduction in genetic load may permit an increased frequency of self-fertilization (Lande and Schemske 1985; Schemske and Lande 1985; but see also Waller 1986; Holsinger 1988b; and the discussion in Barrett and Kohn, Chapter 1).

In contrast to the many studies of genetic variation in rare plants, very little is known about their reproductive biology (Mehrhoff 1983; Kruckeberg and Rabinowitz 1985; Clampitt 1987). However, a few rare taxa have been studied in considerable detail. One of these is *Limnanthes bakeri*, a vernal pool species restricted to a 10-km-long valley in Mendocino County, California. Like its rare congener *Limnanthes macounii*, this taxon is quite distinct, both morphologically and electrophoretically, from other members of the genus (Kesseli and Jain 1984). Reduced protandry (spatial separation of anthers and stigma [Schoen 1982]) in this species results in a much higher level of autofertility than is found in widespread *L. douglasii* (Table 6.2; Kesseli and Jain 1986).

The breeding system of rare *Stephanomeria malheurensis* resembles that of *Limnanthes bakeri*. *Stephanomeria malheurensis*, known from only a single hilltop south of Burns, Oregon, is thought to be derived from widespread *Stephanomeria exigua*

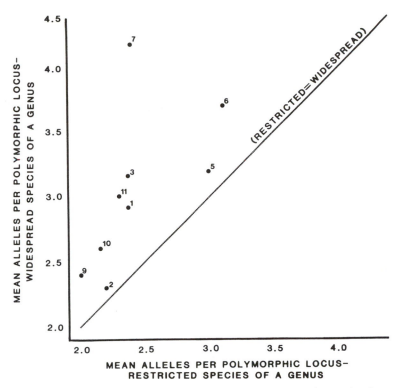

Figure 6.2. A comparison of the mean number of alleles per polymorphic locus for the restricted and widespread species of each genus. If the mean values are identical for the restricted and widespread species, they will fall on the diagonal line. The mean number of alleles per locus is significantly higher for the widespread species (Wilcoxon's signed-rank test; p <0.02). The numbers next to the data points refer to the genera listed in Table 6.1. Note that genus 3 (*Eucalyptus*) was excluded because the authors did not report the number of alleles per locus. Also, genus 8 (*Lisianthius*) was excluded because mean alleles per polymorphic locus could not be calculated for the monomorphic widespread species. (From Karron 1987a)

ssp. *coronaria* (Gottlieb 1973b, 1979). Although the widespread progenitor is an obligate outcrosser, *S. malheurensis* is highly autofertile.

Whereas many rare taxa are self-fertile, at least a few are obligate outcrossers. One such species is *Oenothera organensis,* which is restricted to the Organ Mountains of New Mexico. Flowers of this species have long corollas and are visited by strong-flying hawkmoths (Levin et al. 1979). Despite a low level of electrophoretic polymorphism (Levin et al. 1979), this species has a very well-developed genetic self-incompatibility system (Emerson 1939). Perhaps other factors limit the evolution of self-fertility in some restricted species. For example, if progenitor populations lack self-compatible phenotypes, derivative populations may lack them as well, despite potentially strong natural selection favoring self-fertility.

To test further the predicted patterns concerning breeding systems of rare taxa, I gathered data on levels of (1) pollinator service, (2) self-fertility, and (3) inbreeding

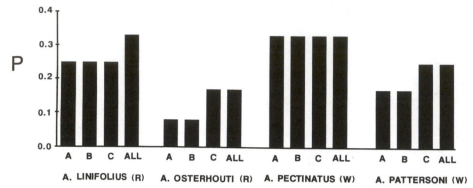

Figure 6.3.   Proportion of 12 loci exhibiting polymorphism (P) in populations of two restricted and two widespread species of *Astragalus*. Twenty-five individuals were sampled in each population. "A," "B," and "C" refer to the three populations of each species. "All" summarizes data for the three populations. (From Karron et al. 1988)

depression in both locally endemic and geographically widespread species of *Astragalus*.

Plants of each species typically produce hundreds of white or pale-colored flowers, with corolla lengths of 15–20 mm (Barneby 1964). The bilaterally symmetrical flowers are usually pollinated by medium to large bees in the genera *Bombus*, *Osmia*, and *Anthophora* (Karron 1987b). These similarities facilitate comparisons of the frequency of pollinator visitation to the rare and widespread congeners. At a site where restricted *A. linifolius* and widespread *A. lonchocarpus* co-occur at equal density, pollinator visits to the wide-ranging taxon were three times as frequent as visits to the rare species (Figures 6.5 and 6.6). This pattern was consistent in two different years.

During the two-day life span of an *A. linifolius* flower, the mean number of visits was less than one. Since geitonogamous pollinations (intraplant pollen transfer) make up the majority of floral visits, many *A. linifolius* flowers did not receive outcross

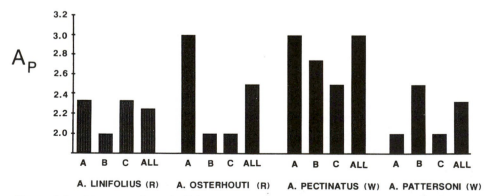

Figure 6.4.   Mean number of alleles per polymorphic locus (A_P) in populations of two restricted and two widespread species of *Astragalus*. Twenty-five individuals were sampled in each population. "A," "B," and "C" refer to the three populations of each species. "All" summarizes data for the three populations. (From Karron et al. 1988)

Table 6.2. Components of Self-Fertilization in Two Species of *Limnanthes*

Component	*L. bakeri* (restricted)	*L. douglassii*[a] (widespread)
Autogamous seed set per flower ($\bar{X}$)	2.11	0.06
Natural seed set per flower ($\bar{X}$)	3.29	3.48
Selfing rate[b]	0.79	0.05

[a]Data for *L. douglassi* population 511.
[b]Selfing rate = (1 − multilocus outcrossing rate).
Source: Modified from Kesseli and Jain (1986).

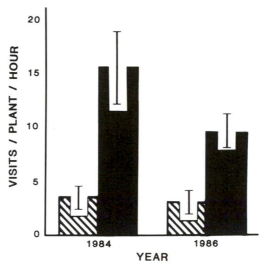

Figure 6.5. Frequency of pollinator visitation to plants of geographically restricted *A. linifolius* (hatched bars) and geographically widespread *A. lonchocarpus* (solid bars). Standard errors of the mean are shown with error bars. (From Karron 1987b)

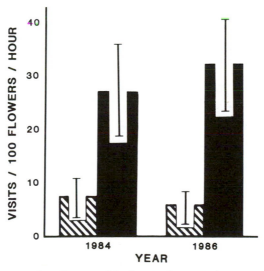

Figure 6.6. Frequency of pollinator visitation to flowers of geographically restricted *A. linifolius* (hatched bars) and geographically widespread *A. lonchocarpus* (solid bars). Standard errors of the mean are shown with error bars. (From Karron 1987b)

pollen. A similar lack of pollinator service was also evident for rare *A. osterhouti* (Karron 1987b).

Given that the restricted species receive few visits from pollen vectors, I performed experimental pollinations to determine whether these taxa are more self-fertile than their widespread congeners. Rare *A. linifolius* will often set fruits in the absence of pollinators, and both *A. linifolius* and *A. osterhouti* are highly self-fertile (Figure 6.7). By contrast, closely related widespread *A. pectinatus* rarely produces fruits when self-pollinated. However, a second, less closely related widespread congener, *A. lonchocarpus*, is self-fertile. Thus these data only partially support the hypothesis that restricted species exhibit higher levels of self-fertility than do their widespread congeners.

Since *A. linifolius* fruits are often produced by selfing, natural selection may reduce the frequency of deleterious recessive alleles. If inbreeding depression results primarily from the effects of such alleles (see the discussion in Barrett and Kohn, Chapter 1), then this species may be expected to exhibit a low level of inbreeding depression. To investigate this prediction, I germinated *A. linifolius* seeds from self-,

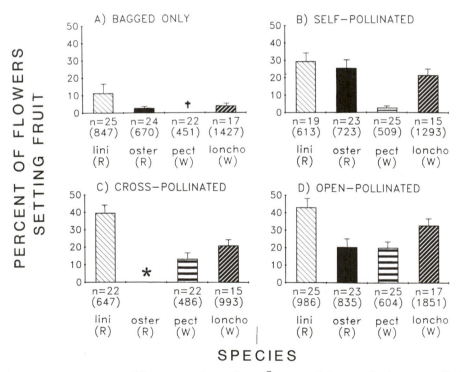

Figure 6.7. Percentage of flowers producing fruits ($\bar{X} \pm 1$ *SE*) following the four controlled pollination treatments indicated (A–D). Pollinations were performed in the field on plants of two restricted (R) and two widespread (W) *Astragalus* species. Abbreviations and symbols: "lini" = *A. linifolius*; "loncho" = *A. lonchocarpus*; "oster" = *A. osterhouti*; "pect" = *A. pectinatus*; *n* = number of plants and (flowers); † = fruit set equals zero; * = cross not attempted. (From Karron 1989)

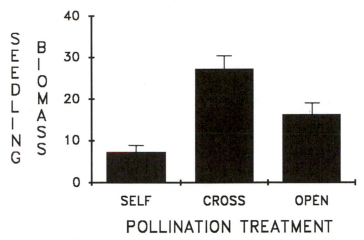

Figure 6.8. Total seedling biomass (in milligrams) for *A. linifolius* seeds resulting from self-, cross-, and open-pollination treatments ($\bar{X} \pm 1$ *SE*). Seedlings were grown in a growth chamber for 60 days. (Modified from Karron 1989)

cross-, and open (natural)-pollination treatments. Seedlings were grown in a growth chamber for two months and then harvested and oven-dried.

The three treatments did not differ in percent germination. Thus there was no evidence for inbreeding depression at this stage of the life cycle. However, 60-day-old seedlings produced by selfing had one-third the dry weight of progeny resulting from outcrossing (Figure 6.8). This dramatic level of inbreeding depression was unexpected, since it is higher than many of the values obtained in studies of widespread species (Barrett and Kohn, Chapter 1).

Seedlings produced by open-pollination are presumably a mixture of self- and cross-fertilization. As might be expected, these progeny were of intermediate fitness. This suggests that inbreeding depression may frequently occur in *A. linifolius* populations. Although these results cannot be readily explained, the genetic load implied by the inbreeding depression is consistent with the moderate level of genetic polymorphism recorded for this species (Figures 6.3 and 6.4). Clearly, additional studies of inbreeding depression in rare taxa are needed.

IMPLICATIONS FOR CONSERVATION OF RARE SPECIES

Since rare taxa exhibit a wide range of life-history characteristics, generalizations concerning their genetic structure and breeding systems should be interpreted with caution. Often the most useful studies are comparisons involving pairs of closely related rare and common congeners.

Locally endemic species tend to have reduced levels of polymorphism (Table 6.1). However, this generalization has several exceptions, as is evident in studies of *Pinus longaeva* (Hiebert and Hamrick 1983), *Astragalus linifolius* (Karron et al. 1988),

Layia discoidea (Gottlieb et al. 1985), and several species of *Orcuttia* (Griggs and Jain 1983).

Some rare species exhibit genetic differentiation between populations. For example, a small population of *Astragalus osterhouti* (Karron et al. 1988) contains alleles that are absent in other, larger populations. Therefore, an attempt should be made to conserve genetic material from as many populations as possible.

It is even more difficult to generalize about the breeding systems of rare species. Some narrow endemics, such as *Limnanthes bakeri* (Kesseli and Jain 1986), will set seed in the absence of pollinators. However, many self-compatible species are not autofertile; they will produce seeds only when an insect visitor manipulates the flower. Finally, some species, such as *Oenothera organensis* (Emerson 1939) and *Layia discoidea* (Gottlieb et al. 1985), are self-incompatible. These taxa require cross-fertilization for seed production.

Whenever possible, conservation biologists should perform experimental pollinations to determine the nature of a rare species' breeding system. If a taxon requires pollinators for full seed set, *in situ* conservation programs should attempt to protect nearby pollinator populations. Two types of human disturbance are particularly detrimental to bee populations:

1. Pesticides used to control insect herbivores may be harmful to bees. Since most bee species have low fecundity, their populations may not recover for many years following spraying (Kevan 1975; Johansen 1977; Tepedino 1979; Thomson and Plowright 1985).
2. Some bee taxa, such as *Bombus,* nest in abandoned rodent burrows (Byron 1980), and grazing by livestock decreases the number of nest sites by causing burrows to collapse (Sugden 1985).

When pollinator populations are inadequate, supplemental pollinations by humans will be useful, although time-consuming. Also, hand-pollination will often be necessary for *ex situ* conservation programs.

Few data are available concerning inbreeding depression in rare plants. However, recent studies of more common species indicate that self-compatible taxa often exhibit moderate levels of inbreeding depression (Barrett and Kohn, Chapter 1). The relative fitness of self- and outcross-pollinated progeny has recently been determined for one rare species: *Astragalus linifolius*. Although this taxon is highly autogamous, seedlings produced by selfing exhibit high levels of inbreeding depression (Karron 1989). This indicates that plant conservation biologists will sometimes need to incorporate breeding protocols into their management programs.

ACKNOWLEDGMENTS

This chapter is dedicated to the memory of Deborah Rabinowitz. I am grateful to Yan Linhart for his advice and encouragement. Field research on *Astragalus* was funded by an NSF Dissertation Improvement Grant (BSR-8514441) and a grant-in-aid of research from Sigma Xi.

7

Sampling Strategies for Genetic Variation in *Ex Situ* Collections of Endangered Plant Species

A. H. D. BROWN and J. D. BRIGGS

Strategies for the sampling of genetic resources should rest on a clear statement of the purposes for which the material is to be used. For *ex situ* collections of rare and endangered species, various purposes have been proposed (e.g., Given 1987). The major actual or potential roles for such collections are to provide material (1) to cultivate for study and display in botanic gardens; (2) to exploit commercially in horticulture, plantations, or forestry, and thereby replace commercial harvesting from natural populations; (3) to reestablish or enrich wild populations; (4) to be a source of single genes or characters for future transfer to domesticated species; and (5) to act as a refuge of last resort against extinction.

The *link* between conservation of endangered species in designated secure sites *in situ,* and in collections *ex situ,* is fundamental to the design of sampling strategies. This is so despite the problems associated with and criticisms made of using *ex situ* material for replanting (Frankel and Soulé 1981; Ashton 1987). In reality, *ex situ* and *in situ* conservation are complementary and, as Falk (1987a) and Given (1987) have argued, should not be viewed as alternatives. This link goes beyond the issue of whether a return to nature is possible or practicable. It influences sampling techniques, informs us about the species' ecology and constraints on population size, suggests uses of the material, and reinforces management and conservation *in situ.*

The first decision to be made is the order of priority of the species in question. In some regions of the world, such as the species-diverse tropical habitats, vast areas are presently subject to widespread destruction (Raven, 1976). There, the lists of endangered species are long and incomplete, and the amount of effort per species is strictly limited. Species sampling for *ex situ* conservation must be highly selective for usefulness and cover as much species diversity as possible.

This chapter, however, focuses on the situation in other regions of the world such as North America, Europe, and Australia, where the number of endangered plant species is not so overwhelming. The task of assembling collections of all such listed species for *ex situ* conservation is a realistic one. Even so, priorities should be assigned, based on the degree of threat, current and potential use, scientific interest, and ease and sense of sampling and conserving.

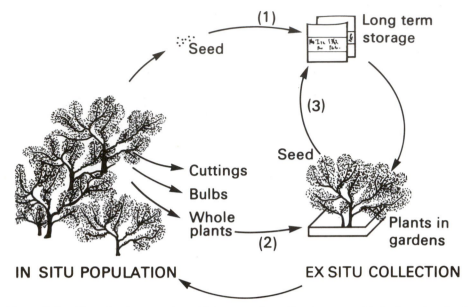

Figure 7.1. Three basic routes by which an *ex situ* collection is established, and its feedback to the source *in situ*.

Ex situ conserved collections are of two kinds. The first is material grown in gardens, either institutional botanic gardens or private gardens and the horticultural trade. The second kind is material in the stored state, commonly as processed and packaged seed in long-term storage, but conceivably as tissue cultures in meristem banks (Hawkes 1987a) and possibly DNA libraries (Ashton 1987). These two kinds of collections, the growing and the stored, differ strikingly in the number of individuals and of species that can be handled.

Figure 7.1 portrays the options in the process of setting up a conserved collection *ex situ*. The source population of the endangered species is the object of *in situ* conservation, and provides material (seeds, cuttings) for the *ex situ* collection, with which it shares a continuing link (signified by the arrows). The two kinds of *ex situ* collections are the plants in gardens and the stored seed. The *ex situ* collection is then set up (1) by sampling seed directly from original sites and processing and storing the natural seed, (2) by growing the source material to maturity as plants in gardens that are replanted as necessary to maintain the species, or (3) by harvesting seed from the mature plants in gardens and processing this seed for long-term storage. This model presently underlies the work of the Center for Plant Conservation

THEORY OF SAMPLING STRATEGIES

The basic sampling variables that the collector must set are the number and location of populations, and the number and type of plants in each population. We consider first the number of individuals to sample from a single population, and then the number and arrangement of populations to be sampled. The choice of individuals at a site is discussed in the next section on sampling practice.

Sample Size at a Site

In many situations, the size of sample may be set by the scarcity of individuals. For example, Table 7.1 gives the frequency distribution of population sizes for 130 populations of 47 species that occur in New South Wales and that Briggs and Leigh (1988) list as endangered. Clearly endangered species are often locally very rare. At the extreme, the population may number only one or two individuals, and the sample would obviously consist of seed and/or slips from every individual.

Assuming that there are more than a few individuals at the site, the next logical sampling objective to achieve would be the number required to ensure that at least one living organism (or two, if self-incompatible) for that site successfully enters the *ex situ* collection. Sample size must obviously allow for poor germinability of seed, poor striking from cuttings, survival after quarantine and other handling damage, and the duplication and sharing of material among institutions before or after storage.

Genetic Variation

Why collect any more than this? Hawkes (1976, 1987a) has argued that with the replacement of the typological concept of a species by the population concept, genetic variation within a population should be collected as an end in itself. He stated that the need is "to capture as large a part as possible of the genetic diversity within species." This is a laudable but idealistic aim, and may be particularly desirable for high-priority species, when the *ex situ* collection can be set up directly in storage (method [1] in Figure 7.1).

Yet, it would be counterproductive if excessive effort went into the sampling of one species, while others became extinct. Further, it would be undesirable to commit excessive resources to processing and storing large collections of an easily collected species to the detriment of the proper documentation, study, and use of the material. This problem has emerged in crop gene banks, where the large size of some collections has deterred their use (Frankel and Brown 1984; Brown 1989a).

Therefore, the collection of genetic variation within species for *ex situ* conservation should be thought of as a means to an end, rather than as an overriding end in itself. Genetic variation is the raw material that will increase the chance of the species sur-

Table 7.1.   Population Sizes for 130 Populations of 47 Endangered Species that Occur in New South Wales

Population size class	No. of populations
1–5	29
6–10	22
11–25	23
26–50	12
51–100	8
101–1000	31
>1000	5

Source: Briggs (unpubl. data).

viving through changed circumstances in the *ex situ* collection. It will increase the chance of using the stored material effectively for replanting or exploitation.

From this viewpoint, it is not possible to recommend a specific proportion of the genetic variation present in a population, or in the species as a whole, that must be represented in the *ex situ* collection. This proportion should vary from case to case. In an extremely rare, high-priority species found only in one highly threatened population, one should aim to sample most of the existing variation. In contrast, a more abundant species in several, more secure and diverse sites requires samples of only a fraction of its genetic diversity. Rather than a fixed fraction, the approach we advocate is to have clearly defined minimum numbers as basic targets, and prudently to exceed these minima for insurance when resources and material permit and likely needs indicate.

How are we to specify genetic variation and measure the effect of sample size on collecting it? There follow three basic ways to do this.

1. A diverse array of different alleles is needed to meet future environmental changes, both physical and biotic (e.g., pathogen pressure). The number of distinct alleles per locus, or *allelic richness,* is the appropriate measure for this concept.
2. *Heterozygosity* (gene diversity) in the progeny produced by the conserved individuals is needed to avoid inbreeding depression. The appropriate measure for this would be the average heterozygosity of the sexual *progeny* produced from the sampled individuals.
3. Many different genotypes or *genets* (Harper 1977)—that is, many different combinations of alleles at all the loci constituting the genome—are needed to give a range of phenotypes. We will call the number of distinct genotypes the *genet richness.* This measure of genetic variation is relevant when individuals are preserved *ex situ* by asexual propagation.

These three measures of genetic diversity are sufficient to consider the sampling problem. Other more complex measures are available and may have a limited use. For example, Sirkkomaa (1983) used the information index or allele entropy to compute the effect of bottleneck effects in specific examples of allele frequency. Marshall and Brown (1983) considered the sample sizes required to retain allele frequencies of forage plant populations within specified intervals of their original values. Such a sampling goal is relevant for population genetic studies but is unjustified for present purposes, and the large numbers required to achieve it are impractical. Ashton (1987) argued that character complexes are more important than individual loci, implying that these complexes should be measured. However, such complexes are either held together through linkage and therefore act as alleles of a supergene, or lost through recombination after hybridization. When such complexes are common components of a population, their sampling theory is well approximated by that of single loci.

Infinite Neutral Allele Model

Plant populations display a great range in levels and patterns of genetic variation, or population genetic structure (Hamrick and Godt 1989). Unfortunately, the structure for any target species is largely unknown in advance of sampling. General guidelines for sampling thus must rest on the analysis of theoretical models, refined if possible with knowledge of breeding system and population distribution. As Marshall and Brown (1975) noted, the infinite, selectively neutral alleles model of Kimura and Crow (1964) is an ideal one for this purpose. The model assumes that all variants are selectively neutral. Each allele arises as a unique mutation and is ultimately lost through the sampling effect of finite population size.

To use this model in the present context, however, does not imply that genetic variation is in fact strictly neutral in nature and of no consequence to variation in fitness. Such a *credo* would deny a genetic dimension to conservation, and would allege that sampling can be done without genetical concepts in mind. The neutral model is a suitable one for the following reasons: First, an extensive theory now exists of the population genetic structure of neutral variation, and of the sampling of such variation.

Second, the neutral theory yields a single parameter family of distributions of allele frequencies. The indexing parameter is θ and equals four times the effective population size times the mutation rate. For a particular value of θ, the "evenness" of the allelic frequencies of the neutral allele model is intermediate. Its frequencies are more even than are the profiles for models that assume a balance between mutation and deleterious selection on the one hand, and less even than the profiles for models that assume heterotic selection on the other.

Thus the model of infinite, selectively neutral alleles is an "average" one, with known sampling theory and is a convenient yardstick. Despite indications that the model endorses sampling strategies that are widely applicable (Marshall and Brown 1975; Sirkkomaa 1983), more research is needed on the following questions: Do other models of allele frequency distributions lead to similar strategies? What kinds of allele frequency distributions are present in target species?

Allelic Richness

Allelic richness, or the number of alleles per locus, is the measure of genetic variation of most use in crop plant genetic conservation (Marshall and Brown 1975). For crop species, *ex situ* collections serve primarily as sources of single characters or individuals genes that plant breeders will transfer into their locally adapted stocks. Arguing that the more *common* alleles in the source population (alleles with frequency in excess of, say, 0.05–0.10) merit priority, Marshall and Brown (1975) proposed that the aim of sampling should be to recover in the sample at least one copy of such alleles with guaranteed probability of, say, 0.90–0.95. To meet this objective, a sample of 50–100 randomly chosen individuals is required. Any further effort should then be spent visiting as many different and diverse sites as possible, rather than collecting larger samples at one site. This strategy is efficient in sampling from various theoretical allelic profiles in the recovery of numbers of distinct alleles (Marshall and Brown 1975).

Despite criticism (see Marshall [1989] for review and citations), the same reasoning and strategy was later extended to forest trees (Brown and Moran 1980), forage plants (Marshall and Brown 1983), and wild species related to crops (Chapman 1989). The case of rare and endangered species is perhaps covered by these extensions. However, it requires particular focus because the constraints on the sampling are quite different. There are larger numbers of endangered species to consider, and the populations of each endangered species are usually few in number and small or highly variable in size. The *ex situ* collection can be set up in several ways (Figure 7.1).

Further issues include the potential trade-off for more interspecific diversity at the expense of intraspecific diversity, and the importance of conserving the source population *in situ* if at all possible.

Perhaps the most contentious issue in extending the formulae for crop plants to endangered species is the ranking of alleles in conservation priority. In particular, some may question whether sampling strategies should focus on common alleles and neglect rare ones. (Note that the word "rare" here refers to the allele frequency in a single population, and not necessarily to a paucity in absolute numbers.) A rare allele is defined as having a frequency less than 0.10. Thus by definition, a population of five adult plants would have no rare alleles.

We contend that rare alleles have been of much greater concern than they deserve for three reasons:

1. Most endangered populations are small in size—much smaller than crop populations—so that allele frequencies are severely truncated anyway.
2. The sampling of crop genetic resources is aimed at providing *alleles* per se for crop improvement. In contrast, the sampling for *ex situ* collections of endangered species is aimed primarily at the preservation of the *species,* principally from threats due to humans. Rare alleles may have some place in plant breeding, but they are likely to have far less a role in conserving endangered species. The current local rarity of such alleles testifies to their insignificant contribution to present adaptedness.
3. The actual value chosen for the minimum frequency has a great effect on sample size. Alleles at or below this frequency demand excessive resources to ensure their sampling; thus their conservation may jeopardize that of other species.

The Logarithmic Relationship

The key question for our purposes is, How does the number of alleles per locus (k) in the sample vary with sample size (n)? Ewens (1972) first gave the sampling theory for the stationary distribution on allele frequencies in the neutral allele model. A simple approximate formula based on his theory gives the relation between n and k as logarithmic (Brown 1989b):

$$k \simeq a + b \log n$$

where a and b are constants depending on the indexing parameter (θ) This parameter incorporates mating system, population structure, and population size, through their effects on the effective population size. Thus it is very important to note that the logarithmic relationship holds regardless of the values of these variables.

To what extent is the logarithmic nature of this relationship dependent on the assumption of equilibrium? Real populations are rarely, if ever, in equilibrium (Barrett and Kohn, Chapter 1). In particular, as emphasized by Wright (1938), population size can have a critical effect on the level and pattern of genetic diversity. In a population initiated from a few individuals, this is known as the *founder effect.* Nei et al. (1975) drew attention to the effect that a recent sharp reduction in population size to a few individuals, or *bottleneck,* could have on allelic richness. Both bottleneck and founder theory are relevant in broad terms to the problem of sampling for conservation, and of reestablishing natural populations if the primary sites become extinct.

Theoretical treatment of the bottleneck effect for the case of infinite neutral alleles followed the paper of Nei et al. (1975). Griffiths (1979) provided an exact formulation of the transient distribution of allele frequencies and showed that in a large sample the expected number of alleles is close to the expected number in the stationary distribution. Watterson (1984) derived formulae to specify the probability distribution, after a bottleneck, of the number of alleles and other measures of genetic structure. Maruyama and Fuerst (1985b) studied the decline in allele numbers in the early phases of a prolonged bottleneck in a previously large population, using the diffusion approximation. The longer-term consequences of maintaining the bottleneck not surprisingly reflect the dramatic loss of alleles that occurs when a population passes through many generations of reduced size.

Figure 7.2 plots some of Griffiths's (1979) data for the expected number of alleles in samples from three kinds of populations. The effect of sampling after a recent bottleneck (case i), compared with stationarity (case ii), is that fewer alleles are recovered. This is as expected, since fewer alleles are present in the population. What is noteworthy is that all three cases suggest an approximately linear relationship between the number of alleles and the logarithm of sample size. The same form of relationship is

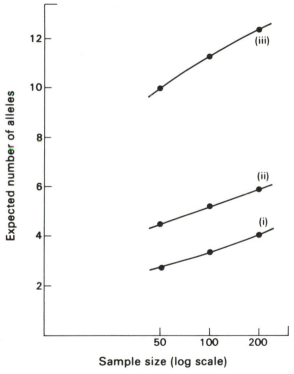

Figure 7.2. The expected number of alleles in a sample of size *n* as a function of log sample size from a population with $\theta = 1.0$. The population is stationary (ii), is 0.4*N* generations from being initially monomorphic (i), or is composed of a large number of equally frequent alleles (iii). (Data from Griffiths 1979)

present in the sampling data of Sirkkomaa (1983) for the proportion of alleles that survive a bottleneck at frequencies >0.05.

This is a very important result if it is shown to hold generally for sampling from real allelic profiles. It argues that there is a strong law of diminishing (genetic) returns on sample size in terms of the allele richness of the sample (Brown 1989b). The first ten organisms randomly sampled from a population are as important as, if not more important than, the additional 90.

Support for the great importance of a logarithmic perspective on population (or sample) sizes comes from another source. Ewens et al. (1987) analyzed a purely demographic model to determine minimum viable population sizes in the face of random catastrophes. They found that the median extinction time for a population subject to such catastrophes depends on the logarithm of its initial size. Thus the major contribution to the postponement of extinction from catastrophes is the first portion of the population conserved.

Thus factors such as local population size, *ex situ* constraints, and/or logarithmic theory argue that an appropriate *minimum target for each population* is ten, rather than 50–100 mentioned above. Yonezawa (1985) has recently suggested that 10 be considered an adequate minimum, if the resources saved can be devoted to sampling other populations.

Heterozygosity

This measure has been frequently considered in the literature on the genetics of animal species conservation (Frankel and Soulé 1981) particularly in captive breeding of large animals (Lacy 1987). Heterozygosity is only partially related to allele richness, and the variation in breeding system among plant species further reduces this relationship. The effect of sample size (n) on the change in heterozygosity (ΔH) from that in the original population (H) is

$$\Delta H \approx - \frac{H}{2n}$$

The inverse relationship indicates that n has to be very small for a real reduction in average heterozygosity to occur. However, the issue is the relation between any decline in heterozygosity and the probability of survival—that is, the inbreeding depression curve. Clearly this will vary greatly among plant species, with inbreeding species likely to show much less effect.

The use of heterozygosity as a measure of genetic variation can be very problematic in plants. One usage is the *expected heterozygosity,* assuming the species is mating at random (i.e., Nei's gene diversity). Another is the *observed heterozygosity.* For the case of endangered plants, we have defined the quantity of interest as the observed heterozygosity in the population bred from the original sample. The value thus involves a mating design that can be varied. For example, one may achieve higher than natural levels of heterozygosity in small samples of the partially cleistogamous *Glycine argyrea* (a rare Australian perennial herb) by deliberate intermating of few individuals or by harvesting legumes from only the open-pollinated chasmogamous flowers (Brown et al. 1986).

Genet Richness

The number of distinct genotypes or genets in a population is a third measure of genetic diversity. In this concept, no account is taken of the nature and extent of genetic differences among the various genotypes (at how many loci, etc.) or of the intensity of testing for genetic differences. Yet the concept is useful in species that are closely self-pollinated or apomictic, or are capable of clonal propagation. One example of the latter is *Eucalyptus pendens*, a rare mallee-type tree species in which several apparently independent trunks are members of the same ancient genet (Moran and Hopper 1987). The sampling of different genets entails watching for phenotypic differences in the field.

The relationship between genet richness and sample size would, for outbreeding nonclonal species, approach linearity because each individual is unique. At first sight, this would argue for much higher sample sizes than were indicated by other measures of diversity. However, if we take a utilitarian view of genet diversity, a strictly limited sample is again justified.

Consider a population of N distinct genets. Assume that some number of the genets (s) are genetically equipped to "survive" both sampling and conserving *ex situ*, and fulfill the purpose of the project. They are the future survivors, and the remainder ($N - s$) would die. The case of $s = 0$ is not included because no sampling strategy would rescue such a population. The problem is to design a sampling strategy that gives a reasonable chance of including in the sample at least one of the survivors. This would achieve the basic minimum objective of preserving the species from extinction. There is no way of knowing a priori which of the currently available genets will be survivors. Suppose that a sample n is drawn for *ex situ* conservation. The probability of a successful outcome for the project is

$$P_s = 1 - \prod_{i=0}^{n-1}\left[1 - \frac{s}{n-i}\right]$$

For large populations ($N \gg n$), the geometric approximation is more useful

$$P_s' = 1 - (1 - \frac{s}{N})^n \qquad \text{when } N \gg n$$

Hypergeometric and geometric expressions are typical in these sorts of problems (Marshall and Brown 1975; Sirkkomaa 1983; Brown 1989a, 1989b). Note that the geometric $\bar{P}'_s$ is less than or equal to P_s, and hence is a convenient minimum.

Table 7.2 gives the probability of success (P_s) for various target populations sizes (N) and viable fractions (s/N) of 0.1 and 0.5. When the ability to survive is rare ($s = 0.1N$), the probability of success decreases (at a decreasing rate) with increasing size of the target population. Figure 7.3 plots the sample size required to be 90% sure ($P_s > 0.9$) of success for viable fractions of 0.1, 0.2, and 0.5. When the majority of genotypes are capable of surviving ($s > 0.5N$), the sample size required is independent of the population size, and depends more on the specified level of guarantee. When the surviving fraction is small, the number required to be 90% sure of success increases with population size.

Table 7.2. Probability of a Successful Rescue (P_s) for a Sample of Size n
from a Population of N Distinct Genets, Giving a Fraction (s/N) that Has the Inherent
Capacity to Survive Sampling and Conservation *Ex Situ*

	$s = 0.1N$, at $N =$				$s = 0.5N$, at $N =$			
n	10	20	50	Large	10	20	50	Large
1	0.10	0.10	0.10	0.10	0.50	0.50	0.50	0.50
2	0.20	0.19	0.19	0.19	0.78	0.76	0.76	0.76
3	0.30	0.28	0.28	0.28	0.92	0.89	0.89	0.89
4	0.40	0.37	0.34	0.34	0.98	0.96	0.95	0.94
5	0.50	0.45	0.41	0.41	1.00	0.98	0.97	0.97

In self-incompatible species, at least two individuals are needed to survive for seed production. The sample sizes required from a large population to ensure this minimum (two or more survivors) at the 90% level are 7, 18, and 38 for s/N values of 0.5, 0.2, and 0.1, respectively. This is approximately double the requirement for self-compatible species (Figure 7.3).

The concept of "survival" or success is intended to include all phases of the conservation effort, from initial sampling of the *in situ* source to final use of the material. Circumstances that may lead to low values of s include unfavorable material, poor seed viability or germination, insufficient cuttings per plant, incorrect timing of collection, incorrect treatment of seeds or cuttings, maintenance of the plants in an environment dissimilar to that at the origin, and so forth.

What values would the parameter s assume? Few, if any, estimates of this variable

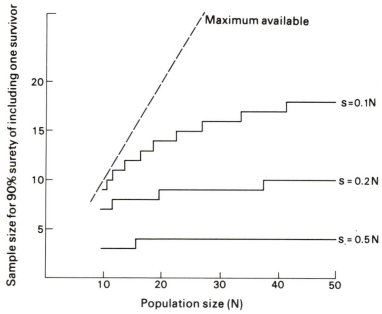

Figure 7.3. Sample size (n genets) from a population of size N genets, required to be 90 percent sure of including at least one survivor, given that s/N ($= 0.1, 0.2, 0.5$) are capable of survival.

in endangered species have been published. We estimated the survival rates to 6–12 months following transplanting of cuttings of 18 rare or threatened species collected in eastern New South Wales and propagated at the Australian National Botanic Gardens, Canberra. The average value of s/n was 0.74, and we assume that this value estimates the value of s/N. This figure is based on samples from a total of 104 individuals from 28 populations (a maximum of 5 individuals had been collected from each). Improvement of this strike rate would follow from handling experience, revisiting of sites, larger samples from individuals, more rapid processing of difficult material, and so forth.

However, the striking of cuttings is only the first phase of the process. The plants must then be maintained in the gardens, the seeds harvested from them, and these seed stored for later use. The whole process may require us to deal with values of s/N as low as 0.1. At the species level, the values in Figure 7.3 suggest that at least 15 individuals *per species* should be sampled. This "species minimum" of 15 would override the minimum requirement of 10 for a population, when only one population exists. For dioecious species, each sex should be sampled separately, with a minimum of 15 from each.

The parameter s/N in practice includes effects that are amenable to managerial control (such as the number of cuttings struck) and those that are not (such as unforeseeable future garden environments). Clearly, sampling and cultural practice should be such as to make s/N as high as practicable, particularly if its value would otherwise fall below 0.1. If, despite the best practices, the inherently surviving proportion of the original population is less than this value, lower levels of surety (P_s) will have to be accepted. Otherwise the excessive resources required by one very refractory species will infringe upon the right of others to survive.

Minimum Sample from Each Single Population

The above discussion of sampling theory and associated issues forms the basis for the following explicit recommendations, where the numbers refer to fully viable material successfully established in the *ex situ* collection.

1. If seed is being collected for direct processing into storage, then collect from 10–50 individuals sufficient seed for a stored sample (ideally amounting to 1000 seed, depending on seed size). The figure of 10 is the number to strive to meet as a "population minimum." That of 50 represents the ideal when seed is easy to collect from such a number of plants.
2. Maximize the diversity within the sample by collecting as many distinct fruit (to increase the number of pollination events and pollen sources), at different times, from different microhabitats.
3. Aim to represent each source plant equally.
4. To maintain options, try to keep the seed from individual plants separate, if this is feasible.
5. The notional overall-figure of 1000 seed is about one-tenth of that suggested by Hawkes (1987a), and is considered a more practical minimum. Samples of fewer than this will need later resampling, or multiplying *ex situ*.
6. If material (cuttings, etc.) is being collected for growing *ex situ*, sample 10

individuals from varying microhabitats or randomly, but include any morphological variant likely to be both genetically based and of definite interest.

Sampling of Several Populations

The next set of decisions by the collector deals with the strategy for multiple population sampling. The sampling of different populations of a species aims to capture a range of its interpopulation diversity. Variation at this level is intermediate in status. On the one hand, differences among *populations* are usually not as significant as differences among *species* or *subspecies*—which are the nominated items for conservation. On the other hand, it is more significant than *intrapopulational* variation, which we have argued is a means to that end. It is particularly important for inbreeding or asexual species because interpopulation variation makes up a larger fraction of their total variation than it does in sexual outbreeders.

Number and Pattern of Populations to Sample

The deployment of resources among populations will be an issue for many endangered species. Table 7.3 gives a classification of 47 such species in southeastern Australia according to the number and maximum distance between their populations. In the majority of cases, more than one population is available for sampling, and such samples should be sought. Indeed, diversity usually can be increased by including more sites with fewer per site. The importance of multisite sampling is that it will include material able to survive in a greater range of environments than the sample from a single site. The need to sample fewer per site when more sites are included may be forced on the collector if material has to be grown before being stored (Figure 7.1, method 3) or is not amenable to storage (method 2).

The examples in Table 7.3 suggest that most endangered species occur in only a few distinct populations. If logistics and the meeting of other targets permit, a reasonable aim is to collect from *all* populations if there are around five or fewer. With larger numbers of populations, it is sensible to employ a clustered arrangement to take account of hierarchical patterns of genetic variation (Marshall and Brown 1975; Hawkes

Table 7.3. The Number of Species Out of 47 Endangered Species in Southeastern Australia with a Recorded Number of Populations per Species Within Classes of Geographical Range

Geographical spread (km)	No. of populations									
	1	2	3	4	5	6	7	8	9	10
<1	22									
1–10		1								
10–50		5	1		1	1				
50–100				1	2					
100–500		1		1		2			2	
500–1000			1		1	1		1		
>1000				1						2
Total	22	7	2	3	4	4	–	1	2	2

Source: Briggs (unpubl. data).

1976, 1987a). The numbers would be a maximum of five clusters with one or two populations per cluster. For two, three, or four clusters, one would sample two or three populations per cluster. The choice of five populations, or five clusters per species as a maximum, rests on the idea that with samples of ten individuals per population, an overall sample of 50 is achieved.

The populations of endangered species often occur in a clustered pattern. For example, the Australian species *Rutidosis leptorhynchoides,* an endangered perennial forb of woodland and grassland, occurs in two highly disjunct areas some 500 km apart (Leigh et al. 1984). The first area is west of Melbourne where five sites remain. There the populations range in size from 11 plants to 300 plants, with a total of about 1000 plants. The total range of occurrences is 40 km. The second area of occurrence is Canberra–Queanbeyan, New South Wales, where five sites with a spread of 12 km remain. There the populations range in size from about 100 plants to a few thousand, with a total approaching 5000 plants. Although this species once occurred further west of Melbourne, it has never been recorded from the 500-km span of country between the Melbourne and Canberra populations. The recommended sample pattern in this case is three populations from the Melbourne area and three from Canberra.

Another practical criterion in the choice of populations relates to the link between *ex situ* and *in situ* conservation. Clearly, sampling preference would go to populations that fit into some long-term conservation plan *in situ.*

Ecotypic Variation

The above discussion has assumed that genetic variation between populations is interchangeable with that within populations. For example, we assume that multisite samples, when regenerated *ex situ,* can give rise to progeny as likely to survive on reintroduction as would a single-site sample. For populations growing at similar latitude, altitude, and aspect, on similar substrate, and broadly within the same community of associated species, this is a reasonable assumption. A transplant experiment could be assumed to succeed. Progeny arising from crosses between populations could be expected to survive and multiply.

However, populations of the same species can diverge into recognizable ecotypes between which gene flow would lead to poorly adapted progeny. Collection and conservation of these ecotypes must take such potential divergence into account. Sometimes explicit scientific or commercial interest in ecotypes of a species, or morphological races, may justify samples from such populations. For example, the marginal, cold-tolerant populations of a rare species related to a crop plant such as the Australian *Glycine latrobeana* are of specific interest. Nominating such an ecotype as an explicit target gives purpose to its conservation (Ashton 1987). However, the sampling of a diverse array of ecotypes in one species may compete with the major effort on other species and must also be considered in that light. Most endangered species will exist in but a few populations, the more marginal of which usually merit no special priority. Such marginal populations may be depauperate of variation owing to founder effects, and difficult to maintain alive *in situ.* Alternatively, they may be broadly adapted genotypes, capable of surviving unusual environments for the species. Until evidence on the comparative adaptability of marginal populations is available in advance of collecting, no bias should operate for or against marginality per se.

Number per Site with Many Sites

Suppose there are stringent limits on the total number of plants that can be grown *ex situ* for any one species. If planting material from several populations is available, should an equal or an unequal number be grown from each population?

This question is analogous to the problem of choosing the entries for a "core collection" from a large germplasm collection of a crop species in which its accessions have been sorted in groups according to origin (Brown 1989a). The theoretical optimum strategy that would maximize the conserved number of distinct alleles per locus, derived from the log relationship above, is as follows:

1. If the populations are *fully differentiated* and share none of their alleles at polymorphic loci, each source population would be represented in proportion to the value of θ in each original population.
2. If the populations are *completely undifferentiated* and genetically identical, any of the source populations can be used in any convenient proportion.

Most situations would fall between these extremes and would require judgment, if possible, assisted by data from genetic markers that estimated the degree of genetic polymorphism.

Repeated Sampling of a Site

Unlike the sampling of crop germplasm for *ex situ* conservation, the *ex situ* samples of most populations of endangered species are ideally intended to ensure the continued existence of the source population. Thus if the population is being protected, it will be available for periodic sampling. The factors that would favor such multiple sampling include the following:

1. The volume of seed required for a stored sample may not be obtainable from a single visit to the site. Repeated sampling in the field will often be easier and cheaper than multiplying in a garden, unless the site is remote and inaccessible. Steps to improve productivity at the site (such as protection from grazing and disease, controlled pollination, and weed control) may be needed. Of course, sampling should not be so intense as to threaten the existence of the source.

2. The success rate with vegetative material or the germinability of the seed may be low, so that a repeated sample, possibly of a different time of year, is needed.

3. The viability of the stored sample may decline to precarious levels. Again, resampling may be easier than regeneration of the old sample in a garden.

4. Temporal genetic differences may exist among different crops, cohorts, or generations at a site. A major cause of such differences in forest trees is variation in male fertility among individuals (Schoen and Stewart 1987). Seed from a single crop can show extreme predominance of a restricted number of pollen sources. However, multiyear collections can lower this bias, since the success of males appears to vary dramatically between years. Alternatively, many species of Myrtaceae and Proteaceae bear fruit for several sequential crops, thus enabling a "multiyear" sample to be obtained in one visit. Another major cause of variation between years could be fluctuating levels of disease or variation in races of a pathogen, which could determine which individuals in any one season produce seed (Burdon 1987). Harvesting in a number of

years would then be expected to yield a total sample more enriched for resistance genes.

SAMPLING PRACTICALITIES

Choice of Individual

Once the population for sampling is delimited and the size of sample decided, the collector must decide how the individuals contributing to the sample will be chosen. The main possibilities are (1) a *simple random* sample to which each plant in the population is equally likely to contribute; (2) a *stratified random* sample, for which the habitat is divided into obviously differing patches (contrasts of aspect, slope, soil, vegetation density, moisture, etc.), and a random sample is taken from within each microhabitat; (3) a *systematic* sample, in which the sampled individuals are equally spaced on a grid or transect; and (4) a *biased* sample, based on variation in appearance (e.g., flower color).

As phenotypic variation for traits like size, height, and seed number is largely nongenetic, either of the two methods of random sampling is preferable to biased sampling for collecting genetic diversity (Marshall and Brown 1975; Hawkes 1987a; Marshall 1989). Systematic sampling has two advantages over strict random sampling: it is easier to carry out, and it spreads the sample over the population. However, it can be biased if variation is periodic in the population. The stratified random technique is probably the best if it is readily executed. Where a plant grows is arguably a better indicator of its genetic adaptation than is its phenotype. If a supplementary biased sample of any variant is taken, it should be numbered and handled separately. Thus this added effort should be made only if the variant is likely to be heritable and of particular interest or value.

Biased sampling may have a role to play at other points in the process, provided that the bias is based on genotype. For example, the number of plants that can be held or replanted in a garden is severely limited. Suppose that the available material has been surveyed for its genetic status through the use of a wide range of genetic markers such as isozymes. The number of alleles detected is an indicator of relative diversity throughout the genome (Brown and Moran 1980; Brown and Burdon 1987) and the level of heterozygosity of recent breeding history. The selection of individuals with maximal heterozygosity and allelic diversity would likely increase the genetic basis of restricted *ex situ* samples and minimize inbreeding.

Type of Propagule Collected

The collector must choose the kind of material to be collected from nature for establishing the *ex situ* sample. The choice is among seed, vegetative cuttings, vegetative propagules (such as bulbs, corms, and tubers), and whole individuals. Many factors impinge on this decision. Seeds and cuttings are the more frequent choices. We first consider, under several headings the features of seeds for sampling, for ready comparison with those of cuttings and other options.

Seed

Availability. The timing of seed collections is crucial for most species. Missions can take place only when fruit is ripe or sufficiently ripe to mature after gathering. On the other hand, delaying a trip risks the loss of better-quality seed through shattering. For example, a delay of two days at shedding can cause the great bulk of seed to be lost for many species of *Acacia.* Multitargeted trips are hard to organize because species differ in maturity date. Variation in maturity among individual plants within and between sites can arise from genetic or environmental causes, and restricts the kinds of plants available for sampling.

Genotypes sampled. The breeding system determines the fidelity of and variation in the seed sample. With uniparental reproduction (selfing or agamospermy), seed will generally be very similar to the sampled parental plants. In contrast, the seed of an outbreeder may be very different from the maternal parent and very diverse, possibly including various inbred fractions. In addition, little control is possible over male fertility variance in outbred source populations. The genetic variation in the sample is both a problem and a benefit to *ex situ* conservation. It introduces an element of unpredictability of performance as well as a contingency for survival in changed conditions. If seed is grown *ex situ* to increase the sample, or later to regenerate it, care must be taken that selection or random genetic drift does not lead to genetic change away from adaptation to survival *in situ* (Given 1987).

Propagation. Techniques must be in place to counter problems concerning after-ripening effects, disease, pest treatment, dormancy requirement, viability in storage, and germinability. Species may have their own peculiar requirements for germination, making uniform handling difficult. Multiplication in gardens gives an opportunity for the sample to become infected with diseases or pests, which inadvertently could be transferred to the *in situ* source (Given 1987).

Storage. As the organ of perpetuation, seed is the optimal choice for storage. Techniques for storage have been closely studied for crop, forage, and forest tree species. In general, the seed of temperate species is readily conserved at about 0°C for prolonged periods after drying to <6% moisture (Hawkes 1987a). Tropical species are less amenable to this method, as their seeds cannot withstand drying and cold storage, and thus rapidly lose viability. Material for long-term storage is immediately available from the field collection. Some monitoring of seed viability is required. Indeed Hawkes (1987a) suggests a viability test every three years, and regeneration of the sample if viability falls below 85%. However, such standards may be possible only for species of the highest priority. Resampling of the original source or indeed another *in situ* population may be a more feasible option than the continual monitoring and regeneration of old samples.

Sample size. The small size of most seeds, the organ of dispersal, enables relatively large numbers of organisms to be sampled, treated, and stored readily. As Hawkes (1987a) emphasizes, it is the use of seed banks that allows us to conserve a portion of the genetic variation of the species. In many species, sufficient seed might be available at the site to form a sample for direct storage while leaving enough for natural regeneration of the population *in situ*. As individual plants within a population usually differ in fecundity, approximately equal numbers of seed per individual should be sampled to avoid unwanted bias or low effective sizes.

Cuttings

Availability. Cuttings can be harvested at most times of the year, although the time of harvest affects the strike rate.

Genotypes sampled. The sampled genotypes are those that have proven survival ability in the wild. They are of known identity with the genotype of the parent plant. They also have the advantage of being replicates of authentic botanical specimens or of possessing some desirable horticultural feature (such as an attractive flower color).

Propagation. Cuttings require speedy processing to retain their viability and need ready availability of space and proven techniques. Woody or perennial species are easier to process than annual herbaceous species; the latter require culture by *in vitro* methods. One advantage of cuttings is that when they readily strike and produce plants, the time to flowering and maturity is often faster than for seedlings, enabling quicker evaluation of the material collected and earlier seed production. Once facilities are in place, the same range of techniques of hormonal treatments can be used with cuttings from a wide range of species. Further, the plants grown from cuttings can be replanted or replicated from further cuttings as needed. They can be closely planted and heavily pruned to keep them small and to encourage softwood production for further cuttings (D. Falk and R. Nicholson pers. commun.).

Storage. From the sampled cuttings, the usual procedure is to establish plants in gardens. Seeds formed in captivity may then be harvested for long-term storage. This requires the culture, flowering, pollination, and fruiting of plants grown in an environment different from their source. Steps are usually needed to prevent unwanted interspecific hybridization in the gardens. Although direct storage of cuttings as meristems in tissue culture may be a technically feasible option, it requires expertise and species-specific technology. It may have a restricted role, as Ashton (1987) suggested, in the conservation of rain forest tree species with recalcitrant seeds.

Sample size. The use of cuttings for conservation places strict limits on the size of samples, mainly because each unit is effectively sampled only after it has been struck and grown into a mature plant. Yet the collector can be sure that an explicit number of distinct naturally occurring individuals has been sampled. In contrast, the number of parents actually contributing to a seed sample is often not known, and is not necessarily related to the total number of seed in the sample.

Bulbs, Corms, and Tubers

For certain species (e.g., families Orchidaceae and Droseraceae) the collection of organs of vegetative reproduction is a feasible alternative strategy. Their sampling features are intermediate between those of seed and those of cuttings, except that they may take more time and effort to collect than do either seeds or cuttings.

Whole Plants

Finally, the collector may consider sampling whole individuals. This technique can destroy the source population. It is inappropriate for a rare and endangered species unless the destruction of the natural source population is certain anyway (Raven 1976). Conservation *in situ* should take clear precedence over the gathering of material for *ex situ* growth or storage. Furthermore, it is usually difficult or impossible to transplant whole individual of deep-rooted perennials from nature to gardens. Transplants of whole individuals require handling as large cuttings.

Collection Records

Many authors have stressed the importance of having adequate records to accompany the samples (e.g., Hawkes 1976, 1987a; Given 1987). Whereas such records are of acknowledged importance in collecting crop plant germplasm, they are even more important for the sampling of endangered species. The records are needed for the continuing link between the sample held *ex situ* and the source population *in situ* (Figure 7.1). When adequate, the records will document the occurrence of the species; enable resampling when needed; contribute evidence on the regulation of numbers and causes of rarity; suggest ways the samples may be grown in gardens, studied, and used in horticulture or breeding; and indicate whether and how reintroduction can be achieved. If sites can be monitored periodically by collaborators, it is possible to build up considerable information on the ecology of the source populations. The U.S. Center for Plant Conservation has designed a field collection data sheet that contains the basic attributes that should accompany samples to be conserved *ex situ*.

CONSERVATION OF RARE AND ENDANGERED SPECIES IN AUSTRALIA

The flora of Australia is both rich and unique. Of the world's 250,000 described species of vascular plants, about 16,000 occur in Australia (Leigh et al. 1984). It is estimated that about one-third more species are awaiting formal description, and some 2% are yet to be found. These last species are presumably rare in frequency and highly restricted or inaccessible in location. Long isolation of the Australian continent has led to a high percentage (ca. 90%) of endemic species. This is a remarkable heritage of global and national importance (Good and Lavarack 1981).

Briggs and Leigh (1988) list 3329 Australian plant species (about 17% of the flora) as either rare or threatened. This fraction is comparable with that for Europe (17%), and higher than that for the United States (10%), South Africa (9%), and New Zealand (9%) (Davis et al. 1986). For Australia, Briggs and Leigh (1988) list 97 species as presumed extinct, 209 as endangered, 784 as vulnerable, 1367 as rare (not threatened), and 872 as poorly known. These categories are broadly in accord with those in *The IUCN Plant Red Data Book* (Lucas and Synge 1978). Thus in Australia, of greatest urgency for both *in situ* and *ex situ* conservation are the 209 (1% of flora) endangered species, and of second priority are the 784 vulnerable species (4%).

More than half the endangered species are perennial shrubs and trees (Table 7.4).

Table 7.4. Life Forms of 209 Endangered Species in Australia

Group	No. of species
Rain forest trees and shrubs	20
Trees	6
Shrubs	110
Perennial herbs	27
Orchids	20
Grasses and sedges	12
Annuals	11
Ferns	2
Aquatics	1

Source: Briggs (unpubl. data).

The large size and long life cycles of such plants place severe constraints on the numbers that can be grown in gardens and on the speed of multiplication for seed samples.

Figure 7.4 shows the number of endangered species in each of 80 regions across the continent. This figure highlights the enormity of the task of traveling to these areas, and sampling and documenting all these species. The clustering in southwestern Australia and along the central east coast will make sampling these areas more efficient. The higher number of endangered species correlates with population density and with

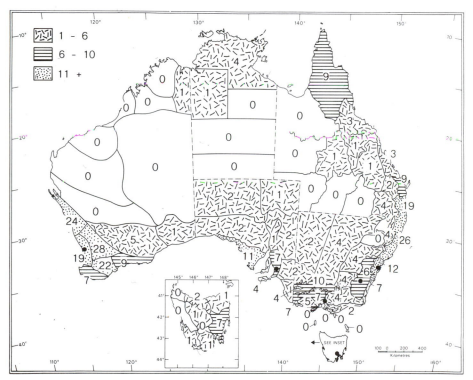

Figure 7.4. Geographical distribution of endangered species in Australia. (From Briggs and Leigh 1988)

regional floristic diversity. It presumably reflects the effects of urban and rural development and temperate-rain-forest destruction. The distributions of vulnerable and of rare species show high numbers of species in the same two regions, but in addition, these categories are prominent in tropical Cape York and Arnhem Land (Leigh et al. 1982a). Despite the widespread location of endangered species, the task of sampling populations of each endangered species is achievable over a five-year period, given the coordinated assistance of botanists in each state.

In Situ Conservation

Of the 209 endangered species, 51 (24%) are known to occur partly or entirely within national parks or proclaimed reserves. Despite this level of security, most of these species generally remain endangered because their conserved populations are very small and are threatened by weeds, herbivores, or unsolved management problems. In five cases, all the known populations are already reserved. Some 390 (50%) of the 784 vulnerable species are known to occur in reserved areas. Again, these species remain threatened because of the small size of these reserved populations. In 44 cases, all the known extant populations are reserved. Away from protected sites, the licensed harvesting of natural stands for flowers and seed to meet the growing local and export demands increase the risk for rare species that are not on lists of protected species (Good and Lavarack 1981).

Steady progress is being made in creating new, often small, protected sites for specific species. This process ideally should result in species being removed from the endangered list and, depending on their security, being transferred to the rare or the vulnerable category. Negotiations for new reserves can be protracted. It is thus important to have propagating material from the prospective site conserved *ex situ* as a safeguard against extinction before reservation is in place. Once a site becomes reserved, *ex situ* collections may be needed to replant it, depending on the nature and extent of past threats.

Ex Situ Collections

At present, the Council of Nature Conservation Ministers (CONCOM) Working Group on Endangered Flora is preparing guidelines for the collection, documentation, and long-term storage of seed of these species. This will allow the organization of a national program aimed at having germplasm of endangered species in storage for research, propagation, and insurance purposes.

In a survey of Australian botanical gardens, Meredith and Richardson (1987) found that only 36% of endangered species were in cultivation. Of the species being grown, some 44% were present at only one institution, 31% at two, and 25% at three or more. Much of this material is likely to stem from samples of only one original plant. Documentation of the origin of these collections and the number of plants sampled is often poor, and the accompanying seed collections from these species are very limited and poorly organized (Meredith pers. commun.). Presently, the endangered species are even more poorly represented in seed collections than they are in grown collections. The populations of virtually all the endangered species will require visiting to obtain adequate samples of satisfactorily documented collections.

CONCLUSIONS

The collector of material for conserving endangered plant species *ex situ* is faced with a wide range of questions. What taxa? When and where to collect? What type of propagule? What type of individual? How many individuals in the sample? How many populations should be sampled? Many of these questions interact.

We have discussed these questions with several principles uppermost in mind. These were that *ex situ* conservation makes most sense when closely linked with *in situ* conservation, that genetic diversity is the basic ingredient for success, but that a strong law of diminishing returns operates so that it is easy to go overboard in effort on one taxon and fail to act on others.

Thus we advocate a "logarithmic" rather than a "linear" view on sampling of genetic diversity. Useful variation increases in proportion to the logarithm of the sample size, rather than the absolute size.

This concept supports a sampling strategy that strives to achieve minimum targets of about 15 individuals per species, or five populations per species and ten individuals per population. These targets are to be adjusted and increased, depending on the availability of material and the desirability of sampling from a wide range of environments. The continuing importance of conserving the source population needs to be stressed as an integral part of the conservation strategy.

ACKNOWLEDGMENTS

The authors are grateful to their colleagues Drs. J. J. Burdon, O. H. Frankel, S. D. Hopper, and J. H. Leigh for discussion and comments on earlier drafts.

III

MANAGEMENT AND ASSESSMENT OF OFF-SITE COLLECTIONS

8

A Comparison of Methods
for Assessing Genetic Variation
in Plant Conservation Biology

BARBARA A. SCHAAL, WESLEY J. LEVERICH,
and STEVEN H. ROGSTAD

The existing levels of genetic diversity and the maintenance of these levels of diversity are major issues in conservation biology. Genetic diversity became an issue when Frankel (1970) postulated that genetic variation is essential for the long-term survival of endangered species. Genetic variation is a necessary prerequisite for any future adaptive change or evolution; presumably, species that lack adequate genetic variation are at greater risk of extinction. In terms of conservation, maintenance of genetic variation is essential if populations are to be successfully reintroduced in the wild or introduced to new habitats. In addition, there are ample data to suggest that genetic variation contributes to the immediate fitness in plant species that maintain fitness by heterosis or heterozygote advantage. Reduction of genetic variation results in a loss of heterozygosity and, consequently, fitness. A concomitant effect of decline in genetic diversity may be the potential random fixation of poorly adapted genotypes following a reduction in population size. Finally, the maintenance of the existing genetic diversity in a species is critical for preserving genetic resources for future applied uses. Recombinant DNA technologies now allow the transfer and expression of genes from one species into another. Such transgenic plants can potentially increase crop yields, and can be used for the production of useful compounds such as pharmaceuticals. An irreplaceable resource of potentially beneficial alleles can be protected by conservation of the genetic diversity of a wide range of plant species.

Plant species that at one time were widespread, were outcrossed, and contained much genetic diversity would be the most sensitive to a decline in genetic variation associated with a restriction in range or population size. Under extreme restriction in population size, the breeding system would move from outcrossing to increasingly inbreeding, and such populations would be subject to inbreeding depression (Charlesworth and Charlesworth 1987; Hedrick 1987). Conversely, those species that experienced long periods of inbreeding in nature would be less likely to show inbreeding depression. In plant species that always have had small, isolated populations and that routinely self-fertilize, consideration of diversity is probably less essential. But even in these species, preservation of the genetic differences that occur among populations may be desirable so that the full range of genetic variation of the species is preserved.

The measurement or estimation of levels of genetic variation within species thus becomes critical to their conservation and management in many cases. Three major types of characters have been used to estimate levels of variation: morphological, allozyme, and, most recently, DNA sequences. Each type of data will be discussed in turn, and its inherent advantages and biases considered. Because morphological and allozyme data have been commonly used, less emphasis will be placed on describing specific techniques. A detailed discussion of DNA methodologies will be presented, since these techniques are less familiar to conservation biologists and there are some recent, relevant discoveries.

MORPHOLOGICAL VARIATION

The most easily obtained assessment of genetic variation is that of measuring morphological or phenotypic variation. Morphology has the advantage of requiring neither breeding nor laboratory studies, and, most importantly, such work can be done directly from field collections. Another distinct advantage of studying morphological variation is that phenotypic characters are often ecologically adaptive. Such morphological variation is often assumed to indicate genotypic variation, local differentiation, or ecotypes. In these cases, variation in phenotype indicates underlying variation in the genome. The classic studies of Clausen et al. (1948) on *Achillea lanulosa* ecotypes show that the morphological differences of ecotypes are often indicative of genetic differences in many other genes, such as physiological ones. Phenotypic plasticity, which is common in plants, is a major modification of this relationship. In many cases, however, using morphology to estimate genetic variation is the most realistic method— as when a rapid estimation of variability is needed or where biochemical surveys are impractical. In such cases, a consideration of morphological variation may be the only practical way to estimate genetic variation.

A good example of how morphological variation can be used to estimate genetic variation and genetic differences among groups occurs in *Clematis fremontii* ssp. *riehlii* (Erickson 1945). This subspecies of *Clematis* occurs only on limestone glades within a small region of the Missouri Ozarks. Erickson found a hierarchical distribution of leaf-shape variation within and among glades (Figure 8.1). He found that areas of relative uniformity of leaf shape occurred within aggregates on a glade; glade populations were divided into these aggregates. Neighboring glades often showed similarity in their leaf shapes; these were termed *clusters,* and clusters could be grouped into a series of glade systems. In this case, the distribution of variation in a single character, leaf shape, corresponded to different genetic groupings, determined by the estimated amount of gene flow among the groups. From these data, Erickson concluded that populations were subdivided genetically. The degree of genetic differentiation would affect the total genome and not just one or two genes. As we will see, a subsequent study of DNA variation in *C. fremontii* showed that Erickson's conclusion was correct; DNA sequences also showed the same population subdivision.

Another somewhat different morphological approach to measuring genetic variation is to analyze quantitative traits. Morphological or phenotypic differentiation, as

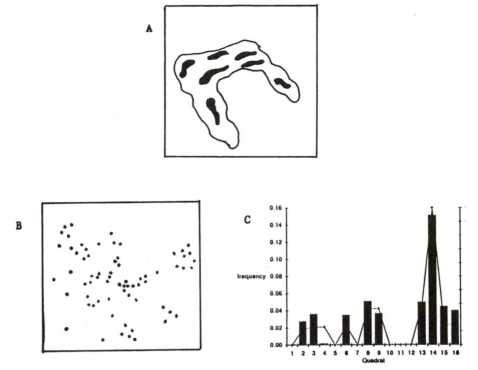

Figure 8.1. *Clematis fremontii* population distribution: (A) the distribution of plants on a glade into morphological aggregates; (B) plants within an aggregate; (C) the distribution of rDNA variants across a glade. (A and B from Erickson 1945; C from Learn and Schaal 1987)

in the case of *Clematis* and *Achillea*, has been documented in many kinds of plant species (e.g., Venable 1984). In many cases, the genetic basis of this variation has been determined from common garden experiments. An alternative morphological approach is a quantitative genetic analysis in which the transmission of morphological traits from parents to offspring of selected crosses is scored. Such studies involve a series of crosses, with the resulting progeny often grown in different environments to determine the degree of environmental versus genetic control of polygenic morphological traits. Quantitative traits are estimated by either a least-squares or a maximum-likelihood procedure (Shaw 1987). Such analyses not only indicate to what degree a phenotypic trait is under genetic or environmental control, but also determine such parameters as narrow-sense heritability (h^2) and genetic covariance between characters (Cov_A), which are important in estimating the potential for evolutionary change within populations. These studies estimate the opportunity for response to selection (Antonovics 1976; Falconer 1981). A quantitative analysis thus can provide information on a critical and central aspect of conservation genetics: how a population will respond to a change in a selective regime that might be associated with a change in environment (see *Holcus* example, below). Relatively few quantitative genetic studies have been conducted on natural populations of native plant species.

A quantitative genetic analysis of morphological characters was conducted in *Hol-*

Table 8.1.   Additive Genetic Variance (V_A) and Heritabilities (h^2) for Two Adjacent Populations of *Holcus lanatus*

Trait	Managed population		Traditional population	
	V_A	h^2	V_A	h^2
Tiller number	186[a]	0.08	−382	−0.17
Tiller dry weight (g)	2.42	0.19	2.96[a]	0.19
Stolon number	−119	−0.29	68.0	0.28
Leaf width (mm)	−0.52	−0.27	−1.10	−0.29
Flowering time (days)	11.8	0.24	−18.1	0.14
Panicle length (mm)	391[a]	0.27	−.087	0.01

[a] $p < .05$.

Source: Billington et al. (1988).

cus lanatus L., an outbreeding perennial grass (Billington et al. 1988; Table 8.1). Two adjacent populations that had been under different management regimes of grazing and cutting were studied. The populations showed significant differences in a number of morphological features in a common garden environment. A polycross experiment revealed that evolutionarily significant genetic divergence had occurred over a relatively short period of time between these two contiguous sites. Moreover, the differences in genetic structure between the two populations were apparent only from the quantitative analysis. Although the populations are adjacent, their potential response to selection would be quite different, even in the same environment. Quantitative differences in life-history features can also be detected among populations of species (e.g., *Impatiens* [Mitchell-Olds 1986]). Such quantitative analyses have the potential to be very important for conservation biology. Not only can genetic variation be measured within populations and differences between populations determined, but such measures can also be used to predict responses to selection. Determining how a population might respond to selection can be extremely important if the species is to be introduced to a new site or habitat. Such studies have several obstacles that currently make them impractical for routine conservation biology studies. They are very time consuming; they require careful experimental design and data analysis; and they cannot be conducted in species that will not readily cross or that have long life spans. Nevertheless, quantitative analyses might be very useful for conserving species that are of special importance, such as a wild relative of a potential crop species.

ALLOZYME VARIATION

In recent years, allozyme electrophoresis has been a predominant technique for examining genetic variation. This technique has greatly expanded our understanding of the genetic processes that occur in plant species, and without it we would have almost no information on the genetic structure of noncultivated species. In spite of its usefulness and widespread application, electrophoresis has some well-known limitations. Furthermore, only genes of a single class, those encoding soluble enzymes, are analyzed, and they often are selected on the basis of the ease of their products' extraction and ability to migrate on a starch gel. Only nucleotide differences in genes that lead to changes in product amino acid composition can be detected, and then detection is usually limited to those changes in amino acid composition that result in a net change in charge of the product molecule. In addition, these genes may not be representative of the genome in general. For instance, there is evidence that many commonly studied allozymes are more variable than other categories of gene products, and that this variability may stem in part from such processes as posttranslational modifications. Interestingly, a molecular study of the F and S alleles of the $Adh-1$ gene in maize showed many differences in sequence, rather than a single base-pair substitution, as had been previously postulated (Sachs et al. 1986).

In spite of these limitations, allozyme analysis often provides the best measure of genetic variation. The technique is relatively straightforward, and methodologies have been developed for many kinds of plants (Hamrick 1989). Mendelian inheritance of allozyme markers has been documented in many species, allowing the mode of inheritance to be inferred in plants that are not amenable to crossing. An allozyme analysis provides an estimate of gene and genotypic frequencies within populations. Such data can be analyzed in numerous ways to measure genetic differentiation, population subdivision (Weir and Cockerham 1984), genetic diversity (Nei 1973), and gene flow (Slatkin and Barton 1989). Because of this repertoire of techniques for analysis, allozyme data are extremely useful in determinations of the genetic structure of species and in comparisons among species. Because a large number of plant species has been analyzed, generalizations may be made about levels of variation and how the levels are influenced by other factors, such as life history, habitat, and breeding systems (Hamrick et al. 1979; Loveless and Hamrick 1984).

In general, allozyme variation does seem to reflect the overall level of variation within a genome. Hamrick (1989) has compared allozyme variation with variation of other traits and, in general, finds good concordance. Table 8.2 shows examples of allozyme data compared with other measures of variation (Hamrick 1989). In most species studied to date, there is often a correspondence between morphological and allozyme data. Many of the species that lack congruence between different variability measures show unusual or rapid rate of evolution. Such a situation occurs in the Ha-

Table 8.2. Concordance of Allozyme Data

Species	Comparison	Correspondence
Avena barbata	Allozyme/morphometric	+
Avena fatua	Allozyme/single genes	+
Hordeum jubatum	Allozyme/morphometric	+
Hordeum spontaneum	Allozyme/morphometric	+
Hordeum vulgare	Allozyme/morphometric/DNA	+
Layia spp.	Allozyme/morphometric	+
Lisianthus	Allozyme/DNA	+
Phlox drummondii	Allozyme/morphometric	+
Pseudotsuga menziesii	Allozyme/morphometric	+
Trifolium hirtum	Allozyme/morphometric	+
Bidens	Allozyme/morphometric	−
Clarkia	Allozyme/morphometric	−
Hordeum murinum	Allozyme/morphometric	−
Hordeum spontaneum	Allozyme/morphometric	−
Pinus contorta	Allozyme/morphometric	−
Silene diclinis	Allozyme/morphometric	−

Source: Hamrick (1989).

waiian *Bidens*. The genus *Bidens* on the Hawaiian Islands has undergone adaptive radiation after initial colonization of the islands, most likely from the Americas. The species differ from one another in morphology and in habitat, suggesting rapid evolution in these parts of the genome (Ganders and Nagata 1984). Yet the species are chromosomally similar (Gillett and Lim 1970), show little allozyme differentiation (Helenurm and Ganders 1985), are interfertile (Ganders and Nagata 1984), and are very similar in their chloroplast DNA (S. O'Kane pers. commun.). From the morphological data, one would infer greater genetic differences among the taxa than is found in the overall genome. The most plausible explanation for the different rates of evolution in *Bidens* is very rapid selection for morphological and physiological genes associated with habitat adaptation during adaptive radiation. Fortunately, such instances are relatively uncommon. Allozymes provide a good, quick method for estimating genetic variation. The method is particularly powerful for comparisons among closely related species that share long periods of evolutionary history. Differences in allozyme variation among evolutionarily similar species most likely reflect differences in population process that can affect total variation. In these species, the relative levels of genetic variation overall can be assessed with high likelihood from allozyme data.

DNA VARIATION

With the development of recombinant DNA technologies, it has become possible to examine variation in DNA sequences in virtually any plant species. An analysis of DNA sequences avoids many of the problems and biases inherit in morphological and allozyme estimates of genetic variation. Because DNA is being directly analyzed, genetic variation can be directly measured as opposed to being estimated from a phenotype. Moreover, virtually any segment of the plant genome can be analyzed. Techniques are available to study sequences that are coding or noncoding, conservative or hypervariable, nuclear or organelle. Different DNA sequences have a range of selective constraints and are differentially affected by molecular mechanisms that generate variation. Thus the levels of nucleotide substitutions vary within different parts of the genome. Some sequences, such as the coding regions of ribosomal DNA, show little variation within a species and may be constant among members of a genus (Schaal and Learn 1988), whereas hypervariable sequences can show genotype-specific variation (Jeffreys et al. 1985). In addition, DNA sequences contain historical information so that phylogenetic affinities can be determined (Templeton 1983). Evolutionary relationships among species are detected by similarities in nucleotide sequences among related species. Thus DNA sequence analysis can be very powerful, and it has the potential to provide information on problems and in species that previously have been intractable. There are, however, some very significant drawbacks to DNA analysis. Laboratory techniques are fairly complicated, time consuming, and relatively expensive, making such analyses impractical in many laboratories. As with allozymes, there are a series of technical pitfalls and potential artifacts. Finally, since only a small segment of the genome is being analyzed at one time, there is the danger of overextending conclusions about variability in one type of sequence to the entire genome.

The techniques for analysis of endonuclease restriction site variation in plants are outlined in the next section, after which we present studies of variation in chloroplast genome and ribosomal DNA. Finally, we discuss our recent work on hypervariable DNA sequences in plants.

Techniques

There are currently two alternative approaches to the analysis of DNA sequences. The first involves direct sequencing of alleles and a direct comparison of the base-pair sequence. Such an approach gives the ultimate information on fine-structure variation and yields extremely powerful data from which to draw evolutionary inferences. Sequencing is very time consuming, and only relatively few individuals can be analyzed, making such an approach not currently practical. However, technical innovations in molecular biology are occurring at such a rapid pace that sequencing of many individuals for population studies will be feasible in the near future. Sequencing of genes directly from DNA that is amplified through a polymerase chain reaction (PCR) has already greatly reduced the time for sequencing.

Currently, most of the data on DNA sequence variation comes from restriction-fragment-length polymorphisms (RFLPs). Here variation is measured by determining the presence or absence of a small sequence (4, 5, or 6 base pairs) of DNA that is specific for a particular restriction endonuclease. An endonuclease cuts double-stranded DNA wherever its specific recognition site occurs (Figure 8.2). The resulting patterns of DNA fragments are compared to determine variation in these restriction sites (Figure 8.2).

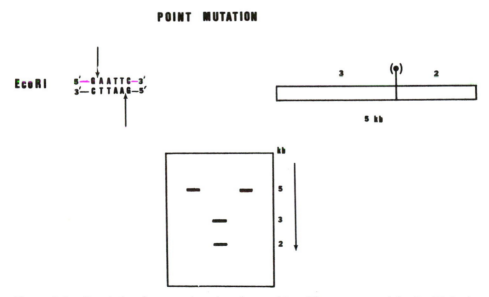

Figure 8.2. Restriction-fragment-length polymorphism. The presence of the *EcoRI* site in the 5-kilobase (kb) fragment cleaves the DNA into two fragments of 2 and 3 kb. Such RFLPs are detected by variation on a gel: some individuals have a single 5-kb fragment, whereas for others the 5-kb fragment is missing, and the 2- and 3-kb fragments are present instead.

To begin such a study, DNA is isolated by one of several available technologies. A commonly used method is the cetyltrimethylammonium bromide (CTAB) extraction procedure of Saghai-Maroof et al. (1984). This is a miniprep method and allows isolation of DNA from many samples within a few hours. An alternative to a chemical isolation is a cesium chloride density-gradient centrifugation, which isolates DNA based on its buoyant density. When total genomic DNA is isolated, all plant genomes are represented, including chloroplast, nuclear, and mitochondrial DNA. Segments of the nuclear and chloroplast genome can be analyzed directly from such DNA isolations. In studies of chloroplast DNA, chloroplasts can be isolated and DNA obtained directly from the chloroplasts. This approach is not always necessary if probes are to be used. Once DNA is isolated, then genomic DNA is digested with one of the restriction endonucleases. There are many different restriction endonucleases available, each with a specific recognition sequence. Many restriction sites within a given DNA sequence can be analyzed by applying a battery of restriction enzymes. After digestion, DNA fragments are electrophoretically separated by molecular weight on an agarose or acrylamide gel. A restriction endonuclease digestion typically yields thousands of fragments, and the DNA, when visualized by ethidium bromide staining, appears as a smear of fragments. The next task is to visualize only those fragments that correspond to the DNA of the particular sequence that is being studied. Such analysis requires the use of a cloned probe.

A cloned probe begins as a cloning vector, such as a virus or plasmid, which has been genetically engineered specifically for cloning of DNA sequences. In preparing the probe, a DNA sequence from another organism, such as a plant, is spliced into the vector DNA. This inserted sequence is the DNA sequence that will be studied for the analysis. Many genes have been cloned and are available for analysis, including sequences such as those for the chloroplast genome, ribosomal DNA, single-copy genes such as *Adh-1*, or hypervariable sequences. In concept, any type of DNA from any species can be cloned and then studied.

The DNA fragments that were separated on a gel are transferred to a nylon or nitrocellulose filter membrane by a Southern blot. The fragments are transferred in the exact pattern of separation on the gel. The cloned probe DNA is radioactively labeled with ^{32}P, and then hybridized to the DNA on the membrane. The labeled probe DNA will hybridize only to those DNA fragments on the membrane that are homologous sequences. The membrane is washed, and only hybridized DNA stays bound to the filter after washing. Those fragments that correspond to the cloned sequences are now radioactively marked and can be visualized by autoradiography. Variation in restriction sites is detected by the fragment patterns (Figure 8.2).

Chloroplast DNA

Much of our initial information on variation in DNA sequences comes from the study of chloroplast DNA (cpDNA)(Palmer 1987). Chloroplast DNA is a circular molecule generally between 100 and 200 kilobases in size. The chloroplast genome is, in general, conservative; the rate of sequence evolution is relatively slow in many plant species. Most of the changes are either small insertions/deletions or base-pair changes. Most of the initial work on chloroplast DNA used interspecific differences to draw phylogenetic inferences. Such studies provided evidence for the origin of *Brassica* species (Palmer et al. 1983) and determined phylogenetic relationships among species of *Lycopersicon* (Palmer and Zamir 1982), *Clarkia* (Sytsma and Gottlieb 1986), and *Lisianthus* (Sytsma and Schaal 1985b). There have been relatively few populational studies of chloroplast DNA. In many species, there is very little intraspecific variation. For example, in *Pennesitum* the chloroplast genome is so conservative that little variation is found, even among members of the genus (Zurawski and Clegg 1987). In *Lupinus texensis,* chloroplast DNA variation was found in only one population of an intensive populational survey (Banks and Birky 1985). As more plant species are being studied, instances of within-species and population variation of cpDNA have been discovered. Levels of variation can be low (*Hordeum vulgare* [Neale et al. 1988]) or can be potentially high, due to hybridization (*Helianthus* [Rieseberg et al. 1988]) or due to somatic variation (as in clonal lineages in *Taraxacum* [King 1989]). Gymnosperms seem to have much more chloroplast DNA variation, and here cpDNA can be used for population studies as in *Pinus* (Wagner et al. 1987).

In most cases, the rate of chloroplast DNA evolution is too slow to provide a good measure of variability within populations or species. In some instances, issues of hybridization or introgression may be of concern to a conservation biologist (Reiseberg, Chapter 11). Here an analysis or chloroplast DNA could be quite useful. In most plant species, the chloroplast genome is uniparentally inherited and passed on as an asexual lineage. This mode of inheritance makes chloroplast DNA an excellent marker for the detection of hybridization and potential introgression, since the molecule is not disrupted by recombination. Even instances of ancient introgression should be detectable by introgressed populations containing another species' chloroplast genome. In a current study of the white oak complex conducted by A. Whittemore, there is strong evidence from chloroplast polymorphisms for introgression in many populations.

Ribosomal DNA

Ribosomal DNA has been used in many studies to examine levels of variation (reviewed in Schaal and Learn 1989). Ribosomal DNA (rDNA) codes for the 17S, 5.8S, and 27S subunits of the ribosome. The sequence is tandemly repeated, with from 3000 to 14,000 copies per genome. The coding sequences are separated by an intergenic spacer region (IGS) that is not transcribed (Long and Dawid 1980). The IGS region has an internal substructure of a variable number of repeating subunits that can range from 125 to over 300 base pairs (Schaal and Learn 1988). Ribosomal DNA has several features that make it suitable for both population and phylogenetic analysis. The coding sequences are conserved and can be used for macroevolutionary studies (Sytsma and Schaal 1985b). The noncoding regions are highly variable, and single plants may contain several variants—up to 20 in some *Vicia faba* individuals (Rogers et al. 1986)—allowing for populational studies.

Currently it is difficult to determine whether rDNA variation is a good reflection of variation within the genome in general. In the genus *Lisianthus,* there is good concordance between patterns of rDNA variation and allozyme variation (Table 8.2). *Clematis fremontii* shows statistically significant local differentiation for rDNA variants within populations (Learn and Schaal 1987) in a pattern similar to that observed for morphological variation (Figure 8.1). Genetic differentiation for rDNA variants is likewise observed for variants of wild barley (Saghai-Maroof et al. 1984). The distribution and apportionment of rDNA variants correspond to population and subspecific boundaries in *Phlox divaricata* (Schaal et al. 1987). On the other hand, some species show little or no variation in rDNA, either in restriction sites or in length variants. Several populations of *Rudbeckia missouriensis* were studied in the same glade habitats as *Clematis*. There were only three variable restriction sites, and these occurred in low

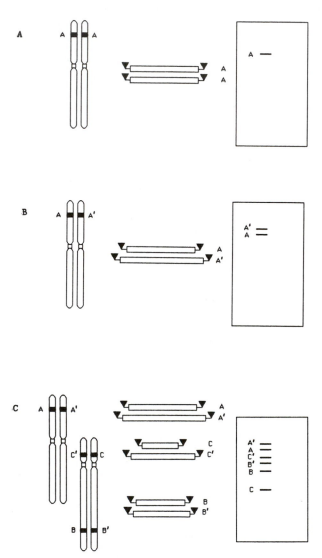

Figure 8.3. Hypervariable DNA sequences: (A) a single locus of a hypervariable sequence. Both chromosomes contain the site, and the individual is homozygous. Only one DNA fragment is seen on a gel. (B) Allelic variation at a site. A' has gained more copies of the consensus sequence, is differentiated by size, and is detected as an additional fragment on a gel. (C) The locus spreads throughout the genome. The same consensus sequence spread to different site within and among chromosomes. Allelic variation occurs at each locus, resulting in a large number of alleles being detected on a gel.

frequency, in marked contrast to the highly variable rDNA of *Clematis* on these geographically isolated sites (King and Schaal 1989). Clearly, more research on population variation on rDNA needs to be conducted before its utility for conservation biology can be assessed.

Hypervariable Sequences

A major advance in the development of suitable genetic markers has been the discovery of hypervariable, or minisatellite, DNA. These sequences can be highly variable in many species of both plants and animals. In several cases, they are genotype specific, yielding "genetic fingerprints." Minisatellite DNA consists of a tandemly repeated core or consensus sequence that in some instances is similar to the Chi replicon sequence in *E. coli* (Jeffreys et al. 1985). These sequences have undergone rapid evolution for the number of both loci and alleles. For example, in humans some minisatellite sequences have dispersed throughout the genome, giving rise to numerous loci with heterozygosities often in excess of 90% (hypervariable loci). Some minisatellite probes can reveal alleles at numerous hypervariable loci in one probing. The resultant individual-specific pattern of fragments is a DNA fingerprint (Jeffreys et al. 1985). (Figure 8.3 shows hypervariable sequences in diagrammatic form.) Minisatellite variation has been found in numerous plant species (Dallas 1988; Rogstad et al. 1988a, 1988b; Ryskov et al. 1988).

Several cultivated and native plant species have been examined for minisatellite variation within and among populations. In some species, the level of variation approaches that necessary for fingerprinting (e.g., *Rubus* [Nybom et al. 1989] and *Malus* [Nybom and Schaal 1990a]). Hypervariable sequences have been used for paternity analysis (Dallas 1988) and for determining the clonal structure of natural populations (Nybom and Schaal 1990b). Moreover, in the few comparisons of variation conducted, it appears that these sequences can be much more variable than allozymes (Nybom and Schaal 1990b). Analysis of these sequences can be potentially very important in conservation biology, since they not only can give an estimation of genetic variation within and among populations, but also can, in some cases, actually determine the genotypes present in a population. The ability to identify genotypes positively is particularly important when current collections are being evaluated. Often records of collections or transfers among gardens are lost, and it is not possible to determine whether two plants represent the same accession. Variable minisatellite sequences can be used to ensure that a diversity of genotypes are preserved in a conservation effort.

CONCLUSIONS

Currently, there is no ideal method for determining genetic variation. Each technique has distinct advantages and disadvantages. Morphological variation can be very easily determined, yet the underlying genetic basis for variation in phenotype is often obscure. Allozymes give an excellent determination of gene and genotype frequencies, but require more extensive laboratory work and are not always indicative of the entire genome. Hypervariable sequences provide very powerful data on genotypic variation, but the technique is complicated enough to limit application.

A conservation biologist must choose which method to use, based on the specific situation for a given plant species. If collecting must be done immediately, then clearly morphology would be the quickest method of making some inference regarding the degree of genetic variation. If the status of a current collection is to be evaluated and laboratory and technical resources are available, then hypervariable sequences can provide the best estimate of genotypic diversity. In many cases, allozymes provide an ideal compromise between data quality and technical accessibility. Allozymes for the most part are a good reflection of genetic variation within the entire genome; they are technically straightforward to analyze, and thus are readily available to a large number of laboratories. Both DNA analysis and quantitative studies seem appropriate for species of special interest, if the species is of particular evolutionary interest or has potential agricultural value.

9

Strategies for Long-Term Management of Germplasm Collections

S. A. EBERHART, E. E. ROOS, and L. E. TOWILL

Through natural evolutionary processes, thousands of plant species have developed. Only a small percentage of these have been selected and utilized as agronomic, horticultural, ornamental, or forestry crops. Hence, most plant genetic resources exist in natural ecosystems according to the principle of "survival of the fittest," with no inventory and no managed preservation. As the world's population continues to expand, the areas required for intensive agriculture and forestry will increase at the expense of habitats for other plant species. Because *in situ* preservation may not be adequate in the future, *ex situ* preservation must be expanded. However, the success of *ex situ* preservation depends on the longevity of the seed under the storage conditions used and the ability to regenerate adequate quantities of high-quality seeds without genetic change when the viability declines. In this chapter, we examine the various strategies available for long-term *ex situ* preservation of seeds and other forms of plant germplasm.

GERMPLASM PRESERVATION OF SEEDS

The most common form of germplasm preservation involves storage of seeds. Seeds are classified as being *orthodox*—that is, capable of retaining viability after being dried to less than about 5–10% moisture content (also termed *desiccation tolerant*)—and *recalcitrant*—that is, losing viability after being dried below a critical limit, usually about 12–30% moisture (also termed *desiccation intolerant*) (Chin and Roberts 1980). Desiccation-intolerant seeds are also termed "short-lived" seeds (Harrington 1972). Although the major thrust of this volume is the storage and preservation of noncrop plant germplasm, the literature contains only a limited number of papers detailing the storage characteristics of these species. If it can be determined that the noncrop species of concern have orthodox seeds, then it can be assumed that the storage characteristics will be similar to those of crop species. Similarly, noncrop species having recalcitrant type seeds will behave as do the common crop plants with recalcitrant-type seeds. Thus the major consideration for long-term *ex situ* seed storage of a given species is the determination of orthodox versus recalcitrant seed behavior. Fortunately, most temperate species have orthodox seeds. In the case of tropical plant species, recalcitrant seeds are more prevalent, although the precise percentage of species with recalcitrant seeds is unknown.

Orthodox Seed Storage

Bewley and Black (1982), Priestley (1986), Roos (1986, 1989), and Toole (1986) have reviewed the topic of seed longevity. Earlier reviews by Barton (1961), Biasutti Owen (1956), Crocker (1938), Harrington (1972), Justice and Bass (1978), and Roberts (1972) all contain excellent information and special insight on this topic. The classic compilation of data on seed longevity by Ewart (1908) is still a primary source of information for many species. Bass (1981) compiled tables of seed longevities for many agricultural and horticultural species stored under various conditions. Claimed longevities of 100 years or more for several species have been compiled and summarized by Roos (1986) (Table 9.1).

Many studies have shown the importance of seed moisture and storage temperature for retaining viability in orthodox seeds (Bass 1980, 1981) (Table 9.2). Longevity also can be affected by environmental factors throughout the growing season during seed

Table 9.1. Seed Longevities of 100 or More Years

Species	Claimed longevity	Comments
Lupinus arcticus S. Wats	10,000 + (??)	[14]C dating of associated rodent bones
"Mummy" wheat and barley	2000–3000 + (??)	Egyptian tombs
Nelumbo nucifera Gaertn.	3075(??)	[14]C dating of associated wood canoe
	120–1080(?)	[14]C dating, geologic and historical records
	237	Museum specimen
Chenopodium album L.	1700(?)	Archeologically dated soil sample
	143	From adobe bricks, positive TZ
Spergula arvensis L.	1700(?)	Archeologically dated soil sample
Various weed species	110–600 + (?)	Archeologically dated soil sample
Cana compacta Rosc.	530–620(?)	[14]C dating of associated materials
Medicago polymorpha L.	200	From adobe bricks, positive TZ
Hordeum leporinum Link	200	From adobe bricks, positive TZ
Malva parviflora L.	183–200	From adobe bricks, positive TZ
Trifolium spp.	193	From adobe bricks, positive TZ
Chenopodium murale L.	183	From adobe bricks, positive TZ
Cassia mutijuga A. Rich.	158	Museum specimen
Albizzia julibrissin Durazz.	147	Museum specimen
Barley and oats	123	Building cornerstone
Hyoscyamus niger L.	116	Museum specimen
Cassia bicapsularis L.	115	Museum specimen
Goodia lotifolia Salisb.	105	Museum specimen
Hovea linearis (Sm.) R. Br.	105	Museum specimen
Malva rotundifolia	100	Buried seed
Verbascum blattaria L.	100	Buried seed
Verbascum thapsus L.	100	Buried seed
Trifolium pratense L.	100	Museum specimen
Lotus uliginosus (L. pendiculatus* Cav.)	100	Museum specimen

Key: ?? = highly questionable; ? = somewhat questionable; TZ = tetrazolium.
Source: Adapted from Roos (1986).

Table 9.2. Effect of Moisture Content (MC) and Temperature on Germination of Orthodox Seeds After 16–19 Years of Storage

Species	MC (%)	Storage temperature (°C)				
		−12	−1	10	21	32
Carthamus tinctoris L.	4	83 (17)[a]	95 (17)	97 (17)	93 (17)	36 (8)
	7	87 (17)	94 (17)	81 (11)	3 (3)	0 (1)
	10	78 (17)	47 (11)	17 (2)	0 (1)	0 (1)
Lactuca sativa L.	4	93 (19)	93 (19)	90 (19)	85 (19)	36 (8)
	7	94 (19)	14 (11)	1 (8)	4 (4)	0 (1)
	10	96 (19)	18 (19)	4 (5)	11 (1)	0 (1)
Sesamum indicum L.	4	93 (17)	93 (17)	93 (17)	92 (17)	2 (17)
	7	91 (17)	23 (17)	39 (5)	0 (2)	0 (1)
	10	28 (1)	30 (1)	6 (2)	0 (1)	0 (1)
Sorghum bicolor (L.) Moench	4	93 (16)	95 (16)	94 (16)	85 (16)	69 (16)
	7	95 (16)	90 (16)	85 (16)	83 (16)	52 (16)
	10	91 (16)	90 (16)	77 (16)	58 (16)	43 (5)
Trifolium incarnatum L.	4	83 (16)	73 (16)	81 (16)	81 (16)	77 (16)
	6	87 (16)	84 (16)	77 (16)	85 (16)	27 (16)
	11	82 (16)	30 (16)	22 (16)	3 (16)	5 (2)

[a]Percent germination (years of storage in parentheses).
Source: Data from Bass (1981).

production as well as during harvesting and drying of the seed before storage (Bass 1980; Roberts and Ellis 1984; Smith 1985). Reducing the seed moisture content to 5–7% and lowering the storage temperature increases longevity for many species, but only limited information is available for seed stored under these conditions, especially at subfreezing temperatures (Roos 1986). Freeze drying has also been used to reduce seed moisture content and improve seed storability (Woodstock 1985; Woodstock et al. 1976, 1983).

Papers have been published on seed storage at cold (<0°C) temperatures, but the storage periods reported usually have been only five to ten years. Roos (1986) listed only two papers that contained data on subfreezing storage of seeds for 25 years. Seeds of *Lobelia cardinalis* L. stored at −5°C and 4.7% seed moisture content had a germination of 64% after 25 years versus an original germination of 73% (Barton 1960). *Hibiscus cannabinus* L. seeds retained their full germination (88%) after 25 years at −12°C and 10.4% seed moisture content (Bass 1981). Rincker (1980, 1981, 1983; Rincker and Maguire 1979) examined the effects of subfreezing storage on germination and forage production of forage grass and legume seed. In one study, 291 seed lots representing seven species were stored for as long as 20 years at −15°C and 60% (RH) (Rincker 1981). In general, seed lots with high initial germination retained high germination potential. For example, 14 lots of birdsfoot trefoil (*Lotus corniculatus* L.) stored for 18 years had 95% germination both before and after storage. Germination declined more quickly for seed lots with lower initial seed germination. For example, 20 lots of bromegrass (*Bromus inermis* Leyss.), with an initial germination of 87%, fell to 55% after 18 years.

In a second study, 260 lots of nine forage species were placed in subfreezing storage directly after harvesting (Rincker 1983). After 20 years, only three seed lots had lost 20% or more of their viability. Upon the retirement of Clarence Rincker from the Agricultural Research Service (ARS), over 3000 seed lots were transferred from Prosser, Washington, to the National Seed Storage Laboratory (NSSL) in Fort Collins, Colorado in order to continue this valuable, long-term seed storage study (Roos et al. 1986). All seed lots were tested for viability and seed moisture content. Analyses were done on the effects of year and location of production on germination (Roos et al. 1987). Whereas location of production appeared to have minimal effect on germination after storage, year of production was found to have a significant effect on storability.

Storing seeds at approximately − 20°C should greatly increase longevity over storage at higher temperatures, thereby reducing the frequency of costly germination tests and regeneration cycles. Approximations based on a number of assumptions suggest that seed from many species may have longevities of about 100 years or more when stored at − 20°C (Roberts 1983; Roberts and Ellis 1984).

Cryogenic storage refers to storage at very low temperatures. Ideally, temperatures less than about − 130°C are preferred because of the absence of liquid water, very low molecular kinetic energies, and extremely slow diffusion (Mazur 1966). Chemical reactions (e.g., metabolism) thus occur very slowly, and storage longevities are postulated to be extremely long and limited by only the buildup of genetic lesions resulting from background irradiation (Ashwood-Smith and Grant 1977). Long periods of storage are desired to reduce problems associated with seed regeneration, including cost of regeneration and risks of losses caused by disease and environmental problems (Roberts 1975). Cryogenic storage is usually in liquid nitrogen (LN, i.e., at − 196°C) or in the vapor above LN (ca. − 150°C to − 180°C) because it is a relatively inexpensive cryogen. Where an adequate cryogen source is available, cryogenic seed storage is cost-competitive with low-temperature (− 20°C) storage (Stanwood and Bass 1981) (Table 9.3). A number of assumptions are used in these calculations, but cryogenic storage is economical when costs associated with regeneration cycles and germination testing are included. Although the costs today would be somewhat different from those in 1981, the overriding effect of increased storage life in LN—particularly for those species characterized by poor storage, for seeds with very high regeneration costs, or for seeds that cannot be replaced—gives LN storage a definite economic advantage.

Orthodox seeds at low moisture levels usually retain the same percentage germination after LN treatment as before; hence, extended seed viability is expected (Stanwood 1985). Some limitations on LN use for orthodox seed may exist and may relate to the seed coat, seed oil content, and seed moisture content. Some orthodox seeds, including temperate fruit and nut seeds (*Corylus* spp., *Juglans* spp., and some *Prunus* spp.), are desiccation tolerant but are sensitive to low temperatures, although the extent of sensitivity and the mechanisms of injury are not known.

Cooling of orthodox seed requires no special equipment. Seed is placed into the vapor phase within a cryotank. Warming of seed from LN storage is not controlled, and samples usually are placed directly at room temperatures after removal from the cryotank.

Recalcitrant Seed Storage

Recalcitrant seeds have short life spans, ranging from a few weeks to several months (Harrington 1972; Roberts 1973; Chin and Roberts 1980). Harrington (1972) compiled a list of plant species with short-lived seeds, many of which we would now call recalcitrant (the term "recalcitrant" had not come into general use at that time). A list of

Table 9.3. Estimated Cost Factors for Mechanical Refrigeration and Liquid Nitrogen Storage for Onion Seeds

Source	100-Year cost (per sample)	
	Mechanical[a]	Liquid nitrogen
Storage		
Refrigerator	$ 1.15	$ 1.49
Inventory	0.15	3.36
Equipment replacement	4.00	0.15
Subtotal	$ 5.30	$ 5.00
Operations		
Utilities ($0.12/year)	12.00	0
Maintenance ($0.20/year)	20.00	0
Viability monitoring ($0.28/year)	28.00[b]	2.80[c]
Liquid nitrogen	0	9.00[d]
Subtotal	$ 60.00	$11.80
Seed replacement		
Four growings at $25	100.00	0
One growing at $25	0	25.00
Subtotal	$100.00	$25.00
100-year cost	$165.30	$41.80
Average yearly cost	$ 1.65	$ 0.42

[a]Assuming standard storage conditions of −18°C and a seed moisture content of 4–7%, storage life of seed is assumed to be approximately 25 years.
[b]Five-year retest frequency.
[c]Based on a reduced germination retest schedule (50-year retest frequency).
[d]Liquid nitrogen cost at $0.42/liter.
Source: Data from Stanwood and Bass (1981).

species suspected of having recalcitrant seeds was compiled by King and Roberts (1980). Three factors may contribute to the short longevity of stored recalcitrant seeds: sensitivity to desiccation, chilling injury, and problems such as microbial contamination and germination during storage that are associated with seeds having a high moisture content.

Recalcitrant seeds are produced by aquatic species, large-seeded species, species native to tropical areas, and some temperate-zone species of trees (e.g., oaks) (King and Roberts 1979; Chin and Roberts 1980). Coconut, cocoa, mango, nutmeg, and rubber plants are examples of economically important species with recalcitrant seed. Adequate tests should be made to determine if seed of a species displays true recalcitrant behavior (Roberts et al. 1984). For example, seeds from citrus, formerly thought to be recalcitrant, were shown to be dormant, requiring a longer period of germination than undried seeds (Roberts 1983).

To retain viability, recalcitrant seeds are stored at as low a temperature as possible under conditions that retain relatively high seed moisture levels and ensure a supply of oxygen for respiration. Seed longevities range from weeks to months under these conditions and are too short for adequate germplasm preservation (Roberts and King 1981). Therefore, these species usually are stored as vegetative plants. If diversity is considerable, many individuals must be conserved, resulting in large space and maintenance costs.

More research is necessary to define the mechanisms of desiccation and chilling injury and to investigate methods of alleviating it. Cryogenic treatments involving cryoprotectants (King and Roberts 1979) or the use of excised embryos (Grout et al. 1984) may prove to be useful, but no procedure has yet been developed for the practical, reproducible cryogenic preservation of recalcitrant seeds. Axillary shoot tip or bud cryopreservation from *in vivo* or *in vitro* plants might be useful for base-collection storage of these species.

Viability Monitoring

Successful seed preservation is dependent on continuous viability monitoring. Current technology requires periodic germination tests. With this destructive method of monitoring, it is important to use minimum seed numbers for each test with intervals of five to ten years between tests, depending on the expected viability of the species. Research is under way that seeks nondestructive methods for viability monitoring as well as reliable techniques to predict longevity for each seed sample.

In general, the ten-year test interval is appropriate when germination exceeds 85% on the most recent viability test for species that can be expected to retain viability for extended periods (e.g., beans, corn, wheat, tobacco, tomato). The five-year interval is probably more appropriate for other materials, especially if the storage characteristics are not known, as would be the case for most wild species. High-quality seed with high initial germination usually stores well for extended periods under the proper storage environment. However, once the deterioration process begins, further decline in viability does not appear to be arrested or prevented by even the best of storage conditions. It can only be slowed down.

The optimum conditions for germination need to be determined for each species, including pregermination treatments to break dormancy when this is a factor, light requirements, day–night temperatures, and seed moisture content. Seed germination testing is costly in terms of manpower and in sacrificing valuable seeds used in viability estimation. At the National Seed Storage Laboratory, the initial germination test uses two replications of 100 seeds, whereas only one replication of 50 seeds is used for the retests. When the seed supply is limited, fewer seeds can be used, with a concomitant loss in sensitivity. When both the seed supply and the viability are low, it becomes a difficult choice whether or not to use up seed in determining the germination percentage. Rather, attempts should be made to rescue those seeds that do germinate and, hopefully, regenerate the stock.

Although the life span of a seed lot can be extended greatly by low-temperature storage, seeds eventually lose the capacity to germinate, and they die. The *accelerated aging* (AA) *test* (Delouche 1965; Delouche and Baskin 1973) was originally developed to predict the relative storability of seed lots. In this test, seeds are placed under high temperatures (40–45°C) and high relative humidities (>90%) for short periods of time (2–5 days) and are then subjected to the standard germination test. Results are compared with those of unaged seeds tested at the same time. Although this test now has become an accepted vigor test (Association of Official Seed Analysts 1983), its original purpose was to predict which seed lots should be carried over to the following year by seed companies. Since this test was developed in Mississippi, it further was assumed that carry-over seed would be subjected to the hot, humid conditions prevailing in that state.

Although the basic principle of the AA test—that seeds having low vigor (see below) will not store as well as those having high vigor—is probably valid, it is not necessarily true that seeds stored under subfreezing temperatures will behave in the same manner. The mechanism for loss in germinability under the two sets of conditions probably is not the same. Evidence for this comes from three observations:

1. Under AA, seed moisture is elevated to a point at which microorganisms can grow readily.
2. Under normal (slow) aging, the percentage of abnormal seedlings increases as germination (percent normal seedlings) drops, whereas under AA this large increase in abnormal seedlings is seldom seen, possibly because the seeds pass quickly through this stage.
3. The frequency of chromosomal aberrations in root tips of germinating seeds, which is normally correlated with germination percentage (Murata et al. 1981), is lower in seeds subjected to AA.

An alternative approach, similar to accelerated aging, to predict which seed lots might deteriorate first in gene banks involves the use of artificial aging techniques (Murata et al. 1980; Roos and Rincker 1982). The significant differences between these two approaches lie in the use of lower seed moisture contents and storage temperatures and longer storage periods. Seed moisture contents typically should be below levels where active fungal growth can occur (<15%), storage temperatures are less than 35°C, and storage periods should be of the order of weeks or months. The objective is to achieve slow deterioration which more closely approximates that of seeds in long-term cold storage. Moore and Roos (1982) compared different artificial aging tests to evaluate the rates of deterioration and statistically analyze the quantal responses. Artificial aging tests that employed high humidities and high storage temperatures (90% RH, 32°C) resulted in high data heterogeneity and concomitant lack of reliability for predicting seed survival. More moderate aging conditions (70% RH, 21°C) resulted in low data heterogeneity. A computer program was developed (Moore et al. 1983) to calculate the deterioration rates and apply the statistical analysis. This program was used by Priestley et al. (1985) to determine differences in seed longevity at the species level.

A mathematical model of seed deterioration has been proposed by Roberts (1972, 1986) and Ellis and Roberts (1981). They have suggested that this "basic viability equation" can be used to predict germination after any storage period at any temperature and moisture content (within reasonable limits). The equation depends on the determination of several constants, including the initial seed quality (specific for each seed lot) and four other constants for the effects of temperature, seed moisture content (determined for each species), and species. Once the constants have been determined for a given species and seed lot, they make predictions as to longevity under any number of different storage conditions. A serious problem inherent in this and in all predictive models is that the equation may be invalid when extrapolated beyond the limits of the data (temperature and seed moisture conditions) used in developing the various constants.

Seed Storage Methodologies

Methods for seed handling, packaging, germination testing, and shipping have been evaluated and somewhat standardized. Orthodox seeds are usually dried to moisture contents of 5–7% and then are sealed in moisture-proof, foil-laminated bags or aluminum cans for storage at $-20°C$ (Justice and Bass 1978; Cromarty et al. 1982; Bass 1984; Ellis et al. 1985a). Care must be taken to ensure that seals are complete and that bags are free of any punctures. Accurate seed-viability tests and methods of breaking dormancy are necessary for evaluating longevity (Justice and Bass 1978; Ellis et al. 1985b).

The recommendation from the International Board for Plant Genetic Resources (IBPGR) for ensuring the retention of maximum diversity within each accession is to regenerate seed whenever germination drops about 15% below the initial germination level (Cromarty et al. 1982). Practical considerations of available labor, space, and time, however, often delay regeneration until viability has dropped 20–50% or even much more from the original value. When the natural population still exists, it may be preferable to re-collect rather than regenerate a new supply from the previous collection.

Mutations, notably accumulations of chromosome damage and point mutations, are associated with the loss of viability during storage. Cells with chromosome change are usually selected against during growth of the seedling (*diplontic selection*), but point mutations may be passed to succeeding generations (Roos 1982; Roberts 1983; Roberts and Ellis 1984). Thus some genetic changes do occur during storage.

CLONAL PRESERVATION

Some germplasm can be maintained only vegetatively. Clones usually are maintained in greenhouses or field plantings, such as orchards or plantations. Field and greenhouse maintenance requires considerable space, and high costs are associated with pruning, disease and weed control, and propagation. Materials held in the field are more susceptible to loss from disease or disaster. Phytosanitary practices may be more stringent in greenhouses and screenhouses, and disease occurrence and spread are thus reduced. Tremendous adaptive characteristics associated with survival under different climatic conditions often exist within a species. Growth of all lines at a single location may place some lines under environmental stress and decrease fecundity.

Plant parts such as cuttings, budwood, root divisions, and tubers can be stored for about one year under refrigeration, and are used for propagation and distribution. The useful storage time varies considerably. Propagules from chilling-sensitive species require storage at warmer temperatures.

In Vitro Preservation

Plants maintained within containers (*in vitro* plants or plantlets) are useful for preservation of a wide array of species (de Langhe 1984; Withers 1984; Withers and Williams 1986; Towill 1988). *In vitro* plants are initiated from meristem tips, buds, or stem tips and are propagated through divisions; thus lines are maintained as clones (Hu and

Wang 1983; Styer and Chin 1983). Shoot multiplication *in vitro* must occur by axillary bud proliferations to avoid selection of somaclonal variants that might arise from adventitious buds produced within the tissue mass. The term *in vitro* may also refer to cells, tissues, or organs held in culture, but their use for germplasm storage is not feasible at the present time, owing to cytological instabilities and the potential for somaclonal variation in regenerated plants (Bayliss 1980; D'Amato 1985).

Storage of *in vitro* plants is under normal growth conditions (such as 20–25°C under light), under reduced temperatures (-3°C to $+12$°C, with or without light), or on a growth-restrictive nutrient medium at ambient or reduced temperatures (Aitken-Christie and Singh 1987). Storage in this manner gives longevities of weeks, months, and, in some cases, years before subculture is needed.

Potential problems with the application of *in vitro* plant preservation include difficulties in culture establishment, micropropagation, rooting, and acclimatization (Stushnoff and Fear 1985; Ziv 1986; Bonga 1987; Hutchinson and Zimmerman 1987; Towill 1988). Many of the details for these stages are genotype specific. Explants from mature woody plants often are more difficult to establish in culture than are those from juvenile plants (Bonga 1987; Hutchinson and Zimmerman 1987). Diverse genotypes may not survive in the same physical and chemical environment. Growth abnormalities may occur (Kevers et al. 1984; Paques and Boxus 1987). There is an uncertainty about the occurrence of somaclonal variants. Adequate characterization of each *in vitro* line (e.g., electrophoretic profiles: isozymes, restriction-fragment-length polymorphisms) must be established and lines periodically examined to avoid identification errors.

In vitro plant preservation has been used mainly for crop species, although some information exists for wild relatives of these species. The methodologies for initiating and maintaining cultures are fairly simple and could be incorporated into programs at botanical gardens. Considerable labor, however, may be required to maintain a collection fastidiously. Experience with *in vitro* maintenance for germplasm preservation, even with crop species, is limited; thus time and cost analyses are lacking. At present, there are inadequate data to provide answers to questions on the extent of duplicate samples, duplicate locations, and handling needs. Until acceptability is gained through experience, *in vitro* storage is now best utilized as an adjunct to *in vivo* storage (van Sloten and Holle 1988).

Cryopreservation

Cryopreservation of meristem tips, shoot tips, or buds is potentially useful for clonal collections, since it provides an inexpensive, space-saving method for long-term storage (Sakai 1984, 1986; Kartha 1985b; Chen and Kartha 1987; Withers 1987). Two general methodologies are used. The first utilizes the ability of some species to cold-acclimate to very low temperatures, whereas the second utilizes the addition of cryoprotectants to enhance the survival of species that are less, or are not, cold hardy.

The extreme cold hardiness that develops in some species is well documented. If twigs of these species are collected in the correct physiological state, they can be cooled at a controlled rate, and then can be preserved in LN (Sakai and Nishiyama 1978; Katano et al. 1983; Hirsh et al. 1985; Sakai 1986; Chen and Kartha 1987; Hirsh 1987; Stushnoff 1987; Tyler et al. 1988). After thawing, excised buds can be grafted to suitable rootstocks to generate a shoot. Alternatively, meristem tips can be dissected from LN-treated buds and cultured to produce a shoot (Moriguchi et al. 1985). Any method must take into account that genotypes within a species differ in hardiness.

Cryopreservation of species that cold-acclimate to a lesser extent or do not acclimate at all requires application of cryoprotectants, such as dimethylsulfoxide (DMSO) or combinations of DMSO with sucrose, glucose, polyethylene glycol, or proline (Finkle et al. 1985; Withers 1987). Buds or shoot tips from field, greenhouse, growth chambers, or *in vitro* plants are excised for treatment. In some cases, cold treatment of the stock plant is beneficial (Caswell et al. 1986; Reed 1988). After excision, the sequence of steps in a cryogenic protocol involve preculture (a few hours to two days) in growth medium with low levels of cryoprotectants; controlled exposure to a defined concentration of cryoprotectant; ice nucleation of the sample just below the freezing point of the solution; controlled cooling (about 0.25 to 1°C/min) to about −35°C to −40°C, followed by transfer directly into LN; rapid warming; and post-thaw treatments prior to culturing to determine viability. The variations on the basic protocol are too numerous to discuss here. Details for several of these steps have been reviewed (Sakai 1984; Kartha 1985a, 1985b; Chen and Kartha 1987; Withers 1987).

A persistent problem in the cryogenic storage of shoot tips using applied cryoprotectants is the maintenance of organization within the shoot tip and regeneration of the shoot without (or with minimal) callus formation or adventitious bud development. With current procedures, shoot tips from herbaceous lines exposed to LN often form a small amount of callus before shoots are regenerated (Towill 1983; Kartha 1985a) or form adventitious shoots at different locations within the tips (Haskins and Kartha 1980). Both probably increase the chance for somaclonal variation. LN-treated buds from hardy or acclimated woody lines develop directly into a shoot, but because of limited studies this cannot be inferred for all species.

The opinion of most cryobiologists is that cryogenic storage is a safe method for retaining viability for many years. Cryogenic storage has been pragmatically used for microbes and for animal cell cultures, semen, and embryos. Storage times reported range from only a few months to a few years. But cryogenic storage of buds and shoot tips from clonal lines requires additional research before practical storage can be initiated. Few laboratories or organizations are engaged in such work, and thus no information is available on storage systems, retrieval, testing procedures, and distribution, or on the number of replicates required per accession and extent of collection duplication.

Although cryopreservation is potentially useful for crop clonal germplasm, it may not be as desirable for wild species. Seed and pollen preservation would generally be more useful for preservation of diversity, particularly for herbaceous species. If clonal maintenance is needed—for example, owing to long juvenility periods—cryopreservation might be justified. But more information is needed, and methods must become simpler and less time consuming for application to many different species.

Pollen Preservation

Pollen is not a conventional vehicle for germplasm preservation, but may be useful for base collections of species that do not produce orthodox seeds (Roberts 1975; Schoenike and Bey 1981). Storage of pollen requires little space. A population of pollen grains collected from genetically different individuals would contain the nuclear genes within that population. Some cytoplasmic genes, however, would be lost.

Pollen, like seed, may be divided into desiccation-tolerant and desiccation-sensitive groups (Towill 1985). Species with recalcitrant seed do not necessarily have desiccation-sensitive pollen. Desiccation-tolerant pollens store best when dried to a low moisture content and when held at low temperatures, analogous to orthodox seed. Dried pollen, such as from some fruit crops of temperate and subtropical climates (Akihama and Omura 1986), survives cryogenic exposure. Desiccation-sensitive pollens (e.g., from many members of the Poaceae) are often short-lived and also lose viability during drying. These pollens, therefore, cannot tolerate temperatures at which the cell water freezes.

Some success has been reported for preserving desiccation-sensitive pollen at low temperatures, notably maize pollen (Barnabas and Rajki 1981), but data are sparse. Careful adjustment of the moisture content to a specific range is crucial (Hoekstra 1986; Kerhoas et al. 1987). Other difficult-to-store pollens, such as that from pearl millet (Hanna et al. 1986) and sugarbeet (Hecker et al. 1986), have survived LN exposure. Information is needed on stability and longevity at subzero temperatures and on the development of handling systems before practical storage is achieved.

Information about storage characteristics of pollen from wild species is fragmentary, existing mainly for some crop relatives and for medicinal and forest species. Storage characteristics may be inferred from systematic relationships with species possessing pollen of known desiccation response. But desiccation sensitivity occurs among species and genera within many families, and thus is not restricted to certain taxa. Studies in pollination biology have produced data that may be useful in predicting the desiccation response for pollen from selected species, but this information, too, is fragmentary.

CONCLUSIONS

With increasing population pressures, *in situ* preservation may not be adequate to preserve the genetic diversity represented by the thousands of plant species that have arisen in the natural evolutionary process. *In situ* preservation must be supplemented by extensive *ex situ* preservation. Long-term storage technologies have been developed and tested for orthodox seeds; these technologies require dry seeds to be stored either in cryotanks with liquid nitrogen or in conventional seed vaults at $-20°C$ in moisture-proof containers. Research must be expanded to identify and develop the technologies for the successful cryopreservation of recalcitrant seed and vegetative parts of plant species that do not produce seeds. Successful long-term preservation is dependent on continuous viability monitoring with re-collection or regeneration whenever viability drops below a minimum level. Development and use of optimum storage conditions will decrease the frequency of viability tests and re-collection or regeneration.

IV

CONSERVATION STRATEGIES FOR GENETIC DIVERSITY

Strategies for Conserving Clinal, Ecotypic, and Disjunct Population Diversity in Widespread Species

CONSTANCE I. MILLAR and WILLIAM J. LIBBY

Why is a chapter on widespread species appearing in a volume on rare species? One answer to that is another question: Why focus on species in the first place? Granted, species do have a unique status. They are more or less closed units genetically, and species extinction signals the end of an evolutionary lineage that may have begun millions of years ago. By contrast, individuals and populations within species are lost and replaced all the time.

If a goal of conservation is to maintain species in the long term, then attention should focus on levels of biological organization within the species. Stripped of their taxonomic identities, organisms cluster naturally into hierarchical units of genetic similarity, with individuals at the lowest level and genetically related groups at the highest level. The species level is part of a continuum of genetic distinctness: diversity and rarity occur at every level, and much of this variation is adaptive. The evolutionary potential and resilience of species depend on the amount and structure of genetic variation at intraspecific levels. Widespread species, by virtue of their size, have complex genetic structures, and cause attention to be focused on these levels. During the early missionary phase of genetic conservation, it was appropriate and effective for the species level to receive the most attention. But now that conservation biology has advanced in its understanding of species and population biology, we recognize that we cannot ignore the unique, rare, and unusual variation that exists at intraspecific levels of most species.

This chapter reviews the genetic architectures of both rare and widespread species, and presents strategies for conservation of diversity within widespread species. Because widespread species by definition comprise many individuals and many populations, conservation strategies differ from those for rare species in that they aim not so much to preserve the entire species, but to sample genetically diverse individuals from some meaningful hierarchy of stands, populations, and races. This chapter draws attention to the inherent value and priority for conserving widespread species and emphasizes that intraspecific variation is important in determining conservation strategies for rare species as well. As forest geneticists, we have chosen as our examples temperate-zone trees. We hope that the stories they provide will be useful for conservation of other taxa.

GENETIC ARCHITECTURE OF RARE VERSUS WIDESPREAD SPECIES

All species, whether rare or common, have a unique genetic profile that defines their hierarchical structure of variation, or genetic architecture of traits (Libby and Critch-field 1987). The existence of intraspecific levels of variation has profound implications for conservation of species. At each level, variation can affect fitness and viability. At the level of individuals, loss of variation (increasing homozygosity) can lead to inbreeding depression, which is often expressed as a severe decline in health (e.g., Libby et al. 1981; Sorensen and Miles 1982; Strauss 1987). At the population level, where variation is often related to adaptation to local environments, reductions in variation or other disruptions of the local gene pool may decrease the chance that a population can adapt to changing conditions. At the species level, loss of diverse populations reduces the potential of the species to respond to environmental changes at regional and global scales. Thus species existence depends to a large part on the genetic health of all the intraspecific levels. Similarly, the appropriate conservation strategy for a species also varies in response to the particular genetic architecture of each species.

Patterns of genetic variation can be grouped into general classes for specific taxonomic groups. Northern Hemisphere conifers have been widely studied, and we summarize some of the patterns of genetic variation that have been described for species in that group.

No or Little Genetic Variation Within Species

This situation is unexpected in most taxonomic groups, but especially in conifers, which typically are highly variable. Red pine (*Pinus resinosa*) is the best known example of this small class. Despite its wide distribution throughout the north central and northeastern United States and adjacent regions in Canada, and a reproductive system that favors the evolution and maintenance of variation, red pine is nearly invariant genetically (Fowler and Morris 1977). This situation is attributed to Pleistocene glaciation, which appears to have reduced red pine to a small area and eliminated variation. Red pine has subsequently recolonized vast areas and, within these areas, appears healthy and aggressive despite loss (presumably) of genetic variation (Barrett and Kohn, Chapter 1; Templeton, Chapter 12).

Species may be invariant in one part of the range and variable in another part. Western redcedar (*Thuja plicata*) grows as a scattered component of mixed conifer forests throughout the Pacific Northwest and British Columbia. In these regions, it appears to have little genetic variation (Copes 1981), perhaps another victim of Pleistocene climate changes. In its southernmost populations of northern California, western redcedar has some genetic variation. These areas may have been refugia for the species during the glacial periods, where variation persisted in moderate-size populations. Unusual environmental heterogeneity in northern California may also contribute to the evolution and retention of variation there.

No Variation Within Populations, Variation Among Populations

Torrey pine (*Pinus torreyana*) exemplifies this small class for conifers. Torrey pine is a rare California species known from only two small populations, one on the mainland near San Diego, and the other on Santa Rosa Island off the California coast, 280 km northwest of San Diego. Genetic studies show this species to have no variation within populations at any of 59 allozyme loci (Ledig and Conkle 1983). There are, however, differences between the populations at two allozyme loci and some variation in monoterpene composition (Zavarin et al. 1967). Thus minor differences between populations, presumably caused by founder events, are the only known genetic variation in the species.

Little Variation Within Populations, Little Variation Among Populations

This class of forest trees is slightly more common than the previous two. Santa Lucia fir (*Abies bracteata*), another rare California conifer that grows scattered in the rugged wilderness of the Santa Lucia Mountains, is an example. There is a modest amount of variation in this species, both within and among populations (Ledig 1987). The differences among populations follow no geographical pattern and are uncorrelated with obvious environmental factors.

These three classes describe patterns of variation that are considered exceptional for the outbreeding, long-lived conifers, but that may be more common for taxonomic groups with different reproductive systems. Geneticists concerned about long-term viability of conifer populations have usually advocated management that avoided inbreeding and that maintained large populations made up of unrelated individuals. These recommendations were aimed to minimize situations of low genetic variation within populations, such as occur naturally in the three classes described above. More commonly, conifers do not tolerate those genetic conditions. That there are so few examples in these classes may testify to this. The species that have survived well despite high homozygosity may be the few winners in an evolutionary game that extirpated many others. Managers have developed techniques to carry populations of normally outbreeding species through an inbreeding bottleneck to a point where high homozygosity is tolerated in the population (Barker and Libby 1974; Libby 1976; Templeton and Read 1983). This may mimic what occurs in the few successful natural instances.

High Variation Within Populations, Little Variation Among Populations

Several widespread conifers exemplify this pattern, either over their entire range or in certain portions of it. Incense-cedar (*Libocedrus decurrens*) is a widespread species common in mid-elevation mixed-conifer forests of California and western Oregon. Variation within populations in the Sierra Nevada portion of the range is relatively high, but only minor genetic differences exist among widely separated populations (Harry 1984, 1987). Low population differentiation but high within-population diversity in western white pine (*Pinus monticola*) is a common pattern throughout the extensive portions of its range in Oregon, Washington, and Idaho. Populations that extend into California, however, are distinct from one another and from the Pacific Northwest populations (Steinhoff et al. 1983). This pattern appears to reflect the relative degree of habitat heterogeneity over western white pine's range—low in the north, high in the south.

High Variation Within Populations, High Variation Among Populations

Conifers in this common situation can be divided into discontinuous and continuous geographical patterns. The difference between these patterns is often only a matter of scale.

Discontinuous

Abrupt genetic differences among populations—that is, large differences at many loci in repeated groupings—occur as a result of natural selection and/or genetic drift. When selection appears to be the primary factor leading to differentiation, as suggested by major differences in environment predictably being associated with particular combinations of genetic variation, the resulting pattern is called *ecotypic*. This may be recognized on a large scale, as in ponderosa pine (*Pinus ponderosa*), where at least four major ecotypes have been described: the Pacific, North Plateau, Rocky Mountain, and Southwestern races (Wells 1964; Smith et al. 1969; Read 1980) (Figure 10.1). The transitions among these races occur in more or less continuous pine populations over hundreds of kilometers. By contrast, on a small scale, bishop pine (*Pinus muricata*) is genetically differentiated into ecotypes that grow on and off pygmy-forest soils in northern California. The two ecotypes meet at abrupt ecotones only 0.5 km wide (Millar 1989).

Discontinuous variation also can occur primarily because of isolation of disjunct populations and genetic drift. In these cases, the pattern of differentiation may parallel the population disjunctions, and similar environments are not predictably associated with particular combinations of genetic variation. Certainly, selection is not suspended, but its role is less obvious, and the term "ecotype" is not used. In Monterey pine (*Pinus radiata*), for example, each of the five disjunct populations has genetically distinct traits (Bannister et al. 1962; Forde 1964; Millar et al. 1988; Moràn et al. 1988). Although selective factors differ among the populations, isolation and genetic drift were probably more important in producing current patterns of population variation.

Continuous

Genetic changes often occur gradually over space rather than abruptly, in which case they are called *clines*. Clinal variation is usually ascribed to gradual changes in natural selection, but migration and genetic drift can also give rise to situations where genetic distance is correlated with geographical distance. At the level of the species range, differences among ecotypes may appear abrupt, whereas at the local level, the transition between races may be clinal (Sorensen 1979; see also Figure 10.1, ponderosa pine). This is common when populations are spatially continuous.

Where important factors of the species' selective environments vary gradually,

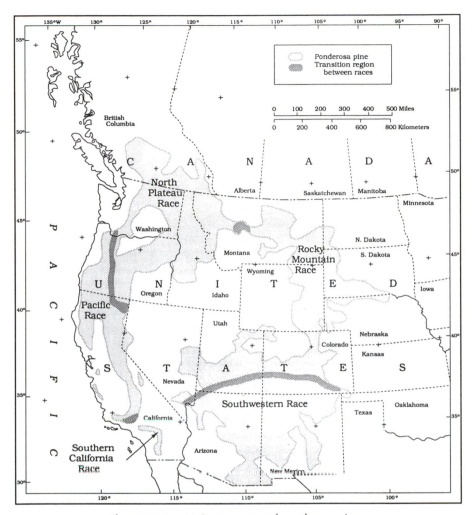

Figure 10.1.   Major ecotypes of ponderosa pine.

such as climate, and when populations are large and more or less continuous, clines are likely to evolve. Many important physiological traits vary clinally in conifers. Most adaptive traits of ponderosa pine and white fir (*Abies concolor*) vary clinally from south to north and from low to high elevations in the Sierra Nevada of California (Callaham and Liddicoet 1961; Conkle 1973; Hamrick 1976; Libby et al. 1980). Average values for second-year height from a common-garden test in white fir vary clinally throughout the species range (Figure 10.2) (Hamrick and Libby 1972). Changes in allozyme frequency in knobcone pine (*Pinus attenuata*) vary clinally along the species' U-shaped distribution (Millar et al. 1988). Many other examples exist.

Clines occur not just in frequencies of genes, but also in levels of diversity within stands. In the southwestern United States, many conifers are most diverse in their more southern populations. Diversity within populations decreases clinally northward (Fins

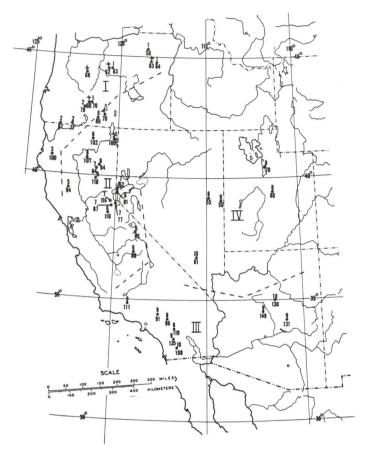

Figure 10.2. Clinical variation in average second-year height (values below black dots) of white fir in common-garden plantations; ecotypic variation in adaxial stomata rows (values above black dots) of white fir. Roman numerals indicate four regional groups indicated by multivariate analyses of nine traits. (From Hamrick and Libby 1972)

and Libby 1982; Steinhoff et al. 1983; Critchfield 1984; Furnier and Adams 1986; Ledig 1987; Millar et al. 1988).

In concluding this section on patterns of variation, we note that different patterns of variation may occur, depending on the traits studied. This discordance has been discussed for adaptive (e.g., survival and growth) traits versus presumably more neutral (e.g., biochemical) traits (Endler 1986; Libby and Critchfield 1987). In many conifers (e.g., lodgepole pine, *Pinus contorta;* Douglas fir, *Pseudotsuga menziesii;* and ponderosa pine), isozyme variation among populations is less than 10%, whereas a much larger proportion of total variation occurs among populations in obviously adaptive traits. For example, in lodgepole pine, 6% of the total isozyme variation is among populations, whereas 38% of total morphometric variation is among populations (Wheeler and Guries 1982). The discordance can occur even among traits within either of those categories. In bishop pine, large differences in allozymes yield one pattern, survival and growth traits another, cone morphology a third, resin canal anatomy a fourth, monoterpene composition a fifth, and reproductive crossing relationships a sixth pattern (reviewed in Millar 1986). Obviously, this is a nightmare for taxonomists and conservationists, but is still entirely within the rules of evolution.

We have enumerated these patterns because on the whole they may occur in rare as well as in widespread species, differing only in geographical scale. Some generalizations are developing about the relative magnitudes of variation in rare versus widespread species. Despite the limitations of using one type of trait to infer a general pattern, enzyme polymorphisms—allozymes—are useful traits for comparing diverse taxa. Hamrick and his colleagues (Chapter 5 and references therein) have surveyed the relationships between allozyme variation and life-history traits for many taxonomic groups. For 28 conifers, endemic and narrowly distributed species had a lower index of variation than widespread species, although this difference was not statistically significant (Hamrick et al. 1981). On a broader taxonomic basis, the influence of geographical range (as well as other life-history traits) on genetic architecture has been compared by using allozyme data from 122 studies involving 91 plant species (Hamrick et al. 1979; Hamrick 1983). For the 91 plant species and four categories of geographical range (endemic, narrow, regional, and widespread), the highest levels of variation occurred in widespread species (Table 10.1). Overall diversity within the species and within individual populations on average was 10% larger in widespread species than in endemic and narrowly distributed species. Species with narrow to regional-size ranges had on average 10% more variability among populations than did endemic species. Of the few comparisons that involved pairs of closely related species with different distribution sizes, no generalizations were possible.

THREATS TO GENETIC DIVERSITY

Aside from strictly biological considerations, widespread species differ from rare species in ways that affect conservation. In forestry and horticulture, many widespread species are economically important, and census numbers of individuals usually increase, not decrease, under management. Intensive use and management of species lead to special conservation problems. The greatest risks are not to the species' existence but to the integrity of the native gene pools (Ledig 1986a, 1988; Millar 1987; Millar and Libby 1989). As such, the effects can extend throughout the species' range, wherever management occurs.

Table 10.1. Relationship of Geographical Range to Allozyme Variability in 91 Species of Seed Plants

Geographical range	No. of studies	H_t	H_s	G_{st}	A_p	P_a
Endemic	10	.275	.208	.200	3.26	.639
Narrow	31	.261	.177	.275	2.96	.609
Regional	38	.238	.154	.312	3.23	.573
Widespread	43	.380	.293	.253	3.70	.702

Key: H_t = total allelic diversity; H_s = mean diversity within populations; G_{st} = proportion of among-population to total diversity; A_p = mean number of alleles per polymorphic locus; P_a = the proportion of the total number of alleles found within each population.
Source: Hamrick (1983).

Some management activities have direct negative (*dysgenic*) impacts on gene pools. Foremost among these is high-grading, whereby the most valuable trees are removed each harvest cycle, and inferior trees are left to reproduce. For example, centuries of such dysgenic selection reduced the gene pools of once vigorous timber-producing forests of *Pinus densiflorus* and *P. thunbergiana* in Japan and Korea (S. K. Hyun pers. commun. 1960), *P. brutia* in Turkey (K. Isik pers. commun. 1977), and *P. roxburghii* in the Himalayas (D. Khurana pers. commun. 1988) to stunted and malformed trees.

Although the dysgenic effects of high-grading are well understood, whenever some trees are left after harvest to regenerate the stand, diversity of their offspring may be affected. Pollen from the remaining trees contributes to the genotypes of future seeds not only in the harvest unit, but in the surrounding forests as well. If only a few trees are left to serve as seed parents (for natural regeneration), then inbreeding and its depression of viability are likely to build up in future generations.

Other actions tend to stir up the gene pool, disrupting rangewide genetic structure by changing allele and genotype frequencies in future generations. The actual effects on diversity and long-term adaptation of forests are usually unknown. *Artificial reforestation*—that is, planting nursery-grown trees back into native habitats—is a common practice of both operational forestry and intensive tree-breeding programs. If not carefully controlled, planting poses a significant threat to gene pools when nonlocal and novel genes are introduced into populations. The potential to disrupt natural genetic architecture increases as more lands are planted. Increasing concern for reforestation and restoration in many temperate forests has prompted enactment of laws requiring that planting follow disturbance of many kinds. If planting is not done with a serious regard for intraspecific variation, the species can be irrevocably affected.

Horticultural use of trees can threaten native gene pools in similar if more localized ways. The origin of trees used for landscaping is usually unknown and nonlocal. Thus when popular species are planted within or near their native ranges, it is unlikely, except for genetically homogeneous or extremely rare species, that they come from the same gene pool as the native population. Interbreeding of planted and native trees occurs, and foreign genes are introduced into native populations, often creating a permanent change. This situation occurs in two of the five populations of Monterey pine native to California, where Monterey pine is widely planted as an ornamental (Libby 1990). The planting stock came from New Zealand landraces, which derived mostly from a third native population several generations ago.

Similarly, it appears that the genetically distinct columnar form of Italian cypress (*Cupressus sempervirens*), which is a widely planted ornamental, has been invading native cypress forests in the Mediterranean region for millennia. Many trees of intermediate crown architecture are the results of interbreeding of the columnar cultivars with the normal broad-crowned forest trees (M. Vidakovic pers. commun. 1980).

In addition to forces that perturb the genetic structure of species, population extirpation is a serious threat to widespread species. The culprits are the same societal forces that threaten rare species—increasing urbanization and conversion to agriculture. In trees, for example, entire populations of MacNab cypress (*Cupressus macnabiana*)—a widespread but patchily distributed Californian species—have been lost because of land conversion. The entire type locality for the species at Whiskeytown, California, was drowned by construction of a reservoir, and a population at Grass

Figure 10.3. Remaining populations of MacNab cypress in Grass Valley, California, after construction of condominiums. (Photo by authors)

Valley, California, has been nearly obliterated by development of a condominium (Figure 10.3). In another example, the gracious valley oak (*Quercus lobata*), once a common symbol of California's Central Valley, has been forced by agricultural expansion into minor riparian refugia plus a few urban parks and estates. These stories are being repeated in most forested regions in the temperate zone.

ETHICAL CONSIDERATIONS

Conservation biology is continually faced with the dilemma of choice: Which taxa are chosen for protection from among the many in need? By limiting the field to the species level and by defining the rarest of rare species and the most endangered among these, there has been a hope that the job could be tackled. But we argue that the species level is only part of a continuum of genetic variation, and that rarity and endangerment occur importantly at levels within species. Species preservation per se may depend on conservation of intraspecific structure. This increases the number of candidates for protection far beyond our hope of even cataloging them.

Obviously, other criteria in addition to rarity and endangerment are needed to guide decisions. Those that have been discussed elsewhere include ecological importance and economic value (Oldfield 1984; Wilson 1988). A genetic criterion should also be added. Taxa would be ranked according to the quality (individual heterozygosity, stand and population diversity, number of hierarchical levels) and quantity of genetic variation they contain as well as to the uniqueness of the taxa, as measured by the genetic distance to the most closely related taxa.

Assigning ranks and assessing genetic value should not, however, follow a straight formula for maximizing diversity. For example, a monophyletic species such as *Ginkgo biloba* would rank higher than a "minor" species in a large section of a large genus such as *Pinus* or *Quercus*. Within species, however, it may not be the genetically distinct outlier populations that rank highest but the core populations from the center of a species range. In this case, within-stand diversity may be qualitatively more important than uniqueness of a population, although we hasten to add that certain types of distinct populations should rank high also (more on this later).

Another criterion is the cause of rarity. Some taxa are naturally rare, whereas others are artificially rare as a result of human actions. Does this situation confer different responsibilities for protection? The answer, we believe, is yes, for two reasons: ethically, we should fix what humans break, and biologically, the taxa that have not evolved under situations of rarity may be less stable than those that are naturally rare. The actual tolerance of naturally widespread species to reductions in population size and concomitant genetic changes depends on many factors, including paleohistory, life history, response to inbreeding, longevity, and phenotypic and developmental plasticity. That many widespread tree species do not tolerate reductions in population size well is indicated empirically by the severe inbreeding depression that occurs when N_e (effective population size) is lowered (e.g., Sorensen 1969; Libby et al. 1981; Sorensen and Miles 1982; Wilcox 1983), and theoretically by such models as genetic transilience (Templeton 1980).

Conservation of diversity within widespread species usually has an objective different from that for rare species. Because species existence is not threatened, it is the evolutionary fitness of the species that is being safeguarded. For rare species, by contrast, the final race against extinction is being fought. In the former case, conservation is the application of preventive medicine, whereas in the latter, it is crisis intervention (Templeton, Chapter 12). And, as in medicine, the benefits reaped by early and thoughtful maintenance are often greater and longer acting than those afforded by treating a battered system suffering a terminal illness. This is not meant to imply that endangered taxa should be given up and all conservation focused on healthy species. On the contrary, it is meant to draw attention to how much can be done and needs to be done before a taxon reaches the critical point. Perhaps this is the greatest challenge to conservation biologists in the long-run: that through our knowledge and influence we not only can rescue species from the brink of extinction, but also can prevent currently healthy taxa from ever becoming endangered.

CONSERVATION STRATEGIES FOR WIDESPREAD SPECIES

In the case of extremely rare plants limited to no more than a few small populations, the goal of conserving the entire species may be possible through habitat protection and *ex situ* preservation. For many other rare but not extremely rare species and for all widespread species, conservation will always involve protecting appropriate samples of the whole species. The goal in this case is to choose the samples such that the important and/or interesting genetic variation in the species is conserved. Such sampling decisions are based on information from genetic architecture studies.

The two main types of sampling are *in situ* habitat protection and *ex situ* germplasm preservation. The pros and cons of each have been widely discussed (Frankel and Hawkes 1975). We take the approach of Falk (1987b) in advocating an integrated approach that involves both methods. Under normal conditions of climatic stability, we would emphasize the values of protecting genetically valuable populations through various levels of *in situ* management. The *ex situ* component would act both as a backup system to safeguard a wider range of variation than could be included in native habitat areas, and as a source of germplasm for research, monitoring, and restoration.

The specter of global climate change, however, threatens the value and security of *in situ* protection areas. If average or extreme temperatures change as fast as predicted, floristic changes will be enormous. On a local scale, there will be widespread population extirpation and vegetation shuffling. *In situ* areas that were originally chosen as prime habitat for specific communities may end up as completely different vegetation types. Although those areas may play new roles in conservation, their intended value as protection areas for specific types may be nil. In this scenario, *ex situ* collecting should play the primary role in sampling and preserving germplasm before it is lost.

We take an uncommitted stand regarding climate change. In the discussions that follow, we assume that both *ex situ* and *in situ* conservation will have value whether climate changes drastically or not. The two components together should be viewed as complementary, and when new information on climate change or other environmental threats becomes available, one or the other component can be favored and expanded.

Geographical Scale

The scale of concern is an important issue in the practical steps that follow. For many widespread species, it is unlikely that one centralized program of genetic conservation will be implemented. Rather, smaller political areas will act on a local level, coordinating, where possible, with other regions. In the following discussions, we assume the scale to be on the order of a single national forest of the U.S. Department of Agriculture Forest Service system. It is important to keep this scale in mind, since the numbers we recommend for conservation are intended to apply to small portions of the species range and not to the whole species.

Index Species

Although we are basically proposing a single-species approach, we consider that some species under consideration may effectively serve as index species for entire ecological communities. Choices among alternative strategies for the individual species will be weighted by their merit in conserving other, nontargeted species. In this discussion, the index species are commercially important tree species and their relatives. This is a realistic choice because trees command a large share of political and fiscal support, because they play central ecological roles in the communities they support, and because in general much more ecological and genetic information is available about them than about many of their associated plants, animals, or microorganisms. Trees are large organisms, with many aspects of their natural (genetic and ecological) history readily accessible. Other index species, such as the spotted owl, peregrine falcon, and black bear (few plant species have been used yet), are chosen for similar criteria.

The Land Component: *In Situ* Management
Genetic Management Systems

Conservation within native habitats protects and perpetuates the integrity of gene pools and of coevolutionary relationships. Individuals, populations, and species interact with one another and with changing environments and thus maintain their evolutionary fitness in ways that are nearly impossible elsewhere. There are, however, many forms of *in situ* management, with conservation objectives having different priorities among them (Salwasser 1987). Some conservationists consider strict reserves, established for the sole purpose of preservation, to be the only form of *in situ* protection. But because only small amounts of land will ever be available for strict reserves, and correspondingly small percents of widespread species will be included in these reserves, we are forced to consider ways in which conservation goals can be incorporated into the management of multiple-use lands.

We propose a composite category of lands called a *genetic management system,* distinct from other management categories because the objectives of genetic conservation would have higher priority over other uses. This category would group together many types of preexisting management classes. One extreme in the category would be strict nature reserves, such as wilderness areas, wild rivers, research natural areas, and other designated nature preserves. Less restricted areas would also be included if their management met certain genetic requirements. Examples include botanical areas and other special interest areas, streamside and riparian buffers, and such areas as spotted owl habitat areas and gene resource management units (GRMUs) (Riggs 1982; Krugman 1984).

For a particular national forest and for a species of interest, the boundaries of preexisting conservation areas would be overlaid on a map of the species distribution. A genetic gap analysis would then be made. This would entail systematically comparing the coverage of preexisting units in the genetic management system with the profile of genetic variation for the species of concern. Where gaps are located (criteria discussed in next section), candidate GRMUs would be sought.

The concept of GRMUs has been proposed and discussed at length both generally (Krugman 1984; Ledig 1988) and for specific cases (Riggs 1982). Authors have used the concept to imply somewhat different types of land management. We define a GRMU as *any designated forest area that meets minimum genetic management objectives.* GRMUs are not exclusive of many other forms of land use. For example, timber harvest could be allowed, although certain restrictions would apply. The main objective is to control the genetic composition of trees in the GRMU so as to maintain, restore, or mimic the natural genetic condition. Natural reforestation is preferred when timber harvest is excluded, but only if the seed and pollen parents are native and if the gene pool of the population has not been compromised (i.e., by high-grading, genetically positive selection, or extreme restriction of population size). If these restrictions do not apply or if planting is required for some other reason, strictly controlled planting is allowed if seeds from a certified native, local, and random sample of trees can be obtained. Species mixes must be maintained in the approximate proportions originally found on the site. To our knowledge, no GRMU has been expressly designated that fully exercises the multiple-use flexibility allowed under this concept.

For many species, it will be best to exclude timber harvest from the GRMU. Many forested areas are considered unharvestable because the lands are too steep, the soil is too erosive, the site is not capable of being replanted, or the growth rates are too poor. If these areas fit into the genetic management system, they can be maintained as de facto reserves because no timber management was planned for them anyway. The serious problem with these areas is that if fire is also suppressed on the GRMU or on surrounding areas, the association of species in the GRMU may change to a different successional type.

In many cases, however, the genetically most appropriate, interesting, and important stands are in designated commercial forests where timber harvest has been a top priority. If timber harvest is permitted on a GRMU, the type of harvest depends on the species and on the management of the surrounding lands. No type of harvest is automatically excluded, as long as a genetic prescription for the regenerating stand is written for the unit. When prescribed, artificial regeneration should be done with seeds obtained from a minimum of 20 randomly selected trees per species growing in the unit and chosen before harvest.

Placement and Design of Units in the Genetic Management System

The two major sampling considerations for conservation of widespread species are geographical placement of units and size of individual units. Many biological and nonbiological factors determine the final numbers. We will consider primarily the genetic ones, while acknowledging that practical concerns limit rather than increase the number and size of areas.

There has been much debate in the conservation community over the relative merits of a few large versus many small *in situ* conservation areas (Jarvinen 1982; Simberloff and Abele 1982; Janzen 1983). Most of the arguments have stemmed from ecological considerations of island biogeographical theory (e.g., Margules et al. 1982). However, the genetic system of the target species also affects the proper allocation and size of reserves. For highly inbreeding species where geographical differentiation is great and within-stand diversity is low (e.g., species of *Avena, Bromus, Elymus,* and *Hordeum* [Hamrick et al. 1979]), many reserves would be better than a few. The reserves could be small, with the major effort aimed at capturing geographical variation with many reserves. For temperate-forest trees, however, which are typically highly outcrossing and where within-stand diversity is high, large reserves would better maintain the important component of within-stand variability. If a few of these were strategically placed on the basis of patterns of genetic variation, much of the variability among populations could also be captured.

The first step in determining the sampling design is—if possible—to complete a study of the genetic architecture of the target species. Such a study can take many forms, from biochemical assays such as restriction-fragment-length polymorphism (RFLP), isozyme, or terpene analyses, to physiological or morphological studies in a common garden. The advantages of the biochemical assays are their speed and ability to yield precise genetic information. Common-garden tests are slow to yield information, and usually do not allow actual genotypic data to be estimated. They do, however, provide information about traits that more directly relate to adaptation than do biochemical analyses.

To determine optimal geographical placement of GRMUs, the following data are useful: magnitude and pattern of genetic variation in the species, amount of variation due to within- and among-stand factors, and geographical factors contributing to variation. These data usually can be obtained from the genetic analyses mentioned above. In addition, the following more specialized data are useful for each sampled region or stand: observed and expected heterozygosities, number of alleles at specific loci, and number of alleles in various frequency classes, including unique alleles.

We suggest that the highest priority for *in situ* conservation systems is to capture the core(s) of variability in the species. By this we mean areas that contain the common or shared genetic variability for a region. For economically important tree species, this area often defines the populations containing the most valued or useful trees as well, and it is important to conserve the base populations from which valued domesticated lines are desired. The number of cores depends on the regional patterns in the species. If a species has racial or regional-level variation (e.g., ponderosa pine, Douglas fir) or smaller-scale ecotypic variation (e.g., ponderosa pine on and off serpentine soils), then each ecotype would have its own core of variation. The percentage of variation that makes up each core depends on the amount of within- and among-stand variability for each species. For a wide-ranging conifer where much genetic variation is found within stands, each core might comprise over 95% of the allozyme variation and perhaps 75% of the adaptive variation in the race or ecotype. Areas that contain this core variation are often in the central portions of species ranges within regions on optimum sites for the species.

Each core of variability should be conserved by a minimum of two independent sites. There are two reasons for this: first, two sites reduce the risk of catastrophic loss (by fire, insects, climate change); and second, two sites capture some of the among-stand variability. To satisfy the second requirement, therefore, the two sites should not be as nearly alike as possible, but should sample some of the site differences within the central core of variation. Such core sites should be large (see below); if two independent sites seriously compromise the size of either one individually, then the single larger core site might be preferable.

As many more units as practical should be added to the genetic management system, but the core areas should be considered essential. For example, in doing a gap analysis, if the preexisting areas sample only the genetic margins or less variable portions of a species range (as is often the case with wilderness areas and other strict reserves [Harris 1984; Ledig 1988]), then the region would be considered insufficiently covered. In adding sites beyond the core, unusual genetic patterns and large discontinuities become important and should be added as possible. Among these, sites with high within-stand variability should be considered first, unless genetically impoverished sites are striking in their complement of unique alleles. The trade-offs between increasing the number of sites at the expense of losing size of individual units is discussed later, but the initial stage of selection should include more candidates than anticipated established areas. In the case of clinal variation or in other cases where no obvious core of variability exists, it would be best to locate two or three areas at significant genetic distance from one another along the main gradient of the cline. Secondary reserves would fill in gaps in the cline as well as any major discontinuities not contained in the primary areas.

After candidates for the genetic management system have been identified for the target species, the value of these sites for other native species (plants, animals, and microorganisms) should be evaluated. On the basis of available information on species occurrence and genetic variation in associated species, the candidate areas that maximize protection of associated native species should be chosen.

Other criteria may also influence selection of final candidate sites. For example, concerns about global climate change suggest that areas with potential escape routes for migration (e.g., whole drainages or at least large elevational spreads) or areas that are ecologically stable despite climate change (e.g., serpentine soils) should be preferred. Management of surrounding lands should also be considered. Whenever an area is situated such that adjacent lands are managed in a way that will compromise the genetic integrity of the GRMU, other areas should be considered. By corollary, any candidates that are adjacent to or in other ways complement other units of the genetic management system should be seriously considered. These areas not only have protected boundaries, but also add value to preexisting reserves by increasing their size and improving their integrity.

The other major sampling question for designing new units in the genetic management system is, How big should individual areas be? The answer depends on whether the goal is to maintain a population size that retains the current levels of diversity or to allow the population to decrease to some acceptable minimum number of individuals that hopefully will still ensure long-term survival of the population. Either way, the genetic integrity of populations depends both on the genetic structure of individuals within the GRMU and on management of adjacent lands.

The genetic factors influencing within-stand dynamics include inbreeding, genetic drift, immigration, and selection and are influenced by the reproductive system of the species. For many outbreeding species, where within-stand variation is normally high, inbreeding caused by reduced population size can cause a severe decline in health and vigor. These species will need relatively large populations to resist the long-term decline of heterozygosity as related individuals mate. The opposite is true for species that tolerate inbreeding. Genetic drift causes a decline in heterozygosity that is negatively correlated to effective population size. The rate at which heterozygosity declines depends not only on population size within the GRMU, but also on immigration from outside the unit. If immigration is reduced either unintentionally due to loss of the species outside the unit or intentionally due to buffer management (see below), drift has a larger effect than if a small amount of immigration is allowed (Crow and Kimura 1970). The drift effect of increasing homozygosity on the fitness of the organisms depends on initial heterozygosity and tolerance to homozygosity.

The uncertainties in these factors suggest that a conservative action would be to establish populations large enough to maintain current levels of heterozygosity. Of particular interest is variability at loci with adaptive effects. Unfortunately, this has proved recalcitrant to estimate. Formulae for calculating effective population sizes required to maintain heterozygosity have been developed for neutral alleles (Crow and Kimura 1970), and these have been applied to allozyme data. Calculation of this size depends on the relation of effective population size to actual census number, mutation rate, and current levels of heterozygosity. Based on allozyme data for ponderosa pine and Douglas fir, the effective population sizes range from 2230 to 910, 640, depending on original heterozygosity and mutation rate (Table 10.2). For alleles with nonzero selective values, the N_e may be larger or smaller, depending on the type of selection on the locus, the intensity of selection, and the mutation–selection relationship.

Genetic variability within GRMUs is affected by conditions outside the GRMU primarily through gene contamination and secondarily by altered selection pressures due to human-caused disturbances. Some geneticists have argued that gene contamination via pollen from ill-adapted stock is not a problem, since maladapted individuals will be selected against in the GRMU. We take the conservative position that any foreign genes should be minimized in the GRMU, since deleterious foreign genes may not be eliminated quickly by selection, and they would cause a gradual decline in fitness. A swamping of deleterious genes, because of massive planting of areas adjacent to nonlocal populations, could cause a catastrophic loss of fitness in the GRMU population. Similarly, deleterious genes that act in late maturity or are deleterious only in rare episodic environmental conditions (droughts, freezes, etc.) will accumulate (Millar and Libby 1989). Either GRMUs must be large enough to effectively buffer populations from external effects (i.e., include their own buffer) or surrounding lands must be managed as a buffer. Either way, we recommend an active program of buffer management.

The size of the buffer depends on the extent of gene flow, which varies by species. In wind-pollinated tree species, viable pollen can disperse tens and even hundreds of kilometers (Lanner 1966; Koski 1970; Millar 1983), although most probably falls in the first hundred meters (Sarvas 1967; Wang et al. 1969). The potential impact from gene contamination depends on the degree of difference in allele frequencies and in local adaptation between the trees in a GRMU and those in the outside forests. In the unlikely event that there is no difference, then gene contamination has no effect, and a buffer against pollen contamination is unnecessary. At the other extreme, if trees near the GRMU have very different allele frequencies, as in plantations with nonlocal stock or highly domesticated trees, then an effective buffer is essential.

Table 10.2. Estimates of Effective Population Sizes (N_e) Needed to Maintain Current Levels of Allozyme Heterozygosities in Populations of Ponderosa Pine and Douglas Fir

Species	Region	H_e	N_e at $\mu = 10^{-5}$	N_e at $\mu = 10^{-7}$
Ponderosa pine	S. California	.082	2230	223,300
Ponderosa pine	N.W. California	.108	3030	302,700
Ponderosa pine	Sierra Nevada	.123	3510	350,600
Ponderosa pine	N.E. California	.157	4660	465,600
Ponderosa pine	N.W. California	.183	5600	560,000
Douglas fir	Sierra Nevada	.190	5860	586,400
Douglas fir	N.W. California	.209	6610	660,600
Douglas fir	N.W. California	.217	6930	692,850
Douglas fir	N.W. California	.229	7430	742,500
Douglas fir	N.W. California	.267	9110	910,600

Note: Estimates are calculated from Crow and Kimura's (1970) equilibrium heterozygosity formula for neutral alleles. Two mutations rates (μ) are given as a range.

Source: Millar (unpubl. data).

In the same way, the impact from altered selection pressures depends on the management of the external lands. If adjacent lands are only slightly modified, as with selection cuts, then a modest buffer suffices. If, however, these lands are completely altered, by clear-cut or conversion to nonforest use, then a wider buffer is necessary to ameliorate the selective effect on the genetic composition of the population in the GRMU.

The management requirements for the buffer resemble those for the GRMU: practice conservative forestry. Allow natural regeneration, or use local seeds. In many regions, there are not enough local seeds to meet all the planting demands (e.g., Riggs 1982). We suggest giving highest priority to the GRMU and the buffer zone when deploying the available certified local seeds. Harvest is allowed in the buffers, but selection cutting is preferred because the trees in the buffers should provide a physical barrier against both an altered microclimate and foreign pollen in the external zone. A loss of some of the resident species because of a shift toward later successional types is preferable to inappropriate populations of those species being planted in the buffer.

If natural regeneration is practiced in the GRMU and the buffer, harvest in the buffer can be timed to occur in the five years or so after harvest in the external zone. This is a good time to cut in the buffer, because then the regenerating trees in the GRMU are not in need of buffer against pollen contamination. Regeneration in the buffer is likely to come from the GRMU and not from outside because the external trees will still be immature. Also, the buffer should grow at the same rate as the outside area, and would guard against contamination as the stands mature.

If natural regeneration is not possible for the buffer and truly local seeds are unavailable, an alternative is to plant the buffer with a well-known, tested, and adapted exotic species (Ledig 1986a). Only those species known to be noninvasive should be chosen. For genetic reasons, a buffer of nonnative species is even better than one of native species because it acts not only as a physical but also as a genetic buffer of edge environment. A nonnative buffer will not become contaminated with genes from the native species, and thus will protect the GRMU longer than would a buffer of native species. Exotic species are easily identified and can be removed if necessary. Exotic species are far more desirable in the buffer than native species of nonlocal origin or even native improved stock. Obviously, the ecological effects of the exotic species on other organisms must be determined.

Once the genetic management system is established, the individual units should be defended and monitored actively. The boundaries of each unit should be clearly marked on the ground and on administrative maps used by forest officers. Opportunities that increase the area of units in the system, or establish new GRMUs, should be sought. If not already available, an ecological survey of the GRMUs and as many of the other units in the system as possible should be conducted. This survey documents baseline information about the areas, and should include a species inventory (plants and animals), soils information, and ecological data about the community types present. Scientific study in the GRMUs should be encouraged, especially pertaining to genetic aspects of both index and associated species. Comparisons between genetic data in strict reserve units and in those GRMUs where timber harvest or replanting is practiced would be useful to guide later management. If possible, reports from research studies should be accessible from a database maintained in a central location for each region.

Gene Banking: *Ex Situ* Management

Ex situ collections serve the essential role of filling the gaps and extending the range of genetic variation protected, and of providing germplasm to damaged habitats. In addition to their protective role if climate changes, *ex situ* collections act as insurance against other, more familiar disasters in the natural environment. Fire, insect epidemic, atmospheric pollution, subminimal viable population size, and changes in regulatory or management policies all threaten populations in natural reserves.

Ex situ collections can also provide important monitoring functions. If collections are made regularly, samples can be genetically analyzed, and changes in genetic composition of the protected stands determined. This would be especially useful if the native populations were thought to be at risk—for example, from gene contamination or inbreeding.

Finally, *ex situ* collections serve as a source of material for scientific research. The information that accumulates from research provides insight into the best conservation use of the collections.

Sampling Design

Many of the same concerns about designing an on-site genetic management system apply to *ex situ* collecting. Genetic information or inferences about genetic structure remain a necessary ingredient for planning.

As a first step, and to serve as backup, *ex situ* collection samples should be drawn from all *in situ* areas. The genetic gap analysis will provide information useful for locating additional collecting areas. These additional sites should be chosen to systematically sample a broader range of genetic variation than was included at *in situ* sites. Special emphasis should be given to genetically distinct populations (especially if they occur in unusual environments or in atypical ecological situations), to small or isolated populations, and to endangered populations. In all cases, efforts should be made to find the least disturbed populations to serve as collection stands. In all cases, detailed records on the collection should be kept, including as much of the following as possible: notes on access, location, and the physical, ecological, and management attributes of the site; number of collection (donor) plants; type and number of propagules; and mapped location of donor plants.

Although the total number of sites chosen for *ex situ* collecting will depend on the genetic structure of the species, in all but the most genetically uniform species there should be many more *ex situ* than *in situ* sites. The more genetically diverse a species is over its range, the more samples should be included. Transition areas between races or ecotypes, and steep gradients along clines should be emphasized. The optimal *ex situ* collections will include replicate samples from genetically similar areas, as well as extensive geographical coverage.

An example of partitioning a species range in forest trees is the practice of establishing seed zones and seed-transfer and collecting guidelines (e.g., Kitzmiller 1976) (Figure 10.4). Forested regions are divided into zones, which in theory include similarly adapted populations. Subzones are delineated to follow clinal or ecotypic contours within each zone. Seeds are collected from multiple stands within each subzone, and guidelines for the number of trees per stand are given.

The number of plants to sample at an area depends on the objectives for the col-

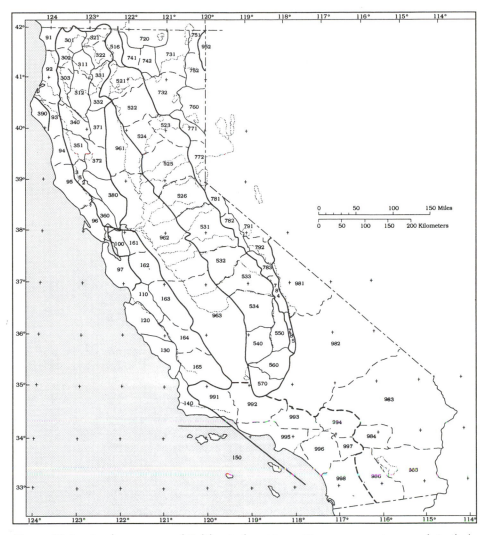

Figure 10.4.   Seed-zone map of California forest trees. Zones represent areas of similarly adapted populations. Seeds for replanting after harvest are drawn from 155-m elevation bands within the same zone as the harvest unit.

lection. Collections may be done to duplicate actual allele frequencies of the native stands or to capture the largest amount of genetic variation without regard for frequencies (Marshall and Brown 1975). Much has been written about determining actual numbers for both situations (e.g., Brown 1975, 1978, 1988; Marshall and Brown 1975; Brown and Moran 1981; Chapman 1984; Yonezawa 1985; Namkoong 1988; Ryder 1988; Crossa 1989). Most calculations require genetic information, especially allele frequencies, within-stand diversities, and numbers of rare and unique alleles per population. For genetically variable species, the numbers of seed parents are usually smaller than those required for *in situ* conservation. This is especially true when the species is outbreeding because the genes in the embryos donated by the pollen parents capture a wide range of local variation. Whenever possible, samples should be kept separate and labeled by parent tree. To reduce relatedness and increase genetic diversity, collection trees should be separated by at least 100 m. This distance depends on pollen flow distances and would be considerably less for species with pollen that does not disperse far.

Storage Considerations

Many kinds of *ex situ* germplasm storage are possible, including living plants in gardens, seeds in cold storage, tissue cultures, and DNA in genomic libraries. The advantages and disadvantages of these have been discussed elsewhere (e.g., Belcher 1980; Ledig 1986a; Bonner 1988, 1989; Millar 1991). We discuss here some special considerations of botanical gardens in maintaining *ex situ* collections.

In all that is done with storage of *ex situ* germplasm, one of the most important is recordkeeping. Proper identification is the first job of recordkeeping. A propagule without a name is useless to genetic conservation. (Although there are biochemical fingerprinting techniques that can help to identify orphan plants, they are expensive and time consuming.) The recordkeeping system should start in the field and track the location and storage conditions of the germplasm, following the germplasm wherever it goes.

The second job of recordkeeping is to maintain information about the performance of the propagules. If plants are grown in gardens or if samples are given to researchers who grow them, as much information as possible on morphology, pathology, growth, and other factors should be kept for that sample. The more information that is available for accessions in a collection, the more valuable the collection becomes. Even casual observations may prove important. All gardens should encourage scientific use of their germplasm, as this fosters an interactive flow of use and information.

Space is usually a problem at botanical gardens. Although many seeds can be stored in small freezers, adequate numbers of all but the smallest whole plants require large planting areas. Furthermore, if genetic information is to be collected from plantations, replication and regular spacing of the plants are required, which also increases the planting space. One option is for gardens to form consortia and to divide the samples of a widespread species among several gardens. Each garden would be responsible for one or several populations, although each population should be replicated in at least two sites. This has the advantages that the populations can be matched to gardens with the best growing conditions for the population and that replication guards against accidental loss at one garden. Furthermore, replication at different gardens can serve as a test of plant performance under various climates. This information may prove useful in the event of major climate change and the need to establish populations in new habitats. If pollination is to be allowed and seeds are to be collected from garden-grown plants, wind pollination can be allowed if only one population has been grown at the garden (but consider the problem noted below). A major disadvantage when the entire collection is not housed together is that relative performance of the populations cannot be compared, since plants from different populations are grown in different environments. Furthermore, dividing the collection makes maintenance, curation, and use more difficult. The benefits, however, may outweigh these problems.

A special problem of gardens is that they may be within gene-flow range of other members of the same species. This is a problem only when seeds are to be collected—

for example, to enhance or replenish stored seed collections—and not if clonal replication of the plants is planned. If the garden is outside the native range of the species, gene contamination can occur owing to planted ornamentals in the vicinity of the garden. The problem is potentially worse if the garden lies within the range of the species, where the native pollen cloud may outcompete the pollen from the planted individuals. This problem may be reduced by bagging female flowers, collecting pollen, and artificially pollinating the plants. This process, however, is time consuming and expensive.

CONCLUSIONS

We propose a five-step plan for conserving widespread species:

1. Study the genetic variation of the index species, or in some other way infer its patterns of genetic variation.
2. Plan and implement *in situ* genetic management systems for local regions (e.g., for trees, areas the size of national forests in the U.S. Forest Service system). Units in the system include strict reserves (e.g., wilderness areas, botanical areas) and multiple-use areas called gene resource management units (GRMUs) where genetic management is the main objective, but some manipulative practices may be allowed. The *in situ* areas should sample the genetic cores of the species for the local region as well as key areas representing important ranges of diversity. The final choice from among candidate units should be decided by the quality of protection that is afforded important associated species.
3. Plan and implement an *ex situ* germplasm conservation program to complement and extend the *in situ* program. *Ex situ* collection sites should be much more extensive and sample more genetically extreme and unusual sites than *in situ* units. Keep the collection viable by replenishing periodically.
4. Monitor genetic diversity over time in both *in situ* units and *ex situ* collections Encourage research use and maintain files of all information on performance of the plants.
5. Actively promote the conservation use of both *in situ* areas and *ex situ* collections. *In situ* areas can be used as habitat for reintroduction or enhancement of endangered species (animals as well as plants). *Ex situ* collections should be offered when restoration of native habitat is needed (after fires, reclamation, etc.).

Conservation biologists have correctly pointed out (Salwasser 1987; Ledig 1988) that since only small percentages of native habitat will ever be fully protected in strict reserves, the crux of the problem lies in wisely managing the remaining lands. Although we also promote conservation-minded management on all wildlands to the extent that is practical, we know there are many practices that by their nature counter the goals of genetic conservation even when they are highly constrained. We do not think that the only conservation strategy is to abolish or severely restrict these practices. In many cases, this would only delay a diversity problem. Rather, an extensive and replicated conservation program such as we propose could release some wildlands for intensive management. This can be envisioned as a trade-off situation: the better the conservation program, the more land that could be made available for other uses. By this, we are not condoning irresponsible or exploitive use of these areas. We are suggesting instead that intensive management of resources is not counter to conservation goals when it is part of a comprehensive program that also includes intensive conservation. With the possibility of relief from excessive environmentalist attack and government regulation, this approach may motivate some previously anticonservation groups to support large conservation programs financially.

11

Hybridization in Rare Plants: Insights from Case Studies in *Cercocarpus* and *Helianthus*

LOREN H. RIESEBERG

It is now evident that hybridization can have both beneficial and harmful consequences for the conservation of biological diversity (Cade 1983). In some instances, hybridization may lead to greater genetic diversity, increased fitness, and adaptation to new environments (Stebbins 1942; Anderson 1949; Harlan 1983). Furthermore, in certain rare cases hybridization may be the only possible way to preserve the germplasm of a rare and endangered taxon. For example, hybridization was used to preserve the alleles of the now extinct dusky seaside sparrow, *Ammospiza maritima nigrescens* (Avise and Nelson 1989).

Hybridization can also lead to the loss of genetic diversity through the genetic assimilation of a smaller population by a larger one (Cade 1983). Genetic assimilation is most likely to be a serious problem for small relict populations or island endemics when they come in contact with a numerically larger or reproductively more successful species. There are many animal species endangered because of this process, including the cutthroat trout (Allendorf and Leary 1988), Pecos pupfish (Echelle and Conner 1989), Mohave chub (Soltz and Naiman 1978), Tecopa pupfish (Soltz and Naiman 1978), yellow-crowned parakeet (Taylor 1975), Seychelles turtle dove (Cade 1983), Mexican duck (Heusmann 1974), and red wolf (Mech 1970). In contrast, there are few known examples of plant species endangered by hybridization despite the great potential for gene flow between close relatives (Anderson 1949; Heiser 1973) and the high frequency of hybridization in island floras (Gillett 1972).

A second possible harmful consequence of hybridization is outbreeding depression. That is, hybrids between genetically differentiated taxa often have reduced fitness (Templeton 1986). Reduction in fitness can occur in first-generation hybrids or may appear only in later generation hybrids or backcrosses. Outbreeding depression usually results from meiotic abnormalities or the disruption of coadapted gene complexes (Dobzhansky 1948, 1970; Sage et al. 1986). Outbreeding depression has been detected in both plants (Price and Waser 1979) and animals (Sage et al. 1986). It has been suggested that certain salmonid fishes may be endangered as a result of this process (Allendorf and Leary 1988), but the possibility of outbreeding depression has generally been ignored in studies of rare plants.

In addition to the general beneficial and harmful consequences of hybridization, hybridization has a direct impact on plant taxonomy and management decisions. Taxonomy and systematics, although sometimes considered among the least important and challenging of the biological disciplines, are critical to conservation biology because they provide the fundamental basis for management decisions (Avise and Nelson 1989). Unfortunately, classical taxonomic methodologies are often inadequate for the precise identification of, and differentiation between, first-generation hybrids, introgressive races, and bona fide hybrid species (Rieseberg et al. 1988). Furthermore, it is presently unclear how these categories should be protected under present laws, such as the U.S. Endangered Species Act (G. Nabhan pers. commun.).

In this chapter, I discuss the evolutionary consequences and management implications of hybridization in three rare plant species: *Cercocarpus traskiae* Eastwood (Rieseberg et al. 1989), *Helianthus paradoxus* Heiser (Rieseberg et al. 1990), and *Helianthus exilis* A. Gray (Rieseberg et al. 1988). It is hoped that these studies will serve as general models for the detection and documentation of hybridization and subsequent recovery strategies.

CERCOCARPUS TRASKIAE

Cercocarpus traskiae, California's rarest tree (Figure 11.1), is confined to a single canyon (Wild Boar Gully) on the southwest side of Santa Catalina Island in Los Angeles County. First collected by B. Trask in 1897 (Thorne 1967), the population has declined from more than 40 individuals to only seven adult plants today.

The decline of *C. traskiae* appears to be directly related to the introduction of large herbivores such as goats, sheep, and pigs to the island during the early nineteenth century (Thorne 1967). Although the Santa Catalina Island Conservancy has made modest efforts to reduce herbivore activity during the past two to three decades, the grazing and rooting activities of goats, pigs, and the recently introduced mule deer remain a serious threat to *C. traskiae*.

Active management of *C. traskiae* began with fencing of two individuals in the late 1970s by the Santa Catalina Island Conservancy (Martin 1984). In 1985, the Conservancy constructed a larger fence around the two adults and around good habitat area approximately 124 m on a side. Only a single seedling was found in 1984, but over 70 seedlings were observed in this area by the fall of 1988. Unfortunately, the establishment of young seedlings outside of the fenced areas appears unlikely, owing to the lack of leaf litter and to the inevitable browsing by herbivores. In addition to the emergence of seedlings, cuttings from the adult trees have been propagated at the Rancho Santa Ana Botanic Garden, and some of these specimens have been returned to test plots on the island.

In addition to herbivore threats, the long-term survival of the species is threatened by limited genetic diversity and by the occurrence of hybridization between *C. traskiae* and the more common *Cercocarpus* species on the island, *C. betuloides* Nutt ex T. and G. var. *blancheae* (C. K. Schneid) Little.

Analysis of 17 isozyme loci by Rieseberg et al. (1989) demonstrated that *C. traskiae* had extremely low levels of genetic variation. Mean values for the proportion of polymorphic loci (P), mean number of alleles per locus (A_p), and mean heterozygosity

Figure 11.1.   (*left*) California's rarest tree, *Cercocarpus traskiae*. (*right*) *Cercocarpus traskiae* leaves (From Rieseberg et al. 1989)

(*H*) of 0.12, 1.12, and 0.04, respectively, are much lower than the mean values of 0.50, 1.85, and 0.15, respectively, reported by Hamrick et al. (1979) for other outcrossing plant species. *Cercocarpus traskiae* is also genetically impoverished relative to most restricted species (Hamrick 1983; Karron 1987a). Nevertheless, levels of genetic variation in *C. traskiae* are only slightly lower than values reported for *C. betuloides* ssp. *blancheae* ($A_p = 1.24$, $P = 0.18$, and $H = 0.05$ [Rieseberg et al. 1989]).

Isozyme evidence was also used to examine the question of hybridization in *Cercocarpus* (Rieseberg et al. 1989). Isozymes are useful for detecting hybridization at the diploid level because the allozymes present in each of the parental species either are or are not found combined in the putative hybrid (Gallez and Gottlieb 1982). Comparison of *C. traskiae* with a nearby population of *C. betuloides* var. *blancheae* revealed the presence of five "pure" adult *C. traskiae* trees, two hybrid trees, and one "pure" *C. betuloides* var. *blancheae* tree in Wild Boar Gully. All seedlings examined were F_1 progeny of two of the "pure" *C. traskiae* trees.

Conservation Considerations

The threat of genetic assimilation, such as that observed in *C. traskiae*, has generally been ignored in most studies of rare plants. This is a serious oversight, given the high potential for gene flow in plants and the high frequency of hybridization in island floras (Gillett 1972). Island plants are extremely susceptible to hybridization and subsequent genetic assimilation because of geographical and edaphic limits to a species range and population size, the general lack of chromosomal sterility barriers among closely related species, invasion and colonization by closely related exotics, and the increasing loss and disturbance of habitat due to human activities (hybridization tends to be much more frequent in recently disturbed areas [Anderson 1949]). For example, in Hawaii, hybrids involving endemic species have been reported for a large number of genera, including *Argryroxiphium* (Carr and Kyhos 1981), *Scaevola* (Gillett 1966), *Lipochaeta* (Sherf 1933a, 1933b), *Railliardia* (Sherf 1934), *Stachytarpheta* (Moldenke 1959), *Argrautia* (Sherf 1944), *Antidesma* (Rock 1913), *Dodonaea* (Leveille 1911), *Dubautia* (Sherf 1933a, 1933b, 1934, 1952, 1956; Carr and Kyhos 1981), *Bidens,* (Sherf 1951, 1952), *Hibiscadelphus* (Baker and Allen 1976), *Nania* (Leveille 1911), and *Viola* (Scottsberg 1939). Hybridization is also common on the California islands (S. Junak pers. commun.), and Liston, Rieseberg, and Mistretta (unpublished) have recently shown that a *Lotus* taxon endemic to San Clemente Island may be threatened by hybridization and introgression. Nonetheless, many more detailed studies of the occurrence and consequences of hybridization in endemic island species are needed before the potential threat of genetic assimilation to the genetic integrity of island floras can be estimated.

In the case of *C. traskiae,* Rieseberg et al. (1989) made the following management suggestions:

1. The single *C. betuloides* var. *blancheae* individual in Wild Boar Gully should be eliminated. Although this individual apparently has not parented any of the *C. traskiae* seedlings observed to date, this is probably because the seedlings occur under trees some distance from the *C. betuloides* var. *blancheae* individual. If any of the three adjacent *C. traskiae* trees flower and produce seedlings, however, the *C. betuloides* var. *blancheae* individual will be a possible parent.
2. Established cuttings representing the five "pure" *C. traskiae* trees should be transplanted to other areas on Santa Catalina where the risk of hybridization is minimal. Thus two or three new populations of this species can be established. The necessary cuttings to initiate this transplant program are already growing at the Rancho Santa Ana Botanic Garden.
3. All nonnative herbivores should be removed from Wild Boar Gully, and a zero population level should be maintained.

The first two management suggestions given for *Cercocarpus traskiae* are generally applicable to most cases of genetic swamping and can be summarized as the following general strategy: eliminate the less desired species from the area of hybridization and/or transplant the rare population to a remote location where the other hybridizing taxon does not occur.

In some cases, it may also be necessary to eliminate all hybrid or introgressive individuals as well as the less desired species. For example, introgressed populations of cutthroat trout are being eliminated by poisoning in order to avoid altering the biological characteristics of this species (Allendorf and Leary 1988). In contrast, the destruction of hybrid *C. traskiae* trees was not recommended by Rieseberg et al. (1989) because limited genetic diversity was deemed to be as severe a threat as hybridization to the long-term preservation of this species. The decision whether or not to eliminate hybrid individuals should be based on the relative severity of the threats of hybridization and limited genetic diversity. The case of *C. traskiae* may be somewhat unique because one-third of the total genetic diversity observed for *C. traskiae* would be lost by destroying the hybrid individuals. It is unlikely that this large a proportion of the total diversity of a species would be lost by eliminating hybrids in most hybridizing rare plant populations.

Although transplanting endangered populations to remote locations may seem to be an ideal solution to problems of hybridization and genetic swamping, this option has many potential difficulties. First, it may be difficult to duplicate the habitat of the original population. For example, *C. traskiae* occurs in soil derived from a rock type (igneous saussurite gabbro) found nowhere else on Santa Catalina Island (Martin 1984). Second, survival rates in introduced populations may be low. Thus the original population may be depleted by repeated attempts to establish new populations. Finally, we know very little about the interspecific interactions that may be necessary for the survival of many rare plant species. For example, an animal-pollinated plant species will not be able to reproduce successfully in the absence of its pollinator.

HELIANTHUS PARADOXUS

Helianthus paradoxus is a rare annual sunflower known only from two populations in Pecos County, Texas; one population in Reeves County, Texas; and a third, possibly extinct population in Dexter, New Mexico (Heiser 1958; Rogers et al. 1982; J. Poole pers. commun.). The species has recently been proposed for protection under the U.S. Endangered Species Act of 1973 (R. Perez pers. commun.). Although restricted to a fairly arid region, *H. paradoxus* occurs in brackish saline waters (12,000–14,000 ppm sodium chloride concentration). Turner (1981) suggested that *H. paradoxus* is simply a hybrid swarm between *H. annuus* L. and *H. petiolaris* Nutt., two closely related annual sunflowers, that was first established in the Fort Stockton area between 1945 and 1947 with irrigation that followed World War II. This hypothesis, Turner felt, could best explain the New Mexico–Texas disjunction. Additional evidence for this hypothesis is the morphological similarity of *H. paradoxus* to some individuals found in known hybrid swarms of *H. annuus* and *H. petiolaris* and the lack of collections of this species from Pecos County prior to 1947. Given these data, Turner suggested that "*Helianthus paradoxus* as a species should be viewed with some skepticism."

Several other authors, in contrast, consider *H. paradoxus* to be a distinctive species (Heiser 1958; Rogers et al. 1982; Chandler et al. 1986). Heiser (1958) noted that although *H. paradoxus* is probably most closely related to *H. annuus* and *H. petiolaris,* the relationship does not appear to be a close one. *Helianthus paradoxus* can be consistently distinguished from both *H. annuus* and *H. petiolaris* by its glabrous pales, phyllaries, leaves, and stem; leaf and phyllary shape; late flowering time (Heiser 1958); and complete lack of sesquiterpene lactones (Spring and Schilling 1989). Heiser (1965) showed that *H. paradoxus,* although highly fertile inter se, is chromosomally distinct from both *H. annuus* and *H. petiolaris.* Finally, *H. paradoxus* is morphologically monomorphic, whereas known hybrid swarms of *H. annuus* and *H. petiolaris* are remarkable for their morphological diversity (Heiser 1947). These data indicate that *H. paradoxus* must be considered a legitimate species. A possible solution to the two very different views of Heiser and Turner is that *H. paradoxus* is a stabilized hybrid derivative of *H. annuus* and *H. petiolaris* (Rieseberg et al. 1990). That is, *H. paradoxus* originated via hybridization some time ago, but is now a fertile vigorous species reproductively isolated from its parental species.

Helianthus annuus and *H. petiolaris* are widespread, polytypic species that are easily distinguished by a number of morphological and chromosomal features (Heiser et al. 1969; Schilling and Heiser 1981; Chandler et al. 1986). The species have similar ranges, occurring commonly in the western United States and less frequently eastward (Heiser et al. 1969). Although both species grow together in a variety of locations and hybrid swarms are common (Heiser 1947), the two taxa are found on different soil types. In general, *H. annuus* is restricted to heavy soils, and *H. petiolaris* to dry, sandy soils (Heiser 1947). Artificial hybridization experiments have revealed that the two species are very different chromosomally (Heiser 1947; Whelan 1978; Ferriera 1980; Chandler et al. 1986), and first-generation hybrids are nearly sterile. Fertility of second-generation plants is highly variable, however, suggesting that it may be possible to overcome the sterility barrier in the F_2 and later generations (Heiser 1947).

To test the hypothesized hybrid origin of *H. paradoxus,* Rieseberg et al. (1990) examined populations of *H. annuus, H. petiolaris,* and *H. paradoxus,* using isozyme, nuclear ribosomal DNA (rDNA), and chloroplast DNA (cpDNA) evidence. Ribosomal DNA variation can be used to detect hybridization directly because rDNA, like allozymes, is additively combined in hybrids (Doyle et al. 1985). Genetic additivity is not expected for cpDNA because cpDNA is generally maternally inherited and nonrecombinant (Sears 1983). However, cpDNA can be used to establish parentage in F_1 hybrids, differentiate between primary and secondary intergradation, and determine polarity in suspected cases of hybrid speciation (Rieseberg et al. 1990).

Their detailed genetic analysis revealed several diagnostic genetic markers differentiating *H. annuus* from *H. petiolaris,* including two isozyme loci, ten rDNA mutations, and three cpDNA mutations. Assuming a recent hybridization event, the expectation is for the hybrid taxon to possess an identical (or nearly identical), maternally inherited chloroplast genome of one parent, and an additive, biparentally inherited allozyme and rDNA pattern from both parents. If *H. paradoxus* was not derived by recent hybridization, allozyme and rDNA additivity would not be expected, and cpDNA analysis would place it with one of its two putative parents but exhibiting a number of autapomorphies, or perhaps even place it as a sister group to the other two species. *Helianthus paradoxus* combined the alleles and rDNA repeat types of its two proposed parents and had no unique alleles. Furthermore, all 30 *H. paradoxus* individuals had the chloroplast genome of *H. annuus,* a condition indicating that *H. paradoxus* cannot be ancestral to *H. petiolaris* and *H. annuus.* These data indicate, therefore, that *H. paradoxus* must have been derived through hybridization. However, because no seed was available from the New Mexico population of *H. paradoxus,* it was not possible to determine whether *H. paradoxus* was the product of more than one hybridization event.

The absence of unique alleles, the presence of a single cpDNA type, and the occurrence of only two small deletions in the rDNA of *H. paradoxus* suggest that *H. paradoxus* is not an ancient taxon. This view is concordant with its limited geographical distribution and lack of morphological variation. In contrast, both *H. annuus* and

H. petiolaris are geographically widespread, morphologically highly variable, and polytypic—characteristics which suggest that they may be more ancient.

It is also interesting to note that *H. paradoxus* is neither morphologically nor chemically intermediate between *H. annuus* and *H. petiolaris.* It lacks all the sesquiterpene lactones present in its relatives (Spring and Schilling 1990); flowers in late fall, in contrast to the summer flowering time of *H. annuus* and *H. petiolaris;* has glabrous leaves, phyllaries, and stems (varying degrees of pubescence are found on the leaves, phyllaries, and stems of its parents); and varies in the direction of neither parent in leaf and phyllary shape. Furthermore, *H. paradoxus* occurs in brackish or saline marshes, a habitat that cannot be considered intermediate between that of *H. annuus* (heavy soils that are wet in the spring but dry by midsummer) and *H. petiolaris* (dry, sandy soils). Thus data from morphology, secondary chemistry, and ecological distribution—three common approaches to the study of hybrid speciation—are misleading in the case of *H. paradoxus.* The discrepancy observed between the molecular data and other data sets is not surprising, since morphological features, chemical characters, and ecological distribution may be under strong selective pressures, whereas molecular markers tend to be neutral.

Although past events regarding the origin of *H. paradoxus* cannot be inferred with certainty, it appears likely that stable fertile types, probably tolerant to saline conditions, were selected from hybrid derivatives of *H. petiolaris* and *H. annuus.* These types differed morphologically and cytologically from the parental types, thus indicating the formation of a new species. Therefore, hybridization not only was responsible for the origin of *H. paradoxus,* but also may have served as a source of new variation (or at least new recombination) for the adaptation of *H. paradoxus* to its unique environment.

Conservation Considerations

The role of hybridization as a source of genetic and taxonomic diversity in a number of plant groups has been the subject of speculation for over half a century (Anderson and Hubricht 1938; Anderson 1949; Stebbins 1950; Lewontin and Birch 1966; Heiser 1973). More recently, a number of thorough electrophoretic studies have detected an increased frequency of rare alleles in hybrid zones (reviewed in Barton and Hewitt 1985). It has been hypothesized that their increase is due to higher mutation rates, intragenic recombination, or relaxed selection. Given the potential importance of natural hybridization as a source of new genetic diversity, natural hybrids and introgressants, where they play no obvious harmful role to an endangered species, should be considered a significant component of genetic diversity within a plant gene pool.

The questions of whether F_1 hybrids, hybrid swarms, introgressive races, or bona fide hybrid species qualify for protection under state or federal endangered species laws are more difficult. These are critical questions because stretching the Endangered Species Act to cover first-generation hybrids, for example, would damage the credibility of the endangered species program and would waste the limited resources of various plant management agencies. In this light, the criteria of Jolly (1989) for protection of hybrid taxa appear appropriate. He suggests protection for the hybrid taxon if (1) its evolution has proceeded beyond the point where crossing of the parental stock could re-create the plant considered, (2) it is taxonomically distinct, and (3) it is sufficiently rare or imperiled. *Helianthus paradoxus* would be protected under these criteria, whereas F_1 hybrids would not.

The possible favorable aspects of hybridization for conserving rare taxa were first noted by Stebbins (1942), who argued that it should be possible to inject new variability and therefore new aggressiveness into rare species through hybridization with more widespread relatives. He further suggested that new hybrid derivatives might be easier to establish in new sites than the original rare species (see also Barrett and Kohn, Chapter 1). Both of these ideas seem problematic, however, given our present knowledge of outbreeding depression (Templeton 1986; Huenneke, Chapter 2). Furthermore, introgression of alien genes into a rare species may result in the homogenization of valuable locally adapted populations and eventual homogenization of the rare species with its more widespread relative. Nevertheless, hybridization may be useful in captive-breeding programs as a last-ditch effort to preserve the germplasm of a rare and endangered taxon. A well-known example is the dusky seaside sparrow, where hybridization was employed because of the lack of living females (Avise and Nelson 1989).

Two factors may threaten the long-term existence of *H. paradoxus*. First, tractor-drawn mowers that scalp Texas highway right-of-ways undoubtedly inflict damage to one of the three known populations of *H. paradoxus,* which grows along Texas Highway 18. Second, both populations in Pecos County, Texas, are threatened by petroleum-production activities in the area that impede water flows.

Both threats listed above are due to human activities. Although it may be possible to convince the Texas Highway Department to refrain from mowing in this area during the summer and early fall of each year, it is unlikely that petroleum-production activities can be limited because they are on private land. This is unfortunate, since the brackish waters of Leon Creek also provide habitat for a rare endemic amphipod (*Gammarus pecos*) and two endangered fishes (*Cyprinidon bovinus* and *Pecos gambusia*) (Echelle et al. 1989). It may be best, therefore, to place large quantities of seed into long-term cryogenic storage and attempt to establish new populations of *H. paradoxus* at less-threatened sites. Both of these options may be possible because *H. paradoxus* is an annual plant, produces large quantities of viable seed, and is represented by several thousand plants in the wild. Furthermore, its natural habitat should be easily duplicated in both Texas and New Mexico.

HELIANTHUS EXILIS

One of the classic cases of hybridization and introgression in plants involves the three annual sunflowers of California: *H. annuus, H. bolanderi,* and *H. exilis* (Heiser 1949). *Helianthus annuus* is a common roadside weed in California, occurring frequently in the Central Valley and in southern California. *Helianthus annuus* was already in the state when the first botanical collections were made and was used by Native Americans for various purposes (Heiser 1949). Because it does not now occur in natural sites in California, it was likely introduced relatively recently by Native Americans (Heiser

1949; Stebbins and Daly 1961). *Helianthus bolanderi* and *H. exilis,* in contrast, are native to California. *Helianthus exilis* is an endangered species; its habitat is poor, moist serpentine soil in the Inner Coastal Range of mountains west of Sacramento. *Helianthus bolanderi* is widespread and weedy, and invades fields, ditches, and road-sides in the Central Valley of California (Heiser 1949). *Helianthus bolanderi* is sym-patric with *H. annuus* in the Central Valley, and the two taxa form extensive hybrid swarms in areas of contact. Using morphological and cytological data, Heiser (1949) hypothesized that *H. bolanderi* developed in California through the introgression of genes from the recently introduced *H. annuus* into *H. exilis.* Although *H. annuus* and *H. bolanderi* do hybridize, and although *H. bolanderi* approaches *H. annuus* in a number of morphological characters, no evidence of gene exchange away from hybrid swarms was documented by Heiser (1949).

Nevertheless, because of the morphological evidence for introgression (Heiser 1949), most authors have considered *H. bolanderi* and *H. exilis* to be closely related introgressive races. Thus *H. exilis* is considered to be a synonym of *H. bolanderi* in recent taxonomic treatments of the annual sunflowers (Heiser 1949, 1978; Heiser et al. 1969; Schilling and Heiser 1981) and has therefore been ignored by the Office of Endangered Species of the U.S. Fish and Wildlife Service. This situation has not changed even in light of morphological studies by Jain and students (Oliveri and Jain 1977), which suggest that *H. exilis* is a unique and threatened species distinct from *H. bolanderi.*

To determine whether *H. bolanderi* actually originated by the introgression of genes from *H. annuus* into *H. exilis,* Rieseberg (1987) analyzed variation in morphol-ogy, flavonoids, and allozymes, as well as restriction-site variation of chloroplast DNA and ribosomal RNA genes (results from the allozyme, cpDNA, and rDNA studies are reported in Reiseberg et al. 1988). Morphological data were in complete agreement with those of Heiser (1949), confirming the morphological intermediacy of *H. bolan-deri* relative to *H. exilis* and *H. annuus.* In contrast, flavonoid, allozyme, rDNA, and cpDNA evidence does not support the proposed introgressive origin of *H. bolanderi.* Although close to 50 taxon-specific markers were used in this study, no evidence of interspecific gene transfer was obtained. In addition, *H. bolanderi* possesses a unique cpDNA, which was outside the range of variation observed for each putative parental species (Figure 11.2). Mean sequence divergence observed between the cpDNAs of *H. bolanderi* and *H. exilis* was 0.30%. This estimate is comparable to sequence di-vergence values observed between closely related species of other plant groups (Riese-berg et al. 1988) and is higher than sequence divergence values observed among most other annual sunflower species (Rieseberg unpubl. data). In fact, this much cpDNA sequence divergence had never been observed within a plant species. This value is also extremely high given the proposed recent divergence of *H. exilis* and *H. bolanderi.* (Heiser [1949] suggests that *H. bolanderi* may have been derived as recently as during the past several hundred years.) In contrast, based on cpDNA divergence estimates, Rieseberg et al. (1988) suggest that the two species may have diverged as long as 3 million years ago. These data indicate, therefore, that *H. bolanderi* was not derived by introgression during the past several hundred years, as hypothesized, but is rela-tively ancient in origin. These data further show that *H. bolanderi* and *H. exilis* are ancient and distinctive species and should be considered as such.

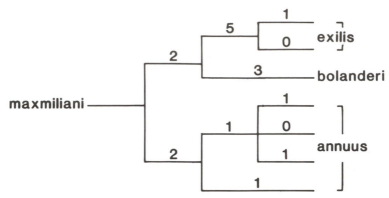

Figure 11.2. Phylogenetic tree of populations of *Helianthus annuus, H. bolanderi,* and *H. exilis* based on 17 restriction-site mutations. The tree is rooted relative to *H. maxmiliani* restriction sites. The number of mutations is given above each branch. (Redrawn from Rieseberg et al. 1988)

Conservation Considerations

This study, along with the previous study of *H. paradoxus,* demonstrates the limits of classical taxonomic methodologies for studying questions of hybridization and introgression in plants. Morphological intermediacy, the criterion used for testing hybridity in most studies, has been misleading in both *H. paradoxus* and *H. bolanderi.* This is not particularly surprising, since it has been known for some time that there are many other explanations for morphological intermediacy (reviewed in Heiser 1973). Nor is this the first time in which a faulty understanding of evolutionary relationships has resulted in misdirected management efforts (Avise and Nelson 1989). In short, when questions of hybridity are raised with respect to rare plants, detailed genetic studies may be necessary for their appropriate identification, classification, and management.

The primary threat to *H. exilis* is extensive mining and development in its native habitat in Napa, Lake, and Colusa counties of north central California. In particular, many of the *H. exilis* populations occur on Homestake Mining lands and are endangered by future mining developments. Members of the California Native Plant Society have shown that 13 of the 64 populations in the Knoxville area of Napa County have already been destroyed (G. Nabhan pers. commun.).

The following management strategies are recommended: (1) recognition of *H. exilis* as a unique and threatened species and subsequent establishment of status survey reporting for this taxon; (2) amelioration of development pressures near known sites; (3) collection of seed for long-term cryogenic storage; and (4) establishment of new populations of *H. exilis* at less-threatened sites. As in the case of *H. paradoxus, H. exilis* is an annual sunflower that produces large quantities of seed. Thus sufficient quantities of seed for both cryogenic storage and the establishment of new populations should be obtained easily. It may be much more difficult to find sites for the establishment of new populations, however, because *H. exilis* is a serpentine endemic, and its natural habitat is rapidly being lost because of development in northern California.

GENERAL DISCUSSION

The problem of hybridization and introgression in rare plants raises questions as to exactly what genetic resources we are trying to preserve and why. Native alleles characteristic of the rare species are not lost via hybridization and subsequent genetic assimilation; they are simply diluted by alleles from the widespread species. Nor are rare plant populations lost through hybridization. The populations remain, although they may come to resemble more closely their more widespread congener or some new compilospecies (Harlan and de Wet 1963). Thus we might take the position of Stebbins (1942) that introduction of new genetic variability in a rare species through hybridization with a more widespread species might actually be beneficial in terms of creating a more aggressive, adaptable species.

There are a number of reasons, however, for fearing the effects of widespread hybridization and introgression in rare plants. First, I can think of no justification for exchanging several distinct rare plant species, each with its own unique growth form and habitat requirements, for a single widespread compilospecies. The advantages of maintaining a number of distinct rare species is clear from an aesthetic viewpoint. From a genetic standpoint, it should be noted that dilution of alleles characteristic of rare plant species may lead to their eventual loss. A second danger of genetic assimilation is the loss of genetically discrete, ecologically specialized plant populations in both widespread and rare species. Finally, widespread hybridization and introgression may result in severe outbreeding depression. Whether outbreeding depression would be a temporary or a long-term problem in hybrid plant populations is unclear.

In summary, hybridization in rare plants may have a number of harmful consequences, including the genetic assimilation of a rare taxon by a numerically larger one, loss of locally adapted populations, and outbreeding depression. These negative consequences have generally been ignored in studies of rare plants, despite their high potential for interspecific gene flow. On the other hand, it is clear that hybridization may be beneficial in certain groups of plants, leading to greater genetic and taxonomic diversity. There are no criteria, however, for the protection of F_1 hybrids, hybrid swarms, introgressive races, or bona fide hybrid species under state or federal endangered-species laws. Both of these problems are exacerbated by the inadequacy of classical taxonomic methodologies for accurate identification of, and differentiation among, the various classes of hybrids listed above. A better understanding of the possible consequences of hybridization in rare plants and the accurate identification of plant hybrids should contribute to the conservation of plant diversity.

ACKNOWLEDGMENTS

This research was supported by NSF Grant BSR-8722643 to LHR. We thank Mike Hanson, Aaron Liston, Scott Zona, Donald Falk, and Kent Holsinger for helpful comments on the manuscript.

12

Off-Site Breeding of Animals and Implications for Plant Conservation Strategies

ALAN R. TEMPLETON

We are living in an era of mass extinction that is due primarily to large-scale habitat destruction caused by human activity. Accordingly, the best remedy to prevent extinction is habitat preservation. Such a strategy preserves not only species that have been targeted because they are endangered, but also the entire ecological community in which the target species lives. This community will often include additional endangered species of which we are not even aware. There is little controversy that this is the ideal strategy for preserving species, but unfortunately habitat preservation is not an option in many cases. The natural habitats of many species have already been completely destroyed, and those of many others have been so reduced in size and so fragmented that the species are in imminent danger of extinction. Even when habitat preservation can be practiced, it frequently requires the reintroduction of propagules. In such cases, off-site breeding of captive or cultivated populations is needed to preserve a species that has gone totally extinct in nature, to provide a backup for habitat preservation efforts, or to serve as a source of propagules for a reintroduction program.

It is hoped that captive populations can ultimately be released into preserved or restored habitats. Hence, the goal of such breeding programs is to maintain the species in captivity until habitat restoration allows its release back into nature. Nevertheless, such restoration could take decades or even centuries. Therefore, captive populations must be managed as a long-term, multigenerational breeding program unless some sort of storage of individuals or gametes is possible (e.g., a seed bank). There has been a general recognition in recent years that the genetic variation present in a species is a valuable biological resource in this era of genetic engineering. Moreover, the restored environments will undoubtedly differ from the original habitats and communities. It is therefore critical that the released populations have sufficient genetic variability to provide adaptive flexibility in an uncertain future. Consequently, it is not enough just to deal with issues of husbandry or horticulture; in addition, we must have genetic management. Species preservation is not just the preservation of a Latin binomial; rather, it is the preservation of an evolutionary lineage consisting of genetically diverse individuals. The loss of a species' genetic diversity represents a type of "partial extinction" (Templeton 1982) that often presages its total extinction.

These general considerations lead to two primary genetic goals in captive management: (1) preserving genetic diversity in the captive population, and (2) altering the genetic composition of the species as little as possible from its initial state. Unfortunately, these two goals are sometimes contradictory. Captivity or cultivation may induce novel selective forces, and populations that respond to these novel selective regimes will be altered from their wild genetic constitution. The capacity to respond to altered selective regimes depends on the level of genetic diversity in the population: the higher the genetic diversity, the greater the capacity to respond to selection. Hence, satisfying the goal of preserving genetic diversity makes it more difficult to satisfy the goal of not altering the species' genetic composition. Consequently, the genetic manager must often balance contradictory goals in the breeding program, as well as deal with practical constraints.

The purpose of this chapter is to explore the lessons learned from captive animal breeding that may be applicable to *ex situ* breeding of endangered plant species. Some fundamental differences between breeding plants and animals are also pointed out that will frequently result in different priorities for managers of plant versus animal species.

PRESERVING GENETIC VARIABILITY

In order to preserve a species' genetic diversity in captivity, it is obviously necessary to carry over much of that diversity from the natural population into the initial captive population. The founding size of the captive population is a critical parameter in determining the extent of this carry-over, with the amount of genetic diversity carried over increasing with increasing initial size. Unfortunately, many captive-breeding programs are established only after the natural population has already been reduced to very few individuals (e.g., the California condors). This is a poor strategy because even if the captive population is founded from all the remaining individuals, much of the species' genetic diversity has been lost by this point. Moreover, with small founding numbers, the genetic manager will also frequently have to deal with additional genetic problems, such as inbreeding depression, as will be discussed in more detail later (see also Barrett and Kohn, Chapter 1). Hence, rather than establishing captive populations as acts of desperation, it is far better to establish captive populations as "insurance policies" when the natural populations are still sufficiently large to contain much genetic diversity (Millar and Libby, Chapter 10). Fortunately, many endangered species of plants still have large populations, even though these numbers are fewer than in historical times. Consequently, in many cases plant populations can be established with an excellent base of genetic diversity, although intelligent sampling of the natural populations is necessary to realize this potential.

Even when the numbers in the founding population are low, a captive management program is worthwhile because much genetic diversity still can be preserved. Nei et al. (1975) showed that populations established from only a single mated pair of individuals can, under the proper circumstances, carry over a sizable portion of the genetic diversity found in the ancestral population. Their studies indicated that founder populations usually preserve the common allelic polymorphisms found in the ancestors and that the primary reduction in genetic diversity is due to the loss of rare alleles. Because common alleles make the overwhelming contribution to the average heterozygosity and polymorphism found in the population, the average heterozygosity and polymorphism are often reduced in only a minor fashion by even a severe founder effect, given that other conditions are satisfied. Among these conditions are that the ancestral population be panmictic (not genetically subdivided) and randomly mating. Together, these conditions ensure that much of the species' overall genetic diversity is present in the form of individual heterozygosity. Thus a few individuals can carry over much genetic diversity. Unfortunately, one or both of these conditions are frequently violated.

For example, my laboratory and the Missouri Conservation Commission have been involved with sampling natural populations of the collared lizard (*Crotaphytus collaris*) for reintroduction on restored habitats in the Missouri Ozarks. These lizards live on Ozark glades: rocky, treeless outcrops usually near the tops of ridges with southerly or southwesterly exposures. These glades provide xeric habitats separated from one another by the predominant oak–hickory forest found in the Ozarks. Hence, the natural habitat of these lizards is highly fragmented. We have performed a variety of genetic surveys on these lizards, including isozyme analysis, restriction mapping of mitochondrial (mtDNA) and nuclear ribosomal (rDNA) DNA, and DNA "fingerprinting." For the isozymes, rDNA, and mtDNA, the common pattern is to find no heterozygosity or genetic diversity at all within a glade (Templeton et al. 1990). However, lizards from different glades, even those separated by just a few hundred yards, often show many fixed genetic differences. Using the higher-resolution technique of DNA fingerprinting, some within-glade diversity is found, but still the vast majority of the genetic diversity is present as between-glade differences. Hence, in this species, most of the genetic diversity is present as between-subpopulation differences, and very little of the genetic diversity is present in the form of individual heterozygosity. A sample of lizards taken from a single glade subpopulation would miss almost all the genetic diversity present in the species regardless of the size of the sample. The best way to preserve this species' genetic diversity is to sample a few animals from many different glades rather than a large number of animals from a few glades.

This example illustrates the critical importance of performing genetic surveys on endangered species. Simply establishing the captive population from a large number of founders does not ensure that genetic diversity is being preserved. Without knowledge of how the species' genetic diversity is divided within and between local populations, it is impossible to design a sampling program that is ensured to preserve a substantial portion of the species' genetic diversity. Basically, the results of genetic surveys are needed to define the stratified random sampling design that best preserves the genetic diversity found in most species (Brown and Briggs, Chapter 7). Information on breeding systems, dispersal mechanisms, habitat diversity, biogeography, and other factors can be used also to formulate the optimal stratified sampling design. However, how a species' genetic diversity is divided within and between local populations or habitats is extremely sensitive to rare events (such as rarely occurring long-distance dispersal), which are not easily detected by field observation. Genetic surveys are often a more accurate and easier method of inferring the pattern of genetic subdivision than are field observations or dispersal studies (Templeton et al. 1990).

Once the founder population has been established, it is also necessary to prevent the loss of the genetic diversity that has been carried over from the natural population. Once again, population size has a great effect. The larger the ultimate captive population size, the more immune it is from the loss of genetic variation through the action of genetic drift. It is therefore optimal to make the captive population as large as is practically possible. However, in achieving this large-as-possible ultimate population size, there are other parameters affecting the preservation of genetic variation.

Nei et al. (1975) showed that the rate of population increase has a pronounced effect on the retention of genetic variation. In particular, the loss of genetic variation decreases as the rate of population growth increases. Consequently, it is best to increase the captive population to its ultimate carrying capacity as rapidly as possible. For example, a captive population of Speke's gazelle (*Gazella spekei*) was founded in 1969 from one male and three females (Templeton and Read 1983). From a detailed analysis of inbreeding depression in this species, Templeton (1987) inferred that the natural population was largely outcrossing and panmictic. Therefore, much of the species' genetic variability would be in the form of individual heterozygosity. The population was also expanded at a rapid rate after its establishment (Templeton and Read 1983, 1984). Because of these two factors, the theory of Nei et al. (1975) predicts that this captive herd should have much genetic variation. To test this prediction, we screened the herd between 1982 and 1985 for genetic variability in rDNA, mtDNA, and 28 protein-coding loci (Templeton et al. 1987). For the nuclear genes, 14% of the loci were polymorphic, a percentage that is typical for large grazing mammals (11.4%) or large mammals in general (13.2%) (Baccus et al. 1983). Hence, despite an extreme founder event and inbreeding, the herd is not depauperate in genetic variability relative to other mammals.

In contrast to the nuclear markers, the captive herd of Speke's gazelles was depauperate in mtDNA variation. The probability of nucleotide polymorphism in the mtDNA was 0.61% for the captive herd (Templeton et al. 1987), whereas for the typical mammal it is 1.58% (Avise and Lansman 1983). This result is not surprising when one takes into account the maternal, haploid nature of mtDNA inheritance in mammals. Because mtDNA is effectively haploid, there is no carry-over into the captive population through individual heterozygosity. The mtDNA in this herd is mimicking the expected results for nuclear DNA in a highly subdivided population. In such a subdivided population, very little of the genetic diversity is present as individual heterozygosity, and hence the amount of genetic diversity carried over into the captive population is much more sensitive to founder number than it would be in a panmictic, outcrossing population. This illustrates the need to invest much more effort and time in establishing a captive population from a genetically subdivided species.

Although the mtDNA variability is depauperate, the expanding size of the herd preserved very well what was carried over from nature. Because there were only three founding females, the maximum number of distinct haplotypes that is possible is only three, and three haplotypes were indeed found, each traceable to a separate founding female (Templeton et al. 1987). Hence, as expected from Nei et al.'s theory, there was little loss of genetic variation after the founder event in this expanding population.

Besides simply expanding the population as rapidly as possible, genetic diversity is best maintained when each founder contributes more or less equally to the population. MacCleur et al. (1986) have also shown that it is critical that this equalization be done from the onset of the captive-breeding program. If any founder has few offspring, the stochasticity associated with Mendelian segregation makes it highly probable that much of the genetic variability present in that founder as heterozygosity will be lost by chance alone. Equalizing the founder contributions in later generations can never undo this damage; once the variation carried over by a particular founder is lost, it is gone forever. The analysis of the Speke's gazelle herd supports this prediction (Templeton et al. 1987). Before the initiation of the genetic management program, the sole male (M3) and one female (F4) were used extensively during the formation of the captive herd, whereas another founding female (F5) was bred very little. Starting in 1979, the founder representation was deliberately equalized (Templeton and Read 1983), but by this point the damage had been done. MacCleur et al. (1986) analyzed the herd and calculated that over 50% of the genetic diversity carried by F5 would already have been lost. By using the pedigree data, Templeton et al. (1987) were able to trace back to the original founders all the allelic diversity present in the herd in 1982 to 1985. Despite a fairly equal founder representation at that time, almost all the genetic variability in the herd traced back to the founding pair M3 and F4, and none traced back to F5. Subsequent work (Georgiadis et al. 1988) did reveal additional genetic variation that was traceable back to F5, but the vast majority of the herd's genetic diversity was derived from M3 and F4 and very little from F5. This example shows that it is critical to equalize founder representation as soon as possible after the establishment of captivity in order to preserve genetic variability. Mistakes made early in the breeding program cannot be corrected unless additional sampling from the natural population is possible.

Once the herd has grown to its carrying capacity, genetic drift can still erode the genetic variation that has survived to that point. The best way of minimizing the impact of genetic drift in a closed population is to subdivide the population into breeding units that have minimal gene flow. Maruyama (1970) showed that a subdivided population loses its global genetic variation at a much smaller rate than an equal-sized panmictic population. To see why, consider the following extreme example. Suppose a population of 100 is maintained in captivity. If we start with a locus with two neural alleles, say A and a, at equal frequency, the population will eventually go to fixation for one allele or the other with equal probability if the population is maintained as a single panmictic unit. The expected time to fixation in this case is 277 generations, using standard population genetic theory (Crow and Kimura 1970). Suppose now that the 100 individuals had been split into four isolated populations, each of size 25. Within each isolate, genetic drift will eventually cause the loss of one allele, and the expected time to fixation is now 69 generations. However, the probability that all four isolates become fixed for the same allele is 0.125. Hence, in the panmictic case the population will lose its allelic variation at this locus with probability 1. In the subdivided case, each isolate will lose its variation four times as fast as the single panmictic population, but with a probability of 0.875 that the four isolates will not all be fixed for the same allele. Thus the subdivided population will preserve indefinitely the genetic diversity at the global-population level with high probability. As this example shows, a subdivided population maintains global genetic diversity far more effectively than an equal-size panmictic population. We have therefore established several subherds of Speke's gazelles at various zoos. Animals are exchanged between subherds only when they are absolutely needed for management purposes. In this way, we minimize gene flow among the herds, and therefore maximize the global herd's ability to retain genetic variation. Such subdivision also has the benefit of buffering the global herd from extinction due to epidemics or accidents, simply by physically separating the subherds.

Unfortunately, there is much confusion about the role of population subdivision on the maintenance of genetic diversity. Because subdivision results in more inbreeding than would be the case for a single, panmictic population, Simberloff (1988) feared that genetic diversity would be eroded. This fear is based on the widespread misconception that inbreeding decreases genetic diversity. It is genetic drift, not inbreeding, that erodes genetic diversity in a closed population. Because genetic drift frequently results in inbreeding, it is easy to confuse the two phenomena. But as supported by both theory (Maruyama 1970) and observation (Templeton et al. 1987), subdividing a herd into small, inbred demes is an excellent strategy for the maintenance of genetic diversity. Hence, inbreeding and erosion of genetic variation do not necessarily go hand in hand.

PREVENTING EVOLUTIONARY CHANGES IN THE CAPTIVE POPULATION

Adaptations to the captive environment may imperil the chances for success after release. Unfortunately, as long as the breeding populations in captivity are genetically variable, there is the danger that they will adapt to the captive environment. Guarding against any inadvertent selection for domestication (e.g., favoring animals as breeders that are more docile) can help, but we simply cannot anticipate or monitor all the ways in which a population can adapt to a captive environment. For example, in 1986 we obtained some collections of *Drosophila hydei* from a natural population on the island of Hawaii (Templeton et al. 1989). A total of 39 isofemale/isomale lines were established from wild-caught individuals. The flies were transferred to fresh food every two to three weeks, the same transfer schedule that we use for *D. mercatorum*, the fly species that we do most of our work on. However, we noticed that *D. hydei* has a longer egg-to-adult developmental time than *D. mercatorum*, so this transfer schedule was inadvertently favoring flies with fast egg-to-adult developmental times. We then changed the transfer schedule to a longer period, but we wanted to see if this inadvertent selection had had any genetic effects during this approximately six-month time period.

The egg-to-adult developmental time in *D. hydei* is negatively correlated with the proportion of 28S rDNA genes bearing an insertion (Franz and Kunz 1981). We therefore estimated the proportion of inserted 28S genes in the founding flies and in their captive-bred descendants six months to a year later. All but one line out of 39 showed a decline in insert proportions, and in most cases this decline was quite dramatic. This response is consistent with the known life-history consequences of this rDNA insert and the nature of the inadvertent selection we were causing for fast development. As this example shows, rapid and drastic evolutionary changes are possible in captivity as an adaptation to the captive-rearing conditions. It is virtually impossible to anticipate what these inadvertent selective forces may be for a given species. As long as we maintain genetically diverse breeding populations in captivity, we will foster conditions that are conducive to captive environment adaptation. Hence, the goals of maintaining genetic diversity and preventing adaptation to the captive environment are sometimes in tension and complicate each other in practice.

Nevertheless, some of the management strategies for preserving genetic variation also reduce the population's response to selection, so these two basic goals are not always in tension. For example, large founding sizes promote genetic diversity, and they also reduce the chances of *genetic transilience*—rapid adaptive shifts induced by the founder event (Templeton 1980). Genetic transilience occurs because although founder events tend to reduce genetic diversity, they also tend to increase the *additive genetic variance*—that component of genetic variation that is responsive to selection. For example, founder events reduce the range of genetic backgrounds on which a given allele is expressed. Hence, epistatic terms that could not be effectively selected for in the ancestral population can be converted into additive variance in the founder population that now respond to selection (Templeton 1980). This prediction has been confirmed empirically by Bryant et al. (1986b), who have shown that additive genetic variances increase after small to intermediate-size founder events in a manner that is consistent with ancestral dominance and epistatic terms being converted into additive variance. Consequently, a founder event increases the additive variance of traits precisely at the time in which the population is experiencing a novel environment (i.e., captivity). This combination favors rapid evolutionary changes resulting in adaptation to captivity. By making the founder size as large as possible, we not only maximize the carry-over of genetic variation, but also minimize the increase in additive variance induced by the founding event. This, in turn, reduces the chances for rapid adaptive shifts after the foundation of the captive population.

The ability of a population to respond to selection is also limited by the variance of reproductive success. By equalizing reproductive success in captivity to maximize the maintenance of genetic diversity, we also reduce the population's capacity to respond to selective forces.

Finally, a population can respond to selection only if there is genetic variation in the breeding population. As previously mentioned, population subdivision is the best strategy for long-term maintenance of genetic variation. Such subdivision also implies that the breeding units—the subpopulations—will eventually have little genetic variation within them. Hence, there will be little response to selection under this breeding design. Unfortunately, there are two factors that can cause serious deviations from this ideal expectation. First, our experience with Speke's gazelle is that the subpopulations at different zoos cannot be maintained in a totally isolated fashion. For example, because of the vagaries of demographic stochasticity, a particular herd may lack a reproductive male. Hence, there is a low level of exchange of animals for management purposes. Moreover, the animals sent to other zoos are from zoos that are producing a surplus of animals—an exact analogue of Wright's (1932) concept of "interdemic selection." This combination of population subdivision, low levels of gene flow, and interdemic selection is optimal for inducing Wright's (1932) shifting balance process for rapid adaptive evolution. Unless gene flow between subdivisions can be kept close to zero, we find once again that the optimal solution for preserving variation can be incompatible with the goal of preventing adaptation to the captive environment.

The second problem with the subdivision strategy deals with the genetic environment associated with captivity. When a captive population adapts to its new environment, it is not just the external environment that is important. Because genes interact with one another, the selective fate of a gene depends critically on its genetic environment, as has been empirically demonstrated (e.g., Templeton et al. 1977; Templeton 1979). For example, basic population genetic theory shows that selection is ineffective against rare, recessive deleterious genes in random-mating populations because such alleles are in a genetic environment (heterozygosity) that masks their deleterious effects. Hence, a large, random-mating population will accumulate many such alleles. In contrast, even rare alleles will frequently be in homozygous condition in an inbreeding population, and selection is therefore much more effective in eliminating such alleles from the population. As this example shows, the genetic environment plays a critical role in determining the selective fate of allelic variation. The genetic environment, in turn, is determined by system of mating, population size, and population subdivision. When a population is brought into captivity, all three of these factors are commonly altered to some extent, resulting in adaptation to a new genetic environment.

If the ancestral population is large and outcrossing, for example, a captive population founded by a small number of individuals will be much more inbred. This altered genetic environment induces a period of intense selection, as alleles are eliminated that do poorly in this more homozygous genetic environment while other alleles are favored that do well. This period of adaptation to the inbred genetic environment is known as an *inbreeding depression*. In our work with captive mammals, such inbreeding depressions are extremely common (Templeton and Read 1983; Ralls et al. 1988). Likewise, if the ancestral population were highly subdivided, a captive population that mixes animals from different subpopulations would also alter the natural genetic environment, thereby inducing an outbreeding depression (Templeton et al. 1986). Such outbreeding depressions have occurred in captive mammalian populations (e.g., owl monkeys [Cicmanec and Campbell 1977; de Boer 1982]) but are much more rarely encountered in mammals than is inbreeding depression (Templeton 1986, 1987; Ralls et al. 1988). Although inbreeding and outbreeding depressions seem to be opposite phenomena, they both have the same underlying cause: the captive population is breeding under conditions that alter its genetic environment, which, in turn, induces a state of maladaptation. If this altered genetic environment persists, the population will either go extinct or adapt to it, as shown for inbreeding depression by studies on Speke's gazelles (Templeton and Read 1984) and for outbreeding depression by studies on *Drosophila* (Annest and Templeton 1978). If the population survives, captivity has selected for a genetic state that differs considerably from the ancestral, natural state.

The best way to avoid adaptation to an altered genetic environment is simply to minimize changes in the genetic environment under captivity. To do this we need to know what that environment is. This knowledge can be acquired by performing genetic surveys on the natural populations to determine basic population structure and/or by making observations on the breeding system and life-history characteristics of the species. Thus genetic surveys on natural populations play a doubly important role in designing captive-breeding programs, since we had previously seen that genetic surveys were also critical for designing the optimal sampling scheme that preserves the genetic diversity found in nature. If genetic surveys on natural populations are not feasible, it is still possible to infer the ancestral genetic environment by monitoring the captive population for inbreeding versus outbreeding depression (Templeton 1987). Fortunately, inbreeding and outbreeding depressions are distinguishable (for details, see Templeton and Read 1984; Templeton et al. 1986), and this information can be used to modify the captive-breeding program in an appropriate manner.

If the natural population turns out to be highly subdivided into local, inbred demes, one can easily satisfy both genetic goals of maintaining diversity and minimizing adaptation to captivity. In this case, one must sample individuals from several local populations in order to carry over much genetic diversity from nature into captivity. One should then maintain the captive population as highly subdivided breeding units, with the founders of each breeding unit ideally coming from the same local natural population. This would prevent any drastic changes in genetic environment, maximize the long-term capacity to maintain genetic diversity, and minimize adaptation to captivity because the subdivided breeding units would have little genetic variation within them. Hence, this is the optimal situation from the genetic manager's point of view.

If the natural population turns out to be highly outcrossed and not subdivided, the management situation is much more difficult. The best way to avoid changing the genetic environment is to maximally avoid inbreeding within the captive population. This requires using the entire captive population as a single breeding unit. However, this strategy also maximizes the long-term loss of genetic diversity in the population and maximizes the potential of the captive population to adapt to the captive environment. Consequently, some compromise between these conflicting outcomes is necessary. The fundamental decision on which strategy to implement depends on the expected length of time the population will remain in captivity and on the carrying capacity of the population in captivity. Basic population genetic theory predicts that neutral genetic diversity is lost at a rate of $1/(2N)$ per generation (Crow and Kimura 1970), where N is the variance effective size (a number that is generally smaller than the census size and approximately equal to the harmonic mean of the number of breeding females and males). Using this neutral rate loss as an index, the breeding program should last no more than $0.6N$ generations in order to preserve at least 75% of the initial genetic diversity carried over into captivity. Moreover, by keeping the breeding program of short duration, the adaptation of the population to its captive conditions is also minimized. These considerations imply that the larger the breeding size (*not* census size) and the smaller the expected duration in captivity, the more optimal becomes the avoidance of inbreeding strategy. If one can maintain only a small number of breeding individuals (as frequently happens with large mammals) or if the anticipated breeding program exceeds $0.6N$ generations, the subdivided strategy should be implemented.

Unfortunately, practical constraints often dictate the breeding strategy much more than theoretical considerations. For example, in our Speke's gazelle breeding program, our founding population consisted of one male and three females. Once the original founders were no longer reproducing, all possible matings in this herd had to be between close relatives. Hence, there was no possibility of avoiding strong inbreeding in this captive population, so we had no choice but to opt for the inbred, subdivided population strategy. The Speke's gazelle breeding program has been portrayed as a deliberate alternative to avoidance of inbreeding (Simberloff 1988), but in actuality it was simply impossible to initiate a program based on avoidance of inbreeding (Templeton and Read 1983).

If practical constraints or long-term considerations lead one to implement an inbred, subdivided-population captive-breeding strategy on an organism that is largely panmictic and outcrossing in nature, then the compromise one has to make is to adapt the captive population to high levels of inbreeding. During the course of this adaptation, the population will suffer from an inbreeding depression. Such inbreeding depressions can be so severe that they may endanger the success of the entire program. Hence, contrary to the usual advice of no artificial selection, it is sometimes necessary to manipulate deliberately the genetic fate of the population simply to ensure its survival under captivity. This was the situation we encountered with Speke's gazelle. The herd was initiated in 1969, but our breeding program did not start until 1979 (Templeton and Read 1983). By the late 1970s, all but one of the original founders were dead, and inbreeding between close relatives could no longer be avoided. An extremely severe inbreeding depression was encountered that poised a serious threat to the survival of the herd. Since inbreeding could not be avoided, we decided to initiate a breeding program that would, via evolution, reduce the inbreeding depression. In this regard, it is critical to note that inbreeding depression is not an inherent attribute of inbreeding; rather, it is a genetically determined response to an altered genetic environment. In this case, much of the deleterious fitness effects are due to recessive, deleterious alleles that tend to accumulate in outcrossing populations. Once those alleles are eliminated from the population, the inbreeding depression will be greatly reduced.

The initial inbreeding depression in the gazelles indicated a genetic load of 7.5 lethal equivalents per founding animal; that is, the level of first-year mortality was the equivalent of that expected by assuming that each import animal was heterozygous for an average of 7.5 recessive genes that when made homozygous would kill the animal within one year after birth. To eliminate these lethal equivalents selectively, we did the following.

1. We gave precedence to using healthy, but inbred, animals as parents. By using inbred animals as parents, we were able to reduce the number of lethal equivalents by one-half in just three years (Templeton and Read 1984).

2. Many lethal equivalents are not actually due to single loci, but to combinations of loci (Dobzhansky and Spassky 1960); hence, it is important to maximize the genetic effectiveness of recombination in the population so that selection can favor the appropriate combinations of genes. The effectiveness of recombination in generating new combinations of genes depends on the level of individual heterozygosity of the parents. Therefore, parental heterozygosity was increased by trying to equalize the founder representation in each parent (note that this is not the same as equalizing founder representation in the entire herd, as previously discussed for the goal of maintaining genetic diversity). Obviously, with intense inbreeding, one cannot sustain high levels of heterozygosity for many generations, but all that is needed is to do so for the first few generations when the selection induced by the inbreeding depression is most intense. We monitored the potential for recombination, by using the *hybridity coefficient,* which measures for each individual the average proportion of its parent's genetic composition derived from different founding animals. We showed that, as expected, hybridity helped to eliminate the deleterious consequences of inbreeding regardless of the levels of inbreeding in the parents (Templeton and Read 1984).

3. The best long-term selective responses are usually obtained when the selective intensities are moderate and not severe. Hence, to the best of our abilities, we avoided matings that would result in offspring with extremely large inbreeding coefficients in favor of more moderate inbreeding coefficients. Because individual heterozygosity could also buffer the deleterious effects of inbreeding, we also chose mating pairs that

would result in offspring bearing genes from as many founders as possible. The potential for individual heterozygosity was monitored by the *ancestry coefficient,* which measures the extent to which an individual has a mixed ancestry in terms of founder individuals. We were able to show that high ancestry coefficients also aided the elimination of inbreeding depression in a manner that was statistically separable from both inbred parents and hybridity (Templeton and Read 1984).

4. Finally, the response to selection is always constrained by genetic variability; therefore, we implemented the recommendations mentioned before to retain genetic diversity in the population (such as expanding the population size as rapidly as possible and equalizing founder representation in the herd). The combined effect of all these procedures was to rapidly reduce the inbreeding depression to the point that it no longer threatened the long-term maintenance of the herd. At that point, we implemented the subdivided-population strategy by establishing subherds at different zoos throughout the United States.

As this example shows, inbreeding depression can be managed. Consequently, when necessary, it is possible to implement the inbred, subdivided-population breeding strategy on an outbred species. This strategy is optimal for long-term maintenance of genetic diversity and for minimizing the potential for adaptation to the captive environment over long periods of time. However, it definitely alters the genetic environment. It is hoped that by preserving genetic diversity, we can ensure that any population released in the future will have the evolutionary flexibility to readapt to its old genetic environment as well as to its restored habitat. To reestablish genetic variation in any future reintroduction, it would be necessary to mix animals from several subherds. It is possible that the different subherds might have evolved incompatible coadapted complexes during captivity, which, in turn, could lead to an outbreeding depression when the animals are mixed. To minimize the outbreeding depression, it is best to perform the initial mixing in captivity and release healthy F_2 or later-generation individuals. The reason for this is that the brunt of the outbreeding depression is borne by the F_2 (Templeton et al. 1986), and after getting through this fitness bottleneck, the absolute fitness of the individuals can increase substantially (Annest and Templeton 1978). Thus by releasing healthy F_2 or later-generation individuals, the release population does not have to deal with the major selective pressures induced by coadaptation during captivity. This recommendation supports the idea of releasing composite mixtures, as suggested by Barrett and Kohn (Chapter 1).

IMPLICATIONS FOR PLANT CONSERVATION

Plants are often better candidates for off-site breeding programs than are animals. Many plants can be maintained as dormant seeds for long periods of time. This alleviates the constraints on founding size and maintenance size, and retards the erosion of genetic variation through drift and adaptation to cultivation simply by reducing the number generations in absolute time. Other plants, such as trees, have very long generation times. Here, real constraints on founding and maintenance size may exist, but at least the compromises that are forced by the expected length of the breeding program are ameliorated because even breeding programs that are anticipated to require centuries may represent only a handful of generations. Once again, this minimizes the importance of drift and selection under cultivation. Consequently, the types of problems discussed in this chapter are most applicable to that subset of plants that must go through an entire generation's life cycle fairly frequently in absolute time.

Even in this subset, the options for plants are generally better than those for animals. In many plants, there is selfing, apomixis, clonal reproduction, and/or extremely subdivided-population structures. In species such as these, much of the genetic diversity is present as among-population (or clones, or cultivars, etc.) differences. Such species are ideal candidates for the subdivided-population-structure strategy of captive breeding, which has the benefits of maintaining long-term genetic diversity and minimizing (but not eliminating) adaptation to cultivated conditions. Moreover, this strategy does not have to change the genetic environment or induce severe inbreeding depressions for plant species with a naturally subdivided-population structure.

However, as pointed out previously, such subdivided-population structures require much more effort and time for initial sampling to ensure that the genetic variation found in nature is carried over to the cultivated population. Moreover, it is difficult to design an optimal sampling scheme without genetic surveys of natural populations in these cases. High-resolution techniques, such as DNA fingerprinting, are ideal for this purpose and have already been successfully applied to plants (Rogstad et al. 1988a, 1989). If genetic surveys are not feasible, one can make a guess as to what the genetic population structure ought to be, based on studies of mating system, life-history characteristics, biogeography, and other factors. Because rare events (such as occasional long-distance dispersal of either seeds or pollen, occasional outbreeding in an otherwise inbreeding species, etc.) can still have a major impact on how genetic variation is distributed within and among local populations, these guesses may sometimes be inaccurate. Given our overall ignorance of the population biology and genetics of rare plants, it is important to perform genetic surveys on several species and correlate the results with information on system of mating, dispersal mechanisms, and biogeography. Only in this manner can we obtain the data needed to ascertain how often inferences based on nongenetic information may be misleading.

One aspect of plant biology that makes plants less ideal candidates than animals for off-site breeding programs is the occurrence of syngameons, which are particularly common in sexually reproducing, outcrossing perennials. *Syngameons* are defined as "the most inclusive unit of interbreeding in a hybridizing species group" (Grant 1981). Although the species within a syngameon are capable of hybridizing, they maintain themselves as distinct evolutionary units in nature (even for time periods measured in the millions of years [e.g., Eckenwalder 1984]) because they are usually ecologically quite distinct. However, it has also been frequently noted that when the natural environment is disturbed, the ecological selective forces that maintain the species within the syngameon frequently disappear and the original species can be lost in a hybrid swarm. Syngameons also occur in animals (Templeton 1989), and hybridization induced by disturbed environments can indeed drive animal species to genotypic extinction through gene flow (Johnston et al. 1988). However, this problem is undoubtedly much more serious in plants than in animals. An off-site breeding program will almost always occur in what can be considered a "disturbed environment"; therefore, there is a great danger that off-site breeding programs for plants could destroy many species through hybridization. Since the hybrids often do very well in the disturbed environments, the performance of a cultivated population may give no indication that serious genetic damage is being done. Hence, the best way to protect against this type of danger is to perform genetic and ecological surveys on natural populations in order to define the species in their natural environment. Once again, we see a critical role for genetic surveys of natural populations in plant conservation.

In summary, plants in many ways are better candidates than animals for off-site breeding programs once such a cultivated or captive population has been established. But much more effort will generally be required for plants before the breeding program can even be initiated, and the sampling strategies will often have to be more complex than those employed for animals.

ACKNOWLEDGMENTS

I wish to thank Donald Falk and Kent Holsinger for their comments on an earlier draft of this manuscript. This work was supported by NIH Grant R01 GM31571.

13

Conservation of Rare and Endangered Plants: Principles and Prospects

KENT E. HOLSINGER and L. D. GOTTLIEB

Plant conservationists concerned with rare and endangered species are faced with a daunting task. A 1988 report from the Center for Plant Conservation, based on a survey of 130 botanical experts, identified 253 plants in the United States that may go extinct within five years and an additional 427 that may go extinct in ten years—a total of 680 species that may go extinct before the year 2000. Three-quarters of these species occur in only five states or territories: California, Florida, Hawaii, Texas, and Puerto Rico. The numbers are even more startling when considered on a global scale. Although fewer than 1000 of the nearly 250,000 species of vascular plants have gone extinct in the past 100 years, nearly 60,000, mostly in the tropics, are thought to be at risk in the next 50 years (Raven 1987).

Because many species have already been reduced to one or two populations with few individuals (Davis et al. 1986), it is essential to determine what kinds of conservation programs are appropriate, what biological information is required to implement those programs, and what criteria should be used to determine the priority ranking of species to be conserved. In this chapter, we review the biology and genetics of rare plant species, drawing heavily from the other contributions to this book, and we propose several practical guidelines to conserve the very large number of plants already at risk in the United States. To provide an example of how plants are currently selected for conservation and how much biological information is used, we have also included a review of the large, well-organized plant conservation program in California.

RARE PLANT CONSERVATION IN CALIFORNIA

The flora of California is the largest and most diverse in the continental United States. Over 5000 species are native to the state, and nearly a third of them are endemic (Raven and Axelrod 1978). As a result of close cooperation among the Nature Conservancy, the California Native Plant Society, and the Endangered Plant program of the California Department of Fish and Game, the California program is one of the most seasoned programs for plant conservation in the United States. As of December

1989, 209 native California plants had been designated as rare, threatened, or endangered by the Fish and Game Commission (a five-member committee appointed by the governor that establishes policy for the Department of Fish and Game), and it is thought that many hundreds of additional taxa also meet the criteria for listing. With the state's many rare plants and its concerned and articulate citizenry, what happens in California is likely to influence plant conservation programs at both the national and the international level.

The original law to protect California's native plants was contained in the 1977 Native Plant Protection Act of the California legislature. It directed the Department of Fish and Game to establish criteria for determining whether a native plant is endangered or rare and authorized the Fish and Game Commission to list native plants in these categories when appropriate. In addition, the act required that permits be obtained for collecting, transporting, or selling such plants, and authorized programs to conserve listed species. A plant was defined as *rare* when "although not presently threatened with extinction, it is in such small numbers throughout its range that it may become endangered if its present environment worsens." A plant was defined as *endangered* when "its prospects of survival and reproduction are in immediate jeopardy from one or more causes."

In 1984, important amendments to the California Endangered Species Act were signed into law. These amendments aligned California's treatment of endangered species with the U.S. Endangered Species Act; provided a formal procedure to petition the Fish and Game Commission to add, delete, or change the status of a listed species; and established a consultation process to deal with potential threats to species. The amendments recognized *threatened* species as a distinct category, defining them as "not presently endangered with extinction, but likely to become endangered in the absence of special protection and management." They also said that it was the "policy of the State to conserve, protect, restore, and enhance any endangered species or any threatened species and its habitat, and that it is the intent of the Legislature, consistent with conserving the species, to acquire lands for habitat of these species."

The basic information in petitions to add, delete, or change the status of a listed plant species concerns the extent and nature of the threat to a species; its abundance, distribution, and habitat; suggested management practices; and its taxonomic disposition. Most of the petitions have been submitted by members of the California Native Plant Society and use information drawn from the Department of Fish and Game's Natural Diversity Data Base, a computerized list of locality records on California's rare plants and animals. Upon acceptance for review, the Department of Fish and Game has one year to evaluate the petition and augment it as appropriate based on the best available information, including consultation with experts and a field evaluation of known sites. A recommendation is then forwarded to the Fish and Game Commission. The department evaluates 10–20 petitions for listing per year, and the commission makes the final decision to approve or deny each petition.

Of the 209 plant taxa listed by the state, 123 are designed endangered, 18 threatened, and 68 rare. About 40% occur only in five or fewer highly restricted populations, and about 60% in fewer than ten populations (Smith and Berg 1988). Of the 101 genera represented, only nine (*Arctostaphylos, Astragalus, Clarkia, Cordylanthus, Dudleya, Eriogonum, Limnanthes, Orcuttia,* and *Sidalcea*) account for 58 (28%) of the taxa. Of the 209 listed plants 55 (26% of the total) are subspecies or varieties, suggesting that they have very close and genetically similar relatives that are not threatened. Of the plants listed, 26% are annuals, compared with 29% of the flora as a whole, but only two tree species (*Cupressus abramsiana* and *Cercocarpus traskiae*) have made the list.

Review of a number of recent petitions and department reports revealed that the primary information in them is the extent and nature of the threat to the species proposed for listing. Thus the reports describe the location and number of local populations, estimate the number of individuals, state whether historically documented populations have been lost, provide habitat descriptions, and list specific causes of population endangerment, which most often are residential-housing development, road building, mining, agricultural activities, off-road-vehicle disturbance, or cattle trampling. The reports also include a brief taxonomic description and occasional references to closely related and morphologically similar species.

The reports present little or no information about the biological attributes of the species. For example, they do not describe population structure, breeding system, genetic variability, ploidy or chromosome number, unusual or interesting morphological or physiological attributes, taxonomic distinctiveness, or possible relationships to agronomically or medicinally important plants. In nearly all cases, the important biological information is not available because thorough studies of proposed plants have not been done, and the Endangered Plant Program does not have the necessary resources, either financial or scientific, to carry out such studies, although it does some research as its funding allows. Consequently, the decision to list a plant taxon as rare, threatened, or endangered appears to be made almost entirely on the basis of the likelihood of its extinction.

It is apparent that the taxa petitioned and reported to the commission are not compared with one another to determine their relative merits for conservation. Indeed, the legal mandate to assign rare, threatened, or endangered status does not require such considerations. But given that nearly 700 plant taxa are said to be at risk of extinction in the United States in the next decade, we believe it is essential to develop biologically based criteria to identify the ones whose loss would be felt most keenly, so that their survival can be guaranteed. Even if the estimate is two or three times too high, the number of plants at risk is still extraordinary. The significance may be best appreciated by noting that the number of plant species in this risk category is about the same as the total number of bird species that breed in North America north of Mexico. Yet there seems to be little discussion of how to identify the most important plants or what criteria should be used to develop priorities to choose among candidates for listing.

The idea that plants proposed for conservation must be ranked in some way and that their rank must reflect biological considerations in addition to the degree of population endangerment was brought home to us by actions of the California Fish and Game Commission at its meeting on October 5, 1989. At this meeting, the commission denied protection to *Oryctes nevadensis,* an annual herb in the Solanaceae that is the only member of its genus. It is so distinct from other members of the family that its generic status has never been seriously questioned (Averett 1979). This plant is known from only six sites in the Owens Valley and a few sites in neighboring Nevada. At the last census in 1986, only 603 individuals could be found. It is severely threatened by cattle trampling because it is actively growing in the spring, when the surrounding vegetation is used for forage, and additional adverse impacts may result from clearing and grading operations associated with the proposed construction of power lines at two of its known sites. Nevertheless, the commission denied it endangered status in California.

The influential *Sacramento Bee,* in a front-page story (December 31, 1989), quoted one of the commissioners who voted against listing *Oryctes nevadensis* as saying, "Here's a very weedy looking plant. And what I'm really wondering is, why are we worrying about it?" Apparently his question went unanswered, but we believe that there are many reasonable and appropriate responses. He could have been told that *Oryctes nevadensis* is related to plants that provide food, stimulant, and horticultural materials (potato, pepper, tomato, eggplant, tobacco, and *Petunia*). As a wild relative of these plants, albeit a distant one, it has the potential to donate useful genes via molecular engineering that might extend the types of environments in which its already valued relatives could be cultivated.

The proposed listing of *Oryctes nevadensis* was brought up again before the commission at its meeting on January 8, 1990. The written record of the meeting shows that testimony regarding the herb's importance and its membership in the Solanaceae was presented, but the motion to list it died for lack of a second. It is likely that political considerations were a significant factor in this case because the few populations are found on land owned by the Los Angeles Department of Water and Power, which opposed the listing. Since species conservation will find frequent opposition when it appears to limit or increase the cost of development or other land use, as in this case, the arguments for conservation must be buttressed with all the biological facts we can muster. More important, very strong cases, like that of *Oryctes nevadensis,* will be harder to reject if we do not treat all candidate species as equals.

SETTING PRIORITIES FOR PLANT CONSERVATION

The most important biological reason to preserve *Oryctes nevadensis* is that the 603 plants in Owens Valley are the sole survivors of an evolutionary lineage that is morphologically highly distinctive. Indeed, this plant has been accorded the taxonomic rank of genus. It seems to us that taxonomic distinctiveness is one of the most significant factors to consider in evaluating the relative importance of a particular taxon for conservation. Monotypic genera or families deserve the highest priority, and endangered species are of more direct concern than endangered intraspecific taxa. Taxonomic distinctiveness is a particularly useful criterion because it is the simplest and most accessible estimator of the extent of evolutionary divergence, itself an index of genetic uniqueness.

Of course, genus and species concepts vary widely from one group of plants to another and may not always be sufficient to include or exclude a particular proposed listing. But within any particular group, ranking criteria are usually consistent, so that comparison of taxonomic rank will quickly reveal whether one proposed plant is more or less distinctive than another. Within the Chenopodiaceae, for example, *Aphanisma blitoides* is the only member of its genus and is a candidate for federal endangered species status. *Atriplex vallicola*, in the same family, is a candidate for the same status, but it is a member of a large and speciose genus with about 100 species worldwide. On this basis, we would suggest that a higher priority be given to efforts to conserve *Aphanisma blitoides* than to efforts to conserve *Atriplex vallicola*.

Nevertheless, we would also ask whether *Atriplex vallicola* has some particular attribute that counterbalances the lesser value we accorded it on taxonomic grounds. For example, it might rank highly on the basis that it is a close relative of an agricultural or horticultural species, that it has an unusual and interesting suite of morphological characteristics, that it is noteworthy for its attractiveness, that it is important in the ecology of a particular community (e.g., as food for foraging animals, as a pollen and nectar source for oligolectic pollinators, or as a colonizer of unusual substrates), or that it is characteristic of an ecologically important habitat.

Thus we suggest that the case for conservation of a particular plant taxon is best made by considering its biology and relative value from a number of standpoints. The point is that we accept the notion that some plants are more significant than others and that criteria should be developed to identify them. In California, these other issues are rarely taken into account, and, in view of the fact that 26% of the 209 listed plant taxa are varieties or subspecies, we worry that attention to minor categories has shifted concern away from other more critical plants. The denial of endangered species status to *Oryctes nevadensis* and the granting of that status to numerous varieties and subspecies in California suggest that the conservation community cannot continue to treat plants proposed for listing as equals. Some species are more critical than others, and we believe that the public will respond favorably when proposals to conserve them are enriched with information about their biology.

Taxonomic distinctiveness is a valuable criterion because we are interested in preserving as many different kinds of organisms as possible, but we are not concerned about asexual sports or individual hybrid derivatives that have never constituted self-reproducing populations. Such entities are only trivially different from their close relatives and should receive little, if any, conservation attention. For example, *Betula murryana*, which has recently attracted substantial attention from plant conservationists, appears to be a fertile octoploid that combines the genomes of the hexaploid *B. allegheniensis* and the tetraploid hybrid *B. × purpusii* (Barnes and Dancik 1988). Only two trees of *B. murryana* are known to exist, and because they grow very near their presumed parents, *B. murryana* almost certainly is nothing more than those two trees that arose where they now grow. It may be worthwhile to save cuttings for horticultural purposes, but we do not think that these two trees are as valuable as species whose populations have been severely reduced by destruction of their habitats. Similarly, *Prunus maritima* var. *gravesii* has attracted a lot of notice, but it appears to be a mutant of the common beach plum and never existed as a natural population. It is known from only a single individual growing on the Connecticut shore (Anderson 1980). *Helianthus paradoxus*, on the other hand, is a stabilized hybrid derivative of *H. petiolaris* and *H. annuus*. It differs substantially from both its parents in flowering time, secondary compound composition, leaf and phyllary shape, and the habitat in which it is found (Rieseberg, Chapter 11), and it is worthy of conservation because it now constitutes a distinct species even though it arose as a result of a hybridization event.

BIOLOGY AND GENETICS OF RARE AND ENDANGERED PLANTS

All plants that are candidates for a conservation program are rare. Rare plants may be locally common but occur in only a few places. They may be scarce where they occur, but geographically widespread. They may be both scarce and geographically restricted (Rabinowitz 1981; Kruckeberg and Rabinowitz 1985). Examples of each type are readily found in the lists of candidate species for threatened or endangered status in the United States. For example, some rare plants are locally common but geographically restricted because their habitat is geographically restricted (e.g., the California serpentine endemics *Layia discoidea* and *Streptanthus niger*). *Plantago cordata,* on the other hand, occurs in a few widely scattered populations along the margins of woodland streams from New York south to northern Florida and west to central Missouri. Its restriction appears to reflect the destruction of most of its previous habitats. Similarly, *Trifolium stoloniferum* is apparently threatened because it is adapted to the disturbed soils characteristic of buffalo wallows, a habitat that has not been available in the eastern United States since the time of European settlement (Bartgis 1985; Campbell et al. 1988). It is now known from only widely scattered localities in West Virginia, Indiana, Kentucky, and Ohio. *Zizania texana,* a close relative of commercial wild rice that was once a troublesome weed in irrigation ditches near San Marcos, Texas, has been reduced to a single population.

In many respects, the biology of rare plants that are locally common is similar to that of widespread congeners. The primary difference is that they are restricted to a particular geographical area or habitat type. For example, Fiedler (1987) found that rare species of *Calochortus* resembled widespread congeners in ecologically important characters, such as mean numbers of seed set per capsule, mean seed weight, germination behavior, and dormancy mechanisms. Their rarity reflects specific adaptation to habitats that are rare. Similarly, the genetic structure of populations of narrowly distributed but locally common species is often similar to that of more widespread relatives. For example, the narrowly endemic *Layia discoidea* and its widespread progenitor *L. glandulosa* are polymorphic at the same 19 of 21 electrophoretic loci examined (Gottlieb et al. 1985).

Karron's survey (Chapter 6) suggests, however, that many rare locally common species are less genetically diverse than their widespread congeners, even if they are locally common. This presumably reflects the relatively small populations (fewer than 10,000 individuals) characteristic of these species. As Barrett and Kohn (Chapter 1) point out, small population size increases the likelihood of population bottlenecks. As a result, rare plants can be expected to have less genetic diversity than common ones. How should this affect the design of plant conservation programs?

Most geneticists agree that genetic diversity is necessary to preserve the long-term evolutionary potential of a species. The question for conservation biologists is whether lowered genetic diversity also has a direct impact on its short-term viability. To put it another way: to what extent might reduced genetic diversity threaten the survival of a rare plant species in the absence of major environmental or habitat disturbances? Huenneke (Chapter 2) and Menges (Chapter 3) suggest that genetic diversity is a critical feature because it contributes directly to the likelihood of persistence and ecological success. However, information about the genetic structure of rare plant populations is based almost entirely on data from electrophoretic surveys of soluble enzymes (Hamrick et al., Chapter 5; Karron, Chapter 6), but these surveys are not expected to reflect the pattern of genetic diversity at loci controlling ecologically significant traits, and it is these traits that are important for determining the success of conservation efforts (Price et al. 1984; Schaal et al., Chapter 8), at least in the short term. In fact, many small populations do maintain genetic differences in ecologically significant characters. For example, Meagher et al. (1978) detected genetic variation for both number of inflorescences and number of rosettes in *Plantago cordata,* even though the populations they studied contained fewer than 1000 individuals.

Populations may also retain polymorphism for genes involved in defense against pathogens or herbivores because of strong frequency-dependent selection, similar to that for self-incompatibility alleles. An individual with a rare self-compatibility allele has a significant reproductive advantage over common ones because it can participate in more matings. Similarly, a rare resistant genotype is likely to have a significant reproductive advantage over common nonresistant ones because it is more likely to survive. This frequency-dependent advantage can lead to the maintenance of a large number of alleles even in a small population because the rarer any allele becomes, the stronger the selective force opposing its loss from the population (Wright 1939, 1960, 1965). For example, the narrowly endemic *Oenothera organensis* is a native of the Organ Mountains in southeastern New Mexico and may comprise fewer than 500 individuals. Although an electrophoretic survey detected polymorphism at only one of 15 loci examined (Levin et al. 1979), Emerson (1939, 1940) identified more than 45 self-incompatibility alleles in this species. Many rare plant species probably maintain considerable amounts of ecologically significant genetic variation. In any event, reduced genetic diversity is unlikely to be a significant threat to their persistence, especially in the relatively short time relevant for conservation—several hundred, not several million, years.

Rarity still has genetic consequences that are worthy of attention. In small populations, genetic drift allows allele frequencies to depart significantly from their *deterministic optimum* (i.e., the point at which population mean fitness is maximized), resulting in lower average individual viability. The magnitude of this effect can be calculated by using standard results from the theory of genetic drift (Crow and Kimura 1970). Suppose $\Phi(q)$ is the stationary distribution of allele frequencies at a single locus, where q is the frequency of a recessive deleterious allele. $\Phi(q)$ is a function of the effective population size (N_e), the mutation rate (μ), and the selection coefficient against recessive homozygotes (s). The probability that a randomly chosen individual in this population will survive is given by the integral of $(1 - sq^2)\Phi(q)$ across all loci.

If we assume independent gene action and independent assortment, this integral can be evaluated numerically. The results are shown in Figure 13.1, for the case when 1000 loci contribute to the effect. (The number of loci affects only the height of the surface, not its shape under these assumptions.) An individual chosen randomly from a population with an effective size of 100 is less than 75% as likely to survive as one chosen from a population with an effective size of 10,000. On the other hand, for an effective population size of 320, the relative survival probability is about 90%, unless selection against the homozygous recessives is extraordinarily weak. Although these

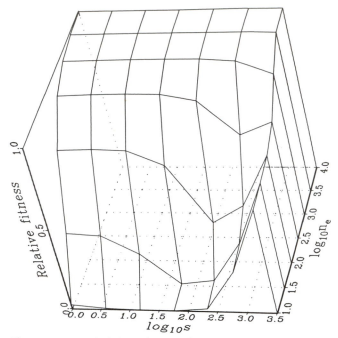

Figure 13.1. The average probability of survival in a finite population relative to that in an infinite population, assuming a mutation rate of 10^{-6}. (s is the selection coefficient against recessive homozygotes; N_e is the effective population size.)

calculations depend on the assumption of completely recessive deleterious alleles, the results are not significantly altered if the deleterious trait is also expressed in the heterozygote or if there is heterozygote advantage.

This result has important implications for conservation because rare and endangered species often have severely reduced distributions and limited numbers, and these changes have come about abruptly. Habitat fragmentation and population decline, in particular, reduce the effective population size, and if the reduction is substantial, the decline in average individual viability could threaten the viability of the entire population. If populations on the order of 10^3–10^6 are required to maintain population viability, as suggested by Menges (Chapter 3), the genetic influences on population viability are likely to be fairly minor. Unfortunately, the effective population size of many rare plants is far smaller than this. One of our primary conservation goals should be to establish self-sustaining populations whenever possible, but doing so may require that ecological manipulations be supplemented with genetic ones—for example, equalization of reproductive success and minimization of inbreeding.

Genetic factors may be less significant for a species whose distribution and abundance has not been historically altered. These species, by virtue of their continued persistence, have already demonstrated an ability to cope with the lower average viability expected in small populations (Huenneke et al. 1986). In the *Calochortus* species already referred to, for example, the widespread *C. albus* has a greater mean seed output than its rare congeners, yet demographic studies of the latter suggest that their populations are stable or increasing (Fiedler 1987). Similarly, Pavlik and Barbour (1988) concluded that populations of *Swallenia alexandrae, Oenothera avita* ssp. *eurekensis,* and *Astragalus lentiginosus,* all endemic to Eureka Dunes in southeastern California (Inyo County), are more endangered by factors extrinsic to the plant populations (e.g., human disturbance, pollinator availability, climatic fluctuations) than by low individual viability. In short, for a species whose distribution and abundance have not been historically altered, demographic considerations rather than genetic considerations will determine short-term population viability.

The idea that demographic considerations may dominate genetic ones in determining population viability is relatively new. Lande (1988a) argued that for wild populations of animals demographic factors will usually be of more importance in determining population viability than will genetic ones. For example, initial plans for the management of old-growth forests in the northwestern United States based viability estimates for populations of the northern spotted owl entirely on genetic considerations. But an analysis of stochastic demography and habitat occupancy (Lande 1988b) suggests that such a narrowly based plan is likely to fail because suitable habitat in the region will be too sparsely distributed to support an ecologically viable population.

Assuming that a population's short-term viability has been ensured, its long-term viability will probably depend, in part, on the amount of genetic variability it retains. But the absolute level of genetic variability is not worrisome for conservation biology. Some suggest that all genetic variation within a species be captured (e.g., Hawkes 1976, 1987a), but as Brown and Briggs (Chapter 7) point out, this is unrealistic. It is also unnecessary. First, many low-frequency alleles are unconditionally deleterious and are maintained only as a result of recurrent mutation. Indeed, analyses of the pattern of nucleotide and amino acid substitution in a variety of genes suggest that most amino acid substitutions are selected against at least in average environments (e.g., Kreitman 1983; Milkman and Crawford 1983; Sawyer et al. 1987; see Nei [1987] for a review). Many low-frequency alleles might actually contribute to genotypes that lower the average viability of individuals.

Second, most adaptively significant variation is contained in alleles found in a moderate to high frequency. Population bottlenecks may lead to the loss of many low-frequency alleles, but the genetic variance that can respond to selection is not so greatly affected. Recent theoretical work by Goodnight (1987, 1988) suggests that the *additive genetic variance*—that portion of the genotypic differences among individuals that can respond to selection—can actually increase as a result of population bottlenecks if there are significant interactions among loci in determining phenotypic characteristics.

Third, low-frequency alleles are likely to be lost in just a few generations. In a population with an effective size of 100, for example, a neutral allele with a frequency of 1% will be lost on average in fewer than 25 generations; one with a frequency of 0.1% will be lost in fewer than five. Thus adaptation to new or different environmental conditions that arise many generations in the future is more likely to result from future genetic variants than from present-day low-frequency ones. In short, maintenance of long-term population viability requires that we attempt to preserve a representative sample of moderate- to high-frequency alleles, whether the population is managed in its natural habitat or samples are collected for off-site preservation (Templeton, Chapter 12).

GUIDELINES FOR PLANT CONSERVATION PROGRAMS

Programs to conserve rare and endangered plant species should take into account goals and management practices that are appropriate for the individual species. In nearly all cases, we will want to preserve existing natural populations, and for many species it will be appropriate to collect seeds or other propagules for off-site conservation to provide insurance against catastrophic events. Occasionally, we may want to establish new populations at a distant site. Some programs may seek to display representative rare and endangered plants in botanical gardens and parks. In fact, management programs that combine all these approaches are the most likely to be successful (Falk, Chapter 14). Because hundreds of plants in the United States are faced with extinction in the next decade or two, we think it is vital to rank them in order of their biological significance by criteria like those described earlier in order to conserve the most critical and interesting plants first. We also believe that it is necessary to have clearly stated goals for each plant taken into conservation. We anticipate different goals for different species, depending on its life form (tree, shrub, herb), its attractiveness, the nature of its native habitat, and its degree of scientific interest.

The average rare and endangered plant species comprises fewer than five populations and perhaps no more than 1000–5000 individuals. The number of populations has in many cases been reduced by loss of habitats, and elimination of the plant is likely if nothing is done to preserve it. We might expect the reasons why a species is rare to have an effect on the management tools selected, but the choice of tools depends primarily on the nature of the species, the degree of population endangerment, and the conservation goals for that species. Preserving plants in their native habitat is clearly a first step, but off-site preservation is also necessary to provide insurance against catastrophic events and to facilitate the possibility of reintroduction in the future when appropriate habitat becomes available.

Mason (1946a, 1946b) and Stebbins (1980) suggested that narrow endemism is usually the result of adaptation to specialized environments that are themselves geographically restricted. Fiedler's (1987) analysis of rarity in *Calochortus,* which we discussed earlier, is consistent with this view. In such cases, the present-day distribution may encompass the entire geographical range that the species has ever occupied. Provided that existing populations of these species are maintaining themselves, they are in little need of direct management intervention because they already cope with the demographic and genetic consequences of local rarity. *Layia discoidea,* for example, has probably always been restricted to serpentine outcrops in the vicinity of New Idria in central California, and its populations seem relatively stable. Monitoring population size and collecting seed for long-term storage are the only conservation activities usually needed for such species, provided that the habitat is protected.

Of course, not all narrow endemics have populations that are self-sustaining. *Amsinckia grandiflora,* for example, is known from a single, small population. It is heterostylous, and the population fluctuates in size dramatically. As a result, the frequency of the two style-morphs also varies considerably. Because only pollinations between different style-morphs are successful, seed set in the population is lowered when one of the morphs is infrequent. Thus the reproductive success of the species varies from year to year, to the extent that direct intervention in the form of hand pollinations between cross-compatible forms, outplanting of greenhouse-grown seed-lings, and exclusion of significant competitors may now be required to prevent its extinction (Huenneke, Chapter 2). Similarly, *Clarkia tembloriensis* ssp. *calientensis* is known from only three small populations southeast of Bakersfield, California (Vasek 1964; Holsinger 1985). Each of the populations consists of only a few hundred indi-viduals, and exotic grasses are threatening them. Outplanting of seedlings and exclu-sion of competitors will ensure the persistence of these populations. In both of these cases, seed placed under long-term storage might make it possible to establish new populations at some time in the future.

Reciprocal transplant experiments of both perennials and annuals have consistently demonstrated that individual survivorship and fecundity are greatest when individuals are replanted in the same microhabitat from which they were collected (e.g., Chapin and Chapin 1981; McGraw and Antonovics 1983; Schmidt and Levin 1985; Silander 1985b). This suggests that microhabitat adaptation will play an important role in de-termining the ecological success of a reintroduction effort. Thus Huenneke (Chapter 2) suggests that reintroductions and reestablishment are best done with seed taken from populations and microhabitats that are ecologically similar to the site where reintro-duction is to occur. Barrett and Kohn (Chapter 1), on the other hand, argue that chances for success are greatest if seed is drawn from a composite cross among the available populations in the off-site collection. They presume that natural selection will weed out unsuccessful genotypes from among the segregating progeny of such hybrid populations. Although we know little about how effective either of these ap-proaches is, it seems likely that the approach Barrett and Kohn advocate will be most applicable when plants are to be established in novel habitats.

Some rare species whose distributions have been significantly altered in historical times are almost certain to require direct intervention to ensure their continued exis-tence. For example, *Plantago cordata* appears to be declining because of the disap-pearance of the stable environments once characteristic of streams in the climax forests of eastern North America (Meagher et al. 1978). Establishment of new populations in undisturbed stream habitats and modification of existing habitats to mitigate the effects of unstable, man-made environments are probably necessary to prevent its extinction. Similarly, it may be possible to save *Trifolium stoloniferum* only by establishing new populations in more open habitats and imposing artificial disturbance regimes that mimic buffalo wallows.

Conservation plans for the maintenance of existing natural populations of an en-dangered species or for the establishment of new populations will depend on detailed study of its ecology. Population viability analyses can pinpoint life-history stages that are particularly critical for persistence, but the results of these analyses will differ from species to species. Because of the large population sizes necessary to buffer popula-tions against demographic stochasticity (Menges, Chapter 3), however, populations large enough to mitigate ecological threats to population viability will mitigate genetic ones as well. When existing populations consist of only a few hundred individuals, however, genetic threats to population viability become important.

Unfortunately, there exist few general guidelines for mitigating genetic threats to population viability. In small populations, the contributions of each adult to the follow-ing generation should be equalized by instituting a controlled crossing program. This will double the effective population size. Similarly, the population size should be increased as rapidly as possible. This will minimize the decrease in average individual viability associated with small population size (Templeton and Read 1984; Templeton, Chapter 12). Beyond these simple suggestions, however, too much depends on the particulars of population structure, breeding system, and population size to make general guidelines useful in the genetic management of breeding populations (Templeton, Chapter 12). Clearly, active management of the genetic structure of an endangered plant species will require an enormous investment of time, money, and expertise. Only a few of the most important species will warrant such heroic efforts.

Nevertheless, conservation of rare and endangered plants may be easier to accomplish than conservation of endangered animal species. Rare and endangered plants are generally more restricted in distribution than are rare and endangered animals, and the ease with which survivorship and establishment in many plant populations can be manipulated permits simpler conservation management than in animals. In addition, seeds of temperate-zone plants are generally amenable to long-term storage. In fact, cryogenic storage may permit such seeds to be stored indefinitely (Eberhart et al., Chapter 9). Seed storage eliminates the loss of genetic diversity that results from reproduction in small off-site populations. Unless the seed storage environment itself is mutagenic or there are genotypic differences in the ability to tolerate long-term storage, neither of which is likely, seed banking provides a way to preserve nearly all of the genetic diversity that was present in the original sample. In short, the problem of genetic conservation for temperate-zone plants is largely reduced to one of obtaining a representative sample of genetic diversity in the off-site collection.

Hamrick et al. (Chapter 5) have shown that life-history characteristics, breeding system, and mode of reproduction explain, at best, only 50% of the variance in genetic diversity. Thus if the design of a sampling program depends on some estimate of the amount and distribution of genetic variation within a species, a preliminary genetic survey will be required. For example, the selection of genetic management units for the forest trees discussed by Millar and Libby (Chapter 10) depends critically on a detailed knowledge of the distribution of genetic variation on a broad geographical scale. As they point out, preliminary surveys of electrophoretic variation, ecotypic variation, and life-history variation will be essential for the design of genetic management units. Such surveys obviously require considerable investments of time, money, and expertise. If similar detailed genetic surveys were required for all the 680 species on the Center for Plant Conservation's list, the chances of saving a substantial proportion of them would be poor. In the 20-year period between 1968 and 1988, fewer than 500 plant species of any kind were subject to the same sort of survey (Hamrick and Godt 1989).

Fortunately, the strategy for obtaining such a representative sample from rare and endangered plants is dominated by one simple fact: rare and endangered plants usually occur in very few populations. Of the 47 endangered species in southeastern Australia, for example, 22 are found in a single population, 16 are found in five or fewer, and none is in more than ten (Brown and Briggs, Chapter 7). Similarly, almost all the 1000 plant species in the Center for Plant Conservation's highest priority ranking for conservation in the United States occur in five or fewer populations, and even for those less threatened species of long-term concern, about 3100 in all, the vast majority are found in fewer than 20 populations. Thus only a few populations need to be sampled to obtain the genetic diversity in such species. An important implication is that genetic surveys will not usually be required for initial conservation efforts because only a few populations are extant. Investigations for management or scientific purposes may require a genetic analysis, but a representative sample of genetic diversity from a few populations can be obtained by simply collecting an appropriate number of plants (Holsinger and Gottlieb 1989).

The capture of genetic diversity within populations is approximately a logarithmic function of the sample size, regardless of the mating system of the plant (Brown and Briggs, Chapter 7). In other words, there is a strong law of diminishing returns associated with increased sampling effort. Increasing the sample size by an order of magnitude will net, on average, only a single additional allele. Thus sample sizes from within a population need not exceed 50–100 maternal parents because this sample size provides a 95% chance of capturing all alleles present in a frequency of greater than 5% (Marshall and Brown 1975). A stratified random-sampling design in which samples are taken from distinctive microhabitats will maximize the chances of sampling unique genotypes. If seeds are being collected, it would be best to collect approximately equal numbers from each maternal parent because the effective population size of the collection will be almost twice the number of plants sampled (Crow and Kimura 1970). The seed collection may be bulked to simplify storage and handling. One disadvantage of seed collections is that little, if any, control on variation in paternal contributions is possible. These contributions can be highly skewed and may vary significantly from year to year (Schoen and Stewart 1987). If only small amounts of seed can be collected in any one year, it may result in an unrepresentative sample of genetic diversity. Thus seed collections from small populations are best made in several different seasons.

If it is desirable to carry out a breeding program in order to maintain genetic diversity, the design of an appropriate crossing scheme is complex. For example, Templeton (Chapter 12) was able to enlarge the herd size of Speke's gazelle from a single male and three females. Detailed pedigree records were used to design a breeding program that simultaneously maximized the number of these ancestors represented within the pedigree of a single individual, equalized their contribution to the population as a whole, and produced moderately high inbreeding coefficients (as opposed to the minimum possible). Such a breeding program requires the involvement of professional geneticists and the investment of substantial resources. Thus it can be implemented for only a few of the highest priority species.

Surprisingly little is known about the overall biology of any individual plant species growing in the wild because scientists often choose plants to study in order to answer particular questions, not to find out all about them as organisms. Geneticists might explore genetic diversity and population structure and seek evidence for natural selection. Ecologists might examine ecotypic variability among populations or how plants interact with different soils or light levels. To make accurate predictions about viability, conservation biologists will have to understand the genetics, ecology, and evolution of each species that they want to conserve. Information on the population structure, breeding system, genetic variability, ploidy or chromosome number, habitat requirements, competitive interactions, life history, and demography will all be necessary.

Clearly, much additional work on the biology and genetics of rare plant species remains to be done. We need to know what fraction of rare plants are rare because their habitats are rare and what fraction are endangered because their habitats have been destroyed. We need detailed analyses of population viability for different kinds of rare plants to determine the extent to which conservation practices appropriate for one species are also appropriate for others. We need studies that examine the usefulness of electrophoretic variation and restriction-fragment-length polymorphisms as markers of genetic changes at ecologically significant loci in small populations. We need to know how differences in mating system or mode of reproduction should affect conservation strategies. We need to determine whether reintroduction and reestablishment efforts are more likely to be successful if seed is taken from populations in a similar habitat than from a composite cross. We need to develop techniques that will allow successful long-term storage of seed from plants that are now regarded as recalcitrant. But we cannot delay conservation efforts until such information is available, because if we did, nearly all the plants that the Center for Plant Conservation identified as being at risk would go extinct first.

Conservation biology is a crisis-oriented science, and its practitioners have to be willing to make a guess about the long-term behavior of a complex system long before its dynamics are completely understood (Soulé and Wilcox 1980; Soulé 1985). Because of the potential for long-term storage of seeds in temperate-zone plants, however, off-site conservation programs can mostly avoid the complexities of designing breeding programs. Similarly, because most rare and endangered species exist in only one or a few populations, obtaining a representative sample of their genetic diversity for off-site collections does not depend on knowledge of population structure. Thus in contrast to the enormous cost and sophistication necessary to deal with the conservation of large vertebrate species, success with temperate-zone plants will be simple to obtain—if we act quickly enough.

ACKNOWLEDGMENTS

We are grateful to Ann Howald and Ken Berg of the Endangered Plant Program of the California Department of Fish and Game for making available to us sample petitions and reports for proposed listings of rare, threatened, and endangered plants. Our conclusions and recommendations, however, reflect our own points of view.

14

Joining Biological and Economic Models for Conserving Plant Genetic Diversity

DONALD A. FALK

Intraspecific genetic diversity is a fundamental element in the modern concept of species. Although not necessarily prominent in the public understanding, the scientific view of species is not as a homogeneous set of organisms, but as a group of evolutionarily related individuals incorporating and expressing wide variety in genetically determined traits. This view of species as biological entities not only tolerates genetic variation, but depends on it; we recognize the importance of genetic variation in reproduction, in ecology, and in evolution and speciation itself (Lewontin 1974). Genetic variation is increasingly recognized as a basic component of population and community structure as well. We would no more consider a single specimen to capture the full functional, ecological, and evolutionary range of a species than we would consider one individual to represent human society.

Conservation of genetic variation as it exists naturally is thus of concern not only to theoretical population biologists. For many species—although by no means all—decreased variability may reduce the viability of offspring, expose populations to stochastic genetic processes (such as drift), and inhibit prospects for evolutionary adaptation and change. Hence, the consequence of failure to preserve the diversity *within* a species may be to reduce its ability to reproduce, function ecologically, and evolve. The preservation of existing diversity within species should thus be considered an important element of successful long-term biological conservation.

Successful and efficient conservation of genetic variability requires two conditions at a minimum: (1) the ability to recognize variation where it exists, and to define biologically adequate levels thereof; and (2) the organization of resources to protect or manage variation at these levels. The former issue lies within the domain of population biology; the latter is essentially a question of resource allocation.

The tools of contemporary population biology can reveal the degree of variation within a population, among populations, and within a species (Hamrick et al., Chapter 5; Karron, Chapter 6; Schaal et al., Chapter 8; Rieseberg, Chapter 11). As with any task of conservation, the biological characteristics of the entity or system to be conserved must always provide the starting point for analysis. Conservationists are frequently in a position of having to act without complete information, but unless we have a clear understanding of what it is we seek to conserve, we risk developing strategies driven by factors external to the objects or systems of concern.

In fact, much of the recent history of conservation consists of the application of increasingly complex and sophisticated models of natural systems to the task of managing and protecting diversity (Soulé 1986; Falk 1990b, 1990c). This applies equally to conservation of individuals, populations, species, communities, or ecosystems—that is, at any level of biological organization. For instance, whether one is determining the number of individuals in a viable population or the acreage required for ongoing community function, the determination of adequacy should always be first derived from biological understanding, and only then subjected to the budgeteer's axe and the biocrat's regulation. Short of preserving "everything, everywhere," the definition of adequacy in conservation must satisfy fundamental biological criteria of function and sustainability a priori.

Conservation, however, has a cost. In some cases, the cost is slight, as in the almost immeasurably small expense a weekend collector incurs to bring home a sample of an odd genetic variant and plant it in a garden. In other cases, costs can run to hundreds of millions of dollars to acquire, protect, and manage whole ecosystems containing thousands of species. To some extent, higher costs are associated with higher levels of biological organization, although the cost of conserving some individual species can be substantial.

The resources available for conservation programs are limited; there is more to do than can realistically be accomplished. The practice of conservation thus falls within the basic definition of an economic activity: "the allocation of scarce resources among competing ends" (Samuelson 1989). Pragmatic considerations demand that we lift ourselves—however unwillingly—from the comfortable confines of theory into the messy, unpredictable, complex *realpolitik* of practicing conservation. And yet, we must do so with full knowledge of the biological consequences of our actions. Otherwise, we risk acting blindly, ignorantly, even counterproductively. This dialectic between theory and practice is fundamental to the field of conservation biology, and to the purpose and structure of this book.

In the end, conservationists face a series of essentially pragmatic decisions, with costs and benefits often imperfectly known: How should a designer of a preserve determine the optimum size of a sanctuary? How many populations should a germplasm collector visit, and how many individuals should be maintained in cultivation? How do we determine the point at which our resources should be turned to a new species or a new site rather than to an extension of an existing project? To what extent (and in what way) should genetic variability and species diversity be incorporated into ecosystem restoration projects?

In seeking to translate the lessons of ecology and evolutionary biology into conservation strategy, we thus encounter a set of questions, some biological and some economic. This unholy union of population biology and microeconomics provides the intellectual framework for this chapter; the objective, however, is to move beyond theory as expeditiously as possible, and to provide guidance for the informed practice of endangered species conservation. To this end, the last section of this chapter is devoted to recommendations for protecting and managing rare plants within a framework of efficient allocation of conservation resources. A more detailed set of guidelines is also offered in the Appendix. It should be noted that these principles may apply equally well to the conservation and management of ecosystems and to sampling for off-site collections of individual species. In each case, we seek to satisfy both the criterion of biological sufficiency and that of efficiency in our use of scarce conservation resources.

THE ENDANGERMENT OF PLANT DIVERSITY IN THE UNITED STATES

Most conservationists recognize that there are more individual populations, sites, and even species than can practically be protected. In fact, from a certain perspective, the practice of conservation consists of deciding which parts of nature to protect and which to leave vulnerable (Lovejoy 1976; Thibodeau 1983). Moreover, in conserving plant diversity, we must be concerned with the inherent variability within species. Thus the number of taxa at risk is only part of the answer; we must also look beneath the surface of the taxon and understand the threats to its internal diversity. In the context of this volume, this may include understanding the erosion of genetic variability within species not yet endangered as a whole. Species diversity, however, provides a useful frame of reference, and has the additional advantage of being the level of organization most commonly referred to in conservation strategies, threat assessments, and legislation (Bramwell et al. 1987).

The Center for Plant Conservation database lists 6723 plant taxa (species, subspecies, and varieties) native to the United States that are of some conservation concern.* This represents approximately one-third of the total flora of the United States of roughly 20,000 taxa (Wagner et al. 1990; Flora of North America Editorial Committee 1992). Of these taxa, however, some are significantly more threatened than others. Approximately 1678 taxa (8.4% of the national total) are known from five or fewer populations, or fewer than 1000 individuals (Nature Conservancy 1990). In at least one of the national databases, 2290 taxa (11.4%) are found at the highest endangerment rank, including hundreds of species that may be within a decade of extinction (Center for Plant Conservation 1988a, 1988b), and 237 species that are listed or proposed as threatened or endangered by the U.S. Fish and Wildlife Service (1990).

The most severely threatened group of species may be those identified in a 1988 survey as facing virtually immediate extinction (Center for Plant Conservation, 1988a). According to this national survey of endangered species botanists, over 780 native U.S. taxa are facing possible extinction within a decade; additional taxa are added to this priority list regularly, as new threats are discovered and as the decline of populations continues. Most of the species are highly restricted geographically; over 600 (77.8%) of the critical U.S. plant species are endemic to a single state. Similar rates of endangerment are found in other temperate, subtropical, and Mediterranean regions of the world (Myers 1979; Davis et al. 1986; Lewin 1986b; Bramwell et al. 1987; see also Brown and Briggs, Chapter 7).

With a relatively small number of extant populations of so many species, we may safely conclude as a starting point that each population or site of an endangered species is of conservation significance. Unfortunately, there is generally little documentation of the rate of loss or decline of populations for most species. In fact, methodologies

---

*This list consists of the superset of taxa included on endangered species lists generated by the U.S. Fish and Wildlife Service (1990), the Nature Conservancy (1990), and the Center for Plant Conservation.

to recognize, quantify, and assess rates of decline are just emerging into the conservation literature. Sutter (1989) has developed a procedure to identify and evaluate species in decline. The method employs an algorithm based on a series of weighted factors for trend information on the number and size of extant populations, the number of populations extirpated, the degree of protection for remaining sites, and habitat quality. In time, methods such as this will allow species that are rare but stable in the wild to be distinguished from those that are more numerous at a given moment but experiencing rapid decline. The primary drawback of such methods is that they require trend information for populations and species over a period of years, data that at present are publicly available and reliable for only a handful of species (Cook and Lyons 1983; Lewin 1986a; Carr 1988; Mehrhoff 1989). Hamrick and Godt (1989 and Chapter 5) have analyzed the available data, and lists of rare plant research citations are provided in Falk (1990a) and in the Bibliography of this volume. Other than these relatively few published sources and the operational databases of organizations (such as the Biological Conservation Database of the Nature Conservancy), population trends are virtually unknown except for a handful of well-studied species. This, in turn, is a consequence of the general lack of long-term monitoring studies for all but a few species (Wildt and Seal 1988). The available data thus permit only broad generalizations regarding the population and genetic structure or trends over time of most rare species.

One of the consequences of this lack of knowledge regarding the genetic composition and stability of and the threats to populations is that the costs of acquiring this information must be factored into conservation budgets. For example, as will be seen below, it is necessary to have some picture of the genetic architecture of a species in order to create a suitable plan to conserve intraspecific diversity (Libby and Critchfield 1987). Short of relying on generalized predictions about the correlates of variability, such as those described by Hamrick et al., Chapter 5, such data are rarely available. It is difficult to conserve diversity efficiently within populations when that diversity has never been assessed; the same principle applies to deciding how many populations should be sampled or protected at the species level.

A FRAMEWORK FOR COST-EFFECTIVE CONSERVATION OF PLANT GENETIC DIVERSITY

The application of microeconomic analysis to the conservation of endangered species may seem a horrifying prospect to some, raising the specter of trying to assign biological—or worse, economic—value to individuals, populations, or species. Fortunately, the laws of microeconomics are more flexible than generally supposed. For instance, in benefit–cost analysis any desired end-product may be considered a "benefit," provided there is some measurable unit (Butlin 1981; Fisher 1981). Economics provides a neutral system of decision making—blind to whether we are saving *Ranunculus*, rain forests, or Rembrandts—to help determine whether resources are being used most effectively. Economics does not define values, goals, or adequacy; it simply provides a conceptual framework and methodology to enable those criteria to be met. In fact, where the needs are great and resources limited, as is demonstrably the case with biodiversity conservation, benefit–cost assessment should be viewed as an essential tool, not a luxury.

Because resource allocation is essentially comparative decision making among competing priorities, it is as important to evaluate what we are *not* doing as well as what we are. Thus in addition to determining whether resources are being used efficiently in a particular case, we must ask what projects have not been funded or undertaken because resources were consumed elsewhere (Marshall and Brown 1975). In economic terms, this means considering not only the direct cost of a project, but also the *opportunity cost* of foreclosed options. This relationship is described in economic terms by production–possibility functions, in which resources are differentially allocated among competing ends. Optimal resource allocation takes place along the "production–possibility frontier," which, in turn, demands that we have some idea of the relative benefits of alternative investments of time and money. Thus as Brown and Briggs (Chapter 7) observe, it is unsatisfactory simply to recommend reserves "as large as possible" or off-site collections of "as many genotypes as possible." Since they set no limits on either size or expenditure, such "guidelines" are in fact no guidelines at all. If resources are to be allocated efficiently, it is essential to determine how much diversity is biologically sufficient in any given case, and then to evaluate the costs of achieving conservation at that scale; otherwise, resources will be consumed by one project, and others will go underfunded. These decisions can be only partially guided by biological considerations, such as the predicted reproductive or ecological effect of reduced population size; judging how to allocate resources invokes other questions of societal values and priorities as well (Norton 1987).

Endangered species can generally be treated through the use of economic scarcity and depletion models (Welch and Miewald 1983). Conventional models describe the exhaustion or sustainable use of renewable and nonrenewable natural resources as factors in production. In this model, genetic resources may be viewed alternatively as "stock" (McInerny 1981), "scarce environmental assets" (Fisher 1981), or "biological capital" (Barbier 1989). However, endangered species differ from most renewable resources in their extreme vulnerability to exhaustion or depletion. Whereas a sustainable harvest rate may be determined for fisheries, forests, or other biological resources, endangered species are generally so close to extinction that they are better treated as nonrenewable resources (see discussion by Holsinger and Gottlieb [Chapter 13] on this point). This distinction becomes clearer when one considers the difference between overall biomass production (which regenerates naturally over years or decades) and the underlying evolutionary gene pool of a species (which, for practical purposes, is limited in extent at any given time). Thus as Butlin (1981), Fisher (1981), and others note, the distinction between renewability and nonrenewability of biotic entities such as endangered species is not clear-cut. Hyman and Stiftel (1988) suggest that endangered species fall into a middle category as a *critical-zone* resource—one that

can be renewed only until a threshold of irreversible depletion has been reached; beyond that level, it may be lost forever. The best example of a critical zone resource is an endangered species, which can expand its population provided that its numbers do not fall below a minimum viable level. However, the minimum viable level is often uncertain. For this reason, a few economists advocate a safe minimum standard approach in dealing with a critical zone resource.

The congruence between critical-zone resource models in economics and minimum viable population theory in biology (Menges, Chapter 3) is hardly incidental. These two homologous models essentially describe the phenomenon of depletion from different perspectives, suggesting the possibility of a unified model joining population biology and microeconomics.

Conserving Multiple Levels of Biological Hierarchy

Consideration of multiple levels of biological organization is an important component of the strategic planning process (Noss and Harris, 1986). Conservation at the ecosystem level, for instance, must clearly be accomplished by protecting habitat. Moreover, as is increasingly evident, the average reserve size required to maintain the full diversity of ecosystem elements and processes is larger than had previously been thought, and substantially greater than all but the largest reserves in the world (Harris 1984; Noss 1987; Wilcove 1988).

Sustainable large-scale preserves and preserve complexes on the order of hundreds of thousands or millions of acres hold the greatest promise for preserving ecosystem diversity. Such reserves, however, will require a commitment of funding several orders of magnitude greater than at present. This will be particularly the case where conservation and commercial interests (mining, forestry, agriculture, housing) are competing for the same units of land, thus tending to drive up considerably the average acquisition cost per acre. Moreover, once acquired, reserves require considerable attention and expenditure for management, including fencing, controlled-burn programs, control of exotic species, seral management, patrolling, and so on. Management thus introduces a long-term operating cost in addition to the initial capital outlay.

Thus it may be highly cost effective to complement large-scale ecosystem conservation with more directed efforts to protect populations or even individuals of particular species. For example, if all that is required is to preserve a few living specimens representing a particular ecotype of a wild crop relative for a plant breeding experiment, then acquisition of an entire ecosystem may be a poor use of scarce resources; this is particularly so if the overall ecosystem or community diversity is unexceptional other than for a few species of interest. In such cases it may be more cost-effective to collect the plants for their local genetic research value. On the other hand, it is virtually axiomatic in conservation biological theory that community and biome levels of diversity must be protected and managed *in situ*. This suggests a principle of matching higher-order levels of biological organization to higher-order (and generally more expensive) conservation tools, and applying other methods to lower levels of hierarchy. This concept underlies the theory of integrated conservation strategies (Falk 1987a, 1990a, 1990b; Edwards 1989).

Even within a given level of biological organization, there are more threatened species, populations, and individuals than can be conserved with the available resources. As Holsinger and Gottlieb (Chapter 13) note, it follows that even here careful choices must be made among competing priorities. As a practical problem in resource allocation, we may assume that we seek the greatest benefit possible at each level of biological hierarchy for a given investment of resources. Although the predicament is widely recognized, there rarely exist uniform criteria to guide the determination of priorities, especially for conservation below the species level. The following two sections approach the problem of setting conservation priorities from the perspective of efficient allocation of scarce resources, at three levels of biological organization: species, populations, and individuals.

Selecting Priority Species for Conservation Efforts

As a rule, conservationists are generally unwilling to permit any species to decline to extinction without efforts to save it. Recent extinctions and near-extinctions such as the California condor, dusky seaside sparrow, and whooping crane illustrate the lengths to which public and private agencies are willing to go in order to prevent the loss of a single species. Although most public attention at the species level is focused on animals, some plants (such as the celebrated case of Furbish's lousewort [*Pedicularis furbishiae*]) have received substantial investments of time and money.

Most conservationists would accept the principle that efforts should be directed to the "most endangered" species. Additionally, other characteristics, such as evolutionary distinctiveness, should also be heavily weighted in selecting target species for conservation (see Holsinger and Gottlieb, Chapter 13). However, criteria for ranking endangerment vary widely, and different systems do not necessarily assign high priority to the same taxa, even species endangered throughout their range. As noted previously, there exist at least four assessments of plant species endangerment in the United States: (1) listings and proposals by the U.S. Fish and Wildlife Service under the Endangered Species Act; (2) Global Ranks as assigned by the Nature Conservancy, based largely on the number and size of populations; (3) the Priority Species assessments of the Center for Plant Conservation, which estimate time to possible extinction; and (4) systems such as those proposed by Sutter (1988), focusing on rates of decline. Each of these systems offers a different perspective on a threatened species, and thus some degree of cross-referencing may be useful. The Center for Plant Conservation (1988b) employs a database sorting algorithm that groups together species found at the highest ranks in the Fish and Wildlife Service, the Nature Conservancy, and the Center for Plant Conservation priority systems. This constitutes an initial screening, which may then be refined by evaluation of the urgency of threats to any particular species within an overall framework for decision making (Center for Plant Conservation 1987). The Biological Conservation Database of the Nature Conservancy (Carr 1988; Jenkins 1989) is similarly designed to assess priorities by use of multiple criteria.

It might be supposed that conservation programs should uniformly be directed toward the most critically threatened species—that is, those reduced to the smallest numbers, declining at the greatest rate, and/or facing the most irreversible threat. Two considerations—one biological, the other economic—suggest that this should not necessarily be the case. On the biological side of the equation, minimum viable population analysis and the effects of close inbreeding (Barrett and Kohn, Chapter 1; Menges, Chapter 3) suggest that some species are, for all practical purposes, too far gone to survive as functioning reproductive, ecological, or evolutionary entities. At least six species in the United States, for instance (several of which are monoecious or obligate outcrossers), are known from only a single individual; examples include *Pritchardia munroii, Arctostaphylos hookeri* ssp. *ravenii,* and *Castilleja uliginosa* (McMahan 1989). In terms of biological hierarchy, the concepts of ecotype, population, and community function are no longer relevant in such cases; there is only the individual, with its particular genotype, to represent the species. Such extreme cases draw attention to economic considerations: Would it be worth incurring the expense—and the opportunity cost—of competing with commercial interests to acquire and then manage the habitat solely for an individual of a species that is no longer a functioning ecological or evolutionary entity? Or, in such cases is it a wiser use of resources to apply a lower-order conservation tool (e.g., a botanical garden, seed bank, or DNA sample) to capture the remaining genetic information, and then concentrate on reintroducing the species on protected land elsewhere? In a practical milieu that requires optimal investment of resources in any given project, the concept of triage will become increasingly important to the conservation of biodiversity.

Sampling Considerations for Populations and Individuals

Having selected a priority species for conservation, the next set of decisions (and the beginning of allocation of significant resources to the case) involves the populations, and individuals within these populations, to be sampled or protected. As noted above, the principles of benefit–cost analysis apply equally well to the acquisition of habitat reserves and to sampling for off-site collections. For purposes of this chapter, attention will be focused on the latter, especially with reference to collecting genetically representative samples of endangered plant species.

Since most populations of a species show some genetic similarity to other populations, any collection of multiple individuals or populations will necessarily entail some repeated sampling of alleles. Moreover, the extent of this repetition will increase with sample size, as a greater proportion of the total gene pool is collected. This relation is due primarily to the inevitable overlap and genetic duplication among samples, which accumulates as the sample grows larger. The problem is thus one of determining an optimal collection size, given the imperative to avoid wasting resources on redundant sampling.

In accord with the neutral-allele hypothesis (Kimura and Crow 1964; Marshall and Brown 1975; Brown and Briggs, Chapter 7), we may make a starting assumption that alleles are of equal significance for the biology of a species, and thus for conservation purposes. We may further assume that the costs of collecting and maintaining individuals within a population, and populations over the range of a species, are also equal. These assumptions may be modified on the basis of specific information regarding a species, but in the absence of such information the neutral-allele and equal-cost models remain useful assumptions (see also discussion of disjunct populations and the depletion function, below).

It is an inherent characteristic of genetic sampling programs that the new genetic content of each successive sample collected is likely to be lower than that of the sample that preceded it. This will always be the case in sampling populations in which the characteristics to be measured through sampling are not unique to individuals within the population. Otherwise, sampling would not be a useful technique, and entire populations would have to be examined. For this reason, we may expect that curves expressing the relationship of genetic diversity conserved by additional effort will be characterized by a negative slope, generally convex toward the origin (Russell and Wilkinson 1979).

The economic implications of this are clear: as the benefits drop off with continued effort, there will come a point at which they are judged to equal the costs of producing them, and the process should be stopped. This may be expressed in the language of the economic theory of production. In our case, "production" is the collection of genetic diversity. A *production function* relates quantities of different inputs (e.g., collecting time and expense) to a given output (e.g., genetic diversity collected). The *marginal product* is the benefit (measured, for instance, in alleles) derived from an extra unit of input (Figure 14.1). It varies according to the level of a given input already allocated to production, and with the current level of other inputs allocated. Although successive increases in the amount of an input assigned to production generally lead to increased output, there usually comes a point beyond which the value of the increased output is less than the value of the input expended to create it (Figure 14.2). Beyond this point, it is uneconomic to increase production by the allocation of more of this resource. This almost universal truth is known as the *law of diminishing marginal returns*.

The explanation for diminishing marginal returns to the conservation of genetic diversity is also attributable to the commonsense technique of sampling first among low-cost populations, then moving to the less accessible, always leaving a relatively small residue of less-accessible populations, the cost of sampling which is much higher. This tendency is comparable to the steadily increasing cost of finding and exploiting fossil fuels as the cheaper reserves are used up, and is known as the *depletion function* of populations (Conrad and Clark 1987). In the case of plants, once the readily accessible populations have been collected, the exploratory effort required for additional populations tends to increase. Pindyck (1978) mapped this relationship for natural resource exploration and demonstrated that costs will generally rise for successive units of resource found (or, to put it another way, that constant input will produce a declining output). For intrinsically rare species, this may be a particularly significant factor in determining the costs of conservation efforts (Lazell 1986). Conventional depletion models for biological resources may eventually be replaced by models em-

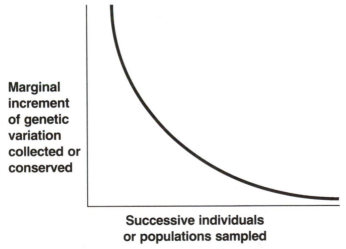

Figure 14.1. Marginal product curve showing the additional genetic content added to the total collection by each successive individual or population sampled. The negative slope is the consequence of the overlap between additional samples and the accumulated diversity in the existing collection.

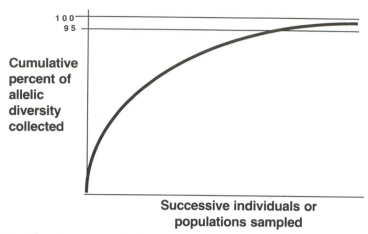

Figure 14.2. Cumulative genetic diversity captured by successive samples of a species, expressed as a percentage of the total allelic diversity within a species. The 95% level represents the proposed minimum viable level of diversity for long-term conservation purposes. The curve represents a declining rate of return for additional samples, as the sum of marginal outputs shown in Figure 14-1.

phasizing cumulative and synergistic degradation effects on natural systems (Barbier 1989). Moreover, the production function can be used as a conceptual tool without direct reference to actual economic value for a gene or a species.

The slope of these curves is determined by the biology of each species—in particular, the extent and partitioning of genetic variability among and within populations. At the population level, following Hamrick et al. (Chapter 5, 1979, 1989; Hamrick 1983), we would expect the marginal utility curve to decline more slowly—reflecting the *greater* marginal benefit of successive populations sampled—in species with a higher index (G_{ST}) of variation among populations. For example, populations of an annual self-fertilizing angiosperm characteristic of early- and midsuccessional stages might follow curve (a) in Figure 14.3. By contrast, a species with life-history characteristics that tend to indicate low variability among populations (e.g., long-lived, outcrossing, wind-pollinated perennial gymnosperms of late-successional stages) might follow curve (b). As Hamrick et al. (Chapter 5) and Holsinger and Gottlieb (Chapter 13) observe, these life-history factors account for only 47% of the observed variation among populations; nonetheless, they do influence the distribution of genetic variation, and should be taken into account in sampling regimes. As Nijkamp (1977) points out, the quantification of biological diversity by indices such as Odum's (1971) species diversity index or standard measures of heterozygosity and polymorphism is essential in applying such models.

As genetic variation in the total collection accumulates, marginal genetic return for each successive sample declines. Nonetheless, owing to the positive slope of the cumulative diversity curve (Figure 14.2), the collection first passes a minimum viable threshold value and eventually reaches a threshold of sufficiency for conservation purposes. Just where this latter point lies is a matter of great conjecture, and undoubtedly varies from species to species; moreover, the judgment cannot be made on biological criteria alone. Following Menges (Chapter 3), Barrett and Kohn (Chapter 1), and Tem-

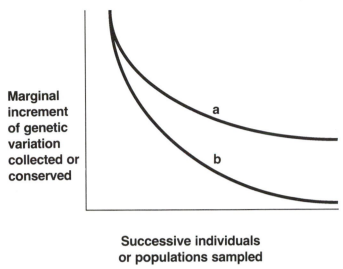

Marginal increment of genetic variation collected or conserved

Successive individuals or populations sampled

Figure 14.3. Schematic representation comparing marginal genetic content of successive collections to two species: (a) an annual self-fertilizing angiosperm characteristic of early- and mid-successional stages; (b) a long-lived outcrossing wind-pollinated perennial gymnosperm of late-successional stage. (After Hamrick et al. [Chapter 5] and Brown and Briggs [Chapter 7])

pleton (Chapter 12), we may suppose that minimum viable population theory provides a suitable definition of a baseline collection, in order to avoid the effects of drift and inbreeding. Few conservationists, however, would be willing to stop at this point, in part because our understanding of species biology remains so incomplete, and in part because the true goal of conservation is to protect the *maximum* amount of natural diversity possible, not to reduce it to a survivable minimum.

In the genetic sampling literature, the objective of capturing alleles occurring at a frequency of 0.05 or greater, with a confidence level of 95%, has been widely accepted (following Marshall and Brown 1975). From a biological perspective, this figure must be regarded as somewhat arbitrary in that there is unlikely to be a true threshold frequency of value for ecological or evolutionary significance. Nonetheless, such figures raise an important issue about the extent of genetic sampling programs. As a matter of probability, alleles that are more common in a population (or species) are more likely to be captured in the early stages of sampling, whereas alleles found at low frequency may be missed because they occur in only a few uncollected individuals. Whether these rare alleles are of reproductive, ecological, or adaptive significance remains a matter of considerable debate. Some authors have argued that rare alleles are of adaptive or evolutionary significance, perhaps representing a reservoir of adaptation to unusual conditions (Huenneke, Chapter 2). An effort to capture such alleles would tend to drive to higher levels the proportion of individuals or populations collected. Others (Brown and Briggs, Chapter 7; Templeton, Chapter 12; Holsinger and Gottlieb, Chapter 13) argue that the low frequency of rare alleles is evidence of their relative reproductive and evolutionary insignificance, representing recent, ephemeral, or declining genetic recombinations. As a conservation matter, this view suggests that rare alleles per se do not merit the additional expense of collecting farther down the "tail" of the curve. Although scientific understanding of the reproductive and adaptive significance of genetic variation, especially in rare species, does not permit a definitive conclusion to be drawn, as a practical matter rare alleles must be considered a relatively low priority. As Figures 14.1 and 14.2 suggest, the additional effort required to sample and maintain low-frequency alleles could be immense, effectively replacing a benefit–cost calculus with an imperative to conserve or collect from every individual in every population. As will be seen, the risk of losing rare alleles would be one of the pragmatic trade-offs implicit in application of the system proposed in this chapter.

A Decision-Making Model for Collecting Strategies

Figure 14.4 illustrates the outcome of these various considerations. Species (a)–(c) are ranked in order of their endangerment, according to the criteria discussed earlier. Successive populations, and individuals within populations, are sampled until a predetermined collection of adequate size (t) is reached. At that point, rather than continuing to invest resources in further collections of species (a), effort is directed to the first sample of species (b), and so on. The result is a constant shifting of species priorities, depending on the biology of the plants and the conservation work already accomplished. Note that for illustrative purposes, the slopes of the curves are shown to vary among species according to life-history characteristics, as discussed above. Thus species (c) requires five collections to reach the desired t threshold, whereas species (a) requires only three.

Several special considerations will influence the application of this model; two are mentioned here briefly. *Disjunct populations* are often thought to be unique biologi-

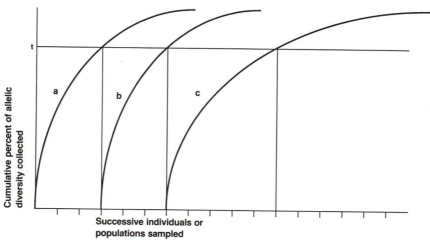

Figure 14.4. A model for conserving species in priority order based on reaching a predetermined level of genetic adequacy (t) in collections. Species (a)–(c) are conserved in succession, with (a) being the most endangered and (c) the least. Successive populations of each species are sampled until 95% of the estimated allelic diversity has been conserved, at which point the conservation effort passes to the next species. For illustrative purposes, the slope of the curve for each species varies according to the discussion of Figure 14-3.

cally, representing either relictual remnants of a species' former range or its extension to the limit of physiological or ecological maxima; they may also represent unique ecotypes (Millar and Libby, Chapter 10; Rieseberg, Chapter 11). For instance, Karron (Chapter 6) has noted alleles in small populations of *Astragalus* that are not found in other populations. By contrast, the findings of Hamrick et al. (Chapter 5)—that endemic species contain less than half the total genetic diversity (H_e) of widespread species—might suggest a lower expected value for isolated populations. However, they also note the significant influence of gene flow on the genetic diversity of populations, indicating that species with limited potential for gene migration among populations may be more highly differentiated at the population level. On the cost side, as Brown and Briggs (Chapter 7) note, the effort and expense of collecting from widely dispersed populations may be significantly greater than the effort and expense of collecting from several clustered populations; this follows Pindyck's (1978) principle concerning exploratory effort. Disjunct populations are thus characterized by higher average collecting costs, but whether the benefits are greater or lower than those for central populations evidently eludes generalization at this point.

A second consideration is the relationship between collecting (or, in the case of habitat protection, acquisition) and maintenance costs. Genetic resource sampling theory tends to focus attention on collecting costs (see, for instance, the estimation by Marshall and Brown [1975] of the number of days required to take field samples). In the long run, however, the total expense of maintaining samples may outweigh the collecting costs by a substantial margin; this has been the experience at the Center for Plant Conservation, and is mirrored in habitat conservation organizations such as the Nature Conservancy, which allocate an increasing budget share to stewardship of protected lands (Nature Conservancy 1988). Cost considerations are certainly among the more compelling reasons for maintenance of collections in seed banks wherever possible (Eberhard et al., Chapter 9; Hawkes 1987a, 1987b). In species for which such methods of maintenance are not feasible, collections must be maintained as growing individuals, with consequent increases in time and expense per specimen. These costs—as well as the risks of mixed pollination in a garden setting—may be reduced by methods such as those proposed by Nicholson (pers. commun.) to maintain collections of woody species in a state of prolonged juvenility and vegetative growth. Holsinger and Gottlieb (Chapter 13) suggest that for outcrossing annual plants, captive effective populations of at least 300 individuals must be maintained to minimize the effects of drift and inbreeding. Although this number exceeds the capacity of most botanical gardens, conservation collections of this magnitude held for periods of up to several years in anticipation of a reintroduction effort might well be practical.

RECOMMENDATIONS FOR OFF-SITE GENETIC COLLECTIONS OF ENDANGERED SPECIES

The general state of conservation biology has progressed substantially in recent years, with the result that earlier efforts to codify genetic sampling procedures (Center for Plant Conservation 1986; Hawkes 1986) may now be improved on considerably. Unfortunately, the total pool of data on the genetics of rare plant species and populations is still discouragingly small. This limitation forces conservation practitioners to continue their reliance on informed speculation rather than direct empirical evidence. Nonetheless, it is a goal of this volume to offer the current generation of genetic collecting standards, in hopes that these general prescriptions will soon be superseded by guidelines specific to particular taxonomic or ecological groups of plants.

Guidelines developed by the Center for Plant Conservation are given in the Appendix. These guidelines are consistent with the contributions found in this volume, and represent an attempt to translate conservation biology theory into practice for endangered plants. The key points of these guidelines are given in the following four sections.

Species

As always, the most critically threatened species should be given priority. However, additional weight should be given to species that represent evolutionarily unique lineages, or that have significant prospects for restoration and recovery. By contrast, some species that are severely depleted below viable levels may best be regarded as terminal cases, with conservation resources directed elsewhere.

Populations

Multiple population sampling appears to be an important component for genetic conservation collecting. This permits existing ecotypic or interpopulational variation to be captured in the sample, especially for disjunct populations. Furthermore, since many rare species are found in a small number of populations (Brown and Briggs, Chapter 7; Nature Conservancy 1990), the additional collecting effort may not be significant.

Current knowledge of the distribution of genetic variation among populations suggests that all populations should be sampled for species known from three or fewer sites, and up to five populations for more widespread taxa (Holsinger and Gottlieb, Chapter 13). Beyond this number, there appear to be significantly lower genetic returns from further sampling of populations.

Individuals

Samples of 10–50 individuals per population appear to capture the more common alleles within acceptable confidence levels. In determinations of sample size, life-history characteristics can be of some use; however, populations of many species have been so disrupted that sample size will be driven more by prospective *use* of the collection for recovery of the species.

Propagules

Samples of ten or more propagules per plant (if possible) will help protect the collection from a loss of genotypes through failure to germinate or grow properly. The need for multiple propagules per individual (or genet) is an inverse function of the anticipated attrition rate. Larger numbers of individuals may be sampled if the attrition rate in cultivation or storage is expected to be high; this may be calculated as the ratio of survivorship (s) to the total collection (N), where the probability of successful maintenance (P_s) $= s/N$ (Brown and Briggs, Chapter 7).

CONCLUSIONS

Faced with an immense problem and with limited resources at their disposal, conservationists must be ever mindful of the need to allocate such resources carefully. This constraint argues strongly for the development of comprehensive conservation strategies that apply the most cost-effective method at each level of biological organization, taking into account long-term management and maintenance costs as well as initial acquisition or collection expenses. Such models will become increasingly powerful tools as conservation biology research reveals more of the pattern of distribution of genetic variation, and as experience with the life-cycle costing of conservation practice accumulates.

One of the strongest arguments for conserving genetic diversity is its latent potential as a usable resource for agriculture or industry. In economic terms, this means assigning an *option value* to benefits placed on a resource, even when there is no economic use at present (Hyman and Stiftel 1988). Conserving critical-zone biological entities such as endangered species can be justified under the rubric of protecting *potentially* useful resources from depletion; the irreversibility of species loss underscores the importance of this argument. If anything, the weight of evidence suggests that the future of agriculture, forestry, and many other industries may rest squarely on our ability to utilize genetic adaptations present in wild species (Prescott-Allen and Prescott-Allen 1986).

The decision to protect (or abandon) an individual organism, a population, or an entire species is part of a larger set of allocative decisions in the field of conservation. Market and planned economies alike weigh the long-term, potential, and often unknown option value of genetic resources against short-term commodity gain. It is hardly necessary to observe that at the present time, short-term resource depletion strategies predominate, although alternative models do exist and are gradually being applied (Barbier 1989). The depletion of "biological capital" is a short-sighted strategy that civilization may not live to regret.

ACKNOWLEDGMENTS

The author wishes to acknowledge the patient assistance of Jennifer A. Klein and Nicole Roth in the preparation of this manuscript, and the insightful and informed commentary of Anthony H. D. Brown, Andrew Cook, Kent E. Holsinger, and Bart Ostro on its contents.

Organizational support was provided by the Center for Plant Conservation and the Arnold Arboretum of Harvard University (Associate Program) during the research. Support for the overall project was provided by the National Science Foundation, the Pew Charitable Trusts, the John D. and Catherine T. MacArthur Foundation, and the Missouri Botanical Garden.

Appendix
Genetic Sampling Guidelines for Conservation Collections of Endangered Plants

CENTER FOR PLANT CONSERVATION

These recommendations are intended to guide individuals and agencies engaged in the creation of genetically representative *conservation collections* of endangered species, in accordance with current scientific literature. The guidelines address a series of pragmatic questions confronted by collectors and program managers in the establishment of conservation collections. These questions focus on the selection of priority species, populations, and individuals from which collections should be made. Such decisions reflect an effort to conserve diversity at four discrete levels of biological organization: species, populations, individuals, and alleles. The guidelines also address situations in which collecting should be spread out over a period of years. Biological considerations affecting each decision are discussed in detail, and a summary and a synopsis in table form are also provided.

Conservation collections are defined as living collections of rare or endangered organisms, established for the purpose of contributing to the survival and recovery of a species. These guidelines are based on principles and data found elsewhere, on the chapters given in this volume, and on the literature cited.

INTRODUCTION TO THE GUIDELINES

The translation of population genetics and demography into guidelines for conservation action is a difficult task. Part of the difficulty lies in the inherent variability of the plants themselves. With enormous variation in life-history characteristics, geographical range, ecological context, and history, rare plant species virtually defy generalization.

This intrinsic difficulty is compounded by the relatively small number of plants understood scientifically to any significant depth. Rare plants, on the whole, are poorly represented in the scientific literature (Prance and Elias 1977; Lewin 1989; Mlot 1989; Falk 1990b, 1990c). Consequently, conservation biologists seeking to develop management guidelines for rare species must rely all too heavily on inference or extrapolation, based on generalized life-history characteristics that show statistically significant correlations with genetic variation. However, as Hamrick and his colleagues (1989 and Chapter 5) have observed, even these factors account for less than half of the observed variation at polymorphic loci among populations, and only 28% of that within populations. A great deal of the observed heterogeneity within and among species remains without clear explanation. Thus if there is indeed any pattern to the distribution of genetic variation in plant populations, we are seeing only a part of it at the present time.

In the face of this uncertainty, conservation practitioners nonetheless require some guidance in creating genetically representative conservation collections. Earlier recommendations (including those of Marshall and Brown 1975; Hawkes 1976; Center for Plant Conservation 1986) were based on interpretation of the available literature. These efforts represented an important evolutionary step in establishing a scientific basis for plant collections. Understanding of the genetics of plant populations has now progressed sufficiently in recent years to permit a more refined statement of guidelines for cost-efficient and effective genetic conservation programs. Hamrick and Godt's (1989) survey of the allozyme variation literature, for example, reviews 653 studies in 449 species, nearly four times the size of their early samples. It is hoped that as understanding of population biology continues to expand, the guidelines proposed here will in turn be superseded by further refinements.

The collector or curator of rare plants is faced with an essentially practical series of questions: Which species should be given priority for collection? How many sites and how many individuals should be sampled? How much material should be taken from each individual? In answering such questions, we make the difficult transition from theoretical population biology to the realm of praxis. A great deal of generalization is required in order to make this transition successfully; for this reason, we join Hamrick et al. (Chapter 5) and Holsinger and Gottlieb (Chapter 13) in advising caution when interpreting these guidelines for any given species. Nonetheless, an attempt has been made here to offer practical guidelines for conservationists, based as directly as possible on the current literature in population biology, demography, and genetics. These findings are described briefly in the following sections, and are summarized in Table A.2.

The insightful reader will note that the five questions posed below constitute a natural hierarchy. That is, they correspond in turn to four important *target levels of biological organization and variation: species, populations (and ecotypes), individuals, and alleles,* considered over the dimension of time. This correspondence is hardly coincidental; it is an axiom of integrated conservation strategies that diversity at multiple levels of hierarchy be taken into account in planning (Falk 1990b, 1990c). Thus by addressing diversity at several levels simultaneously—rather than, for instance, simply representing a "species" by a few specimen plants without regard to genetic variation—the collection becomes more useful, scientifically valid, and applicable.

The collector's work is guided at each level by certain *key considerations.* For instance, at the population level, sampling design is influenced most heavily by the degree of genetic difference among populations, as well as by the specific history (including its founders) of a given site (Barrett and Kohn, Chapter 1). Likewise, the amount of material collected per individual is driven overwhelmingly by the survival ratio of propagules and, to a lesser extent, by the intended use of the collection. These considerations are summarized in Table A.2.

For each of the four primary practical dimensions of a collection, a *recommended range* is shown, representing the best available interpretation of the current literature. A series of *correlate factors* that influence whether a collection should fall toward the low or the high end of the recommended range is also given. As noted above, the literature does not permit these factors to be supplied with mathematical precision; the development of a master sampling algorithm must await further research. However, notwithstanding the caveats of Hamrick et al. (Chapter 5), we can use these correlate factors as indicators of the *relative distribution* of diversity, if not its quantity. The most heavily weighted factors are indicated in Table A.2 by italics, and are shown toward the upper or lower ends of the collecting spectrum (largely following the weights assigned by Hamrick and Godt 1989).

THE FIVE SAMPLING DECISIONS

1. Which Species Should Be Collected?

From a conservation perspective, the *degree of endangerment* still represents the best single indicator of priority for collection. The concept of endangerment, however, is not necessarily easy to define or determine. For instance, Holsinger and Gottlieb (Chapter 13) observe that although all endangered plants are rare, not all rare plants are endangered (cf. Rabinowitz 1981). Examples include naturally rare or sparsely distributed alpine species, certain widely distributed edaphic endemics, and many trees of tropical distribution (Bawa and Ashton, Chapter 4). Any of these may be considered in stable condition genetically, if protected from disturbance. However, any species with limited range or a small number of individuals can be considered at risk in a broader sense, including major changes in land use or in basic climatic and environmental conditions.

Furthermore, as these examples illustrate, an important distinction may be made between *natural and anthropogenic rarity*. The former may well have ecological or reproductive adaptations to persistence in small populations that the latter may lack. These may include adaptations to levels of close inbreeding that would expose lethal recessive alleles in species adapted to gene flow within a larger pool (Barrett and Kohn, Chapter 1; Menges, Chapter 3). By contrast, some species that are relatively more widespread at a given moment may be experiencing rapid decline in the number, size, and health of populations. Projecting population trends in such species may help to identify problems as they emerge; conservation intervention may then be able to protect a larger and more diverse gene pool than would remain if decline were to proceed to the brink of extinction (Millar and Libby, Chapter 10).

In addition, several ranking systems exist for determining the degree of endangerment, with different criteria and frequently different results for any given species. Using rare plants of the United States as an illustration, major classification schemes and databases are maintained by the Center for Plant Conservation, the Nature Conservancy, the U.S. Fish and Wildlife Service, and the International Union for the Conservation of Nature/World Conservation Monitoring Centre. It is not uncommon for a plant considered highly endangered according to one list to be low-ranked or even absent from one or more of the other lists. For example, only 184 (25%) of the U.S.

rare plants thought to be facing extinction within a decade according to a recent survey by the Center for Plant Conservation are currently listed or proposed as endangered or threatened by the U.S. Fish and Wildlife Service under the Endangered Species Act.

As Holsinger and Gottlieb (Chapter 13) note, taxa representing *unique evolutionary lineages* are worthy of special consideration. The degree of genetic distinctiveness among such species is high, making them intrinsically important objectives from the perspective of evolutionary diversity. Monotypic genera and families should thus be accorded high priority for conservation efforts (see also Center for Plant Conservation 1987).

The objectives of a conservation program may also influence the selection of species to some degree. Where the *potential for reintroduction or research* exists, collections may be instrumental in furthering overall conservation goals for a species. This follows naturally from the notion that such collections are part of "survival pathways" for a species, following an integrated model (Falk 1990b, 1990c; Brown and Briggs, Chapter 7). Thus species with the potential to be incorporated into a biological management program may be given high priority for inclusion in a collection, owing to their capacity for more complete recovery.

As Prescott-Allen and Prescott-Allen (1986) and others have observed, native plant species represent an important—and largely untapped—source of genetic variation with potential human utility. Preliminary data from the Center for Plant Conservation indicate that a majority of rare U.S. plant taxa are congeners of species of economic significance in agriculture, forestry, industrial products, pharmaceuticals, or horticulture. Recent advances in biotechnology have made these genetic adaptations increasingly accessible for research and product applications (Martin et al. 1989). From a strategic perspective, this potential represents a bridge between the corporate research and conservation communities, and thus an important means of building support for biological conservation. A close evolutionary relationship to economically important plants is thus a valid additional criterion in selecting taxa for inclusion in a conservation program.

Finally, the feasibility of establishing a species in protective cultivation should be taken into account. There would be little point in collecting propagules from a wild population for a conservation collection if the material could not be successfully maintained over a period adequate to meet the objectives of the program. For instance, while many endangered species or boreal and temperate regions of North America are considered orthodox to standard low-temperature seed conditions, species from the Mediterranean, subtropical, and tropical areas may possess no cold dormancy mechanism, and thus may be recalcitrant (R. Smith and D. Price, pers. commun.); this topic is treated more extensively by Eberhart, Roos, and Towill (Chapter 9). Establishing a conservation collection involves, by definition, removing some living material from a naturally occurring population of an endangered species. Furthermore, many rare species appear threatened due to low seed set or other reproductive failure in the wild, suggesting that removal of seed can have a deleterious effect on population size and dynamics (Wiens et al. 1989; N. Weedon and W. Brumback, pers. commun.). Thus the probability of the successful establishment of a conservation collection in propagated or stored form should be considered a nontrivial consideration that determines which species should be attempted. Low survivability of propagules is also considered below as a factor in determining the collection size of a species.

Recommendation: Give a species high selection priority if it is characterized by one of the following:

1. High degree of endangerment
2. Rapid decline in population numbers or size, due to anthropogenic causes
3. Taxonomic and evolutionary uniqueness
4. Potential for biological management and recovery
5. Potential to serve as a source of genetic material with utilitarian or economic applications
6. Probability of successful establishment in cultivation or maintenance in storage conditions

2. How Many Populations Should Be Sampled per Species?

In constructing a population sampling regime, the key biological consideration is *degree of genetic difference among populations*. Since populations are by definition separated from one another by some distance, it follows that there will be a significant travel expense and time commitment associated with each additional population sampled. Moreover, if separate lineages are to be maintained in conservation collections, each additional population sampled involves a commitment to additional recordkeeping and curatorial measures. Thus the number of populations to be sampled is a nontrivial consideration from the perspective of costs and long-term management, and should be weighed carefully (Falk, Chapter 14).

Roughly 1680 native U.S. plant taxa are known from five or fewer populations (Center for Plant Conservation 1988b) constituting approximately 73% of the 2290 species considered priorities by the Center for Plant Conservation. This corresponds to Brown and Briggs's observation for the southeastern Australian flora that over 80% of endangered species are known from five or fewer populations. Thus in regard to species that qualify as high priorities according to the criteria discussed above, the collector, in practical terms, is faced with a sampling decision among relatively few sites.

As Hamrick and Godt (1989) note, most rare or geographically restricted species partition their diversity in essentially the same manner as do more widespread and diverse species. Trees may constitute an exception to this rule; Hamrick and colleagues (pers. commun.) have found recently that geographical range does influence the distribution of genetic variation among populations of woody plants. On the whole, however, although the overall *amount* of genetic variation within restricted species may be smaller, the *distribution* of variation may be inferred by general rules, and sampling programs constructed accordingly.

Roughly half of loci in plant species are polymorphic (although endemic and narrowly distributed species show a slightly lower figure). Of the polymorphic portion of the genome, Hamrick and Godt (1989) calculate that approximately 22% of total diversity is distributed among populations ($G_{ST} = 0.224$). Variation among populations thus represents a significant fraction of the genetic diversity within species, and should be taken into account in sampling programs. Moreover, this variability among populations, although not quantitatively large, may be qualitatively significant in terms of evolutionary adaptation. Populations may develop unique local ecotypic characteristics due to microhabitat variation, and these, in turn, may represent important stages in the process of evolutionary adaptation and speciation.

On the other hand, as discussed by Falk (Chapter 14), the marginal genetic "benefit" of each additional population sampled declines rapidly after the first few populations (Brown and Briggs, Chapter 7; Holsinger and Gottlieb, Chapter 13). The population sampling question is thus essentially one of balance between the potential biological significance of variation among populations, and the cost and declining benefit of each additional site to be sampled.

In addressing this balance, the Center for Plant Conservation recommends that *all populations be sampled for species occurring in three or fewer locations,* and that three to five populations be sampled among more widespread species. Since few critically endangered species are found from a large number of populations, it will rarely be necessary to sample more than five populations. Brown and Briggs (Chapter 7) and Holsinger and Gottlieb (Chapter 13) consider this to be a typical "rule of thumb" maximum even for widespread taxa. This population sampling range is summarized in Table A.1.

Where a judgment must be made in regard to the number of populations to be sampled, the key consideration is the *degree of genetic differentiation among populations.* Variation among populations is influenced by a variety of factors (listed below) pertaining to the history and demography of the site and to the life-history characteristics of the species. Species with limited potential for gene migration manifest high levels of ecotypic variation and significant indices of diversity among populations. Well-documented and stable ecotypes may be particularly worth sampling and representing in a collection, especially where future reintroduction is contemplated.

Life-history characteristics are thus useful parameters at this level of biological organization, along with certain biogeographical considerations (such as isolation of populations) that tend to influence gene flow. Other things being equal, species that are self-fertilizing herbaceous annuals show higher levels of differentiation among populations than do wind-pollinated, outcrossing, long-lived woody perennials. Other life-history characteristics having significant weight in influencing genetic partitioning include seed dispersal mechanisms and (to lesser a degree) successional stage, regional distribution, and taxonomic group.

In the case of endangered species, the selection of populations may be driven by nonbiological factors, such as the presence of *For sale* signs or bulldozers on the site. Other things being equal, it makes good conservation sense to include in the sampling

Table A.1. The Recommended Number of Populations to Be Included in a Rare Plant Sampling Program for Capturing Genetic Diversity at the Population Level

No. of extant populations	No. of populations sampled
1	1
2	2
3	3
4	3–4
5	3–5
>5	4–5

Note: See text for discussion.

program a population that is in danger of extirpation, regardless of the type of threat. For example, a population of tallgrass prairie species that is disappearing because of fire suppression and shrub invasion of the community may be an excellent candidate for collection, since with proper management the community may later be restored on-site or established elsewhere.

The potential for integrated conservation management offers an alternative rationale for including certain populations in a conservation collection—for example, populations found within protected areas, parks, or species sanctuaries where the responsible agency is interested in long-term biological management. In contrast to unprotected, at-risk sites, populations on protected land may have a likelihood of managed survival in the wild. The trade-off thus falls between protecting a critically threatened gene pool in danger of imminent destruction or protecting a gene pool with better prospects for long-term survival owing to more complete site protection and management. Although such choices should be made on a case-by-case basis, off-site collections should still give priority to populations in immediate danger of destruction, partly because the conservation collection may become the *only* means of conserving those genes. Integrated conservation potential, however, represents an important alternative strategy, with a greater possibility for "success stories" and complete biological recovery.

Recommendation: Sample up to five populations of each species, following the guidelines in Table A.1. More populations should be sampled from species with lower potential for gene flow (and consequently higher diversity) among populations. Such species have the following characteristics, listed approximately in descending order of significance.

1. Imminent destruction of populations
2. High observed ecotypic or site variation
3. Isolated populations
4. Potential for biological management and recovery
5. Recent or anthropogenic rarity
6. Self-fertilization
7. Herbaceous annual or short-lived perennial
8. Early- to mid-successional stage
9. Gravity-, explosively, or animal-dispersed seed
10. Dicot or monocot
11. Temperate–tropical distribution

3. How Many Individuals Should Be Sampled per Population?

The purpose of sampling multiple individuals within a population is to capture the significant fraction of diversity that, in most species, is present in any single population. As Hamrick and Godt (1989; Hamrick et al., Chapter 5) have observed, an average of 78% of the total allozyme variation at polymorphic loci in plants generally is found within populations. This fraction, in combination with the 50% of loci that are monomorphic, means that the great majority of total genetic content of most plant species is present in a single "typical" population. Thus within-population sampling has a different strategic objective from sampling among multiple populations; the latter attempts to capture geographical and ecotypic variation, a quantitatively small portion of the overall genome, but of potential adaptive and evolutionary significance. By contrast, sampling design *within* populations seeks to capture the basic genetic blueprint of the species with a reasonable degree of accuracy.

As Brown and Briggs (Chapter 7) have shown, *the allelic content of a sample is proportional to the log of both population and sample size.* As a consequence, large samples from each population are not required to efficiently capture the primary allelic diversity. Figure 14.1, as discussed by Falk (Chapter 14) and Brown and Briggs (Chapter 7), shows that the great majority of genes in a species are captured by the first few samples. Because of the logarithmic nature of the allele–sample curve, sampling programs would have to be extended dramatically in order to capture alleles occurring at extremely low frequencies.

As a result, the law of diminishing returns applies to sampling programs within populations as well as to those among populations. This function derives ultimately from the inescapable genetic similarity among individuals of a species; the probability of encountering a new (uncollected) allele declines with each successive individual added to the collection. Moreover, monomorphic and common polymorphic alleles are captured early in the sampling sequence. This leaves only those alleles that are infrequent in the population (found at a frequency of 5% or less), unsampled once the collection has attained a reasonable size. This *declining marginal genetic benefit* function has a close homology with microeconomic utility and product curves: as the collection builds, each subsequent individual (or population) is successively less likely to contain anything new. These points are discussed in detail by Falk (Chapter 14).

As a corollary to Question 3, **should the number of individuals sampled vary with the size of the population?** Intuitively, one would expect that larger populations should be represented by samples larger than those of small populations. Recall, however, that genetic diversity is itself proportional to the log of population size. Thus differences in population size at the level of an order of magnitude would be necessary before sample size would be significantly affected. More importantly, however, consider the particular problem at hand of creating a conservation collection of an extremely rare species. Many such species exist in very small populations, whether by natural distribution or by anthropogenic reduction. In the latter case, the few remaining plants may represent a small fraction of the original genepool, which has been both reduced numerically and fragmented geographically. Thus in order to avoid unintentional exacerbation of demographic or genetic effects discussed elsewhere in this volume (Barrett and Kohn, Chapter 1; Menges, Chapter 3; Karron, Chapter 6), it is prudent to take a larger proportional sample of the entire population from a rare species. For the expense of a few additional seeds or cuttings, the conservationist can avoid subjecting the collection to an artificial founder effect.

Some authors (Yonezawa 1985) have suggested that sample sizes from as few as ten individuals per population will balance the criteria of sufficiency and efficiency. Brown and Briggs (Chapter 7) recommend increasing the upper limit somewhat, to take account of the survivability of the propagules and some reproductive characteristics of the species. Self-incompatibility, for example, may require a substantially larger sample size to achieve the minimum effective breeding population size.

The Center for Plant Conservation recommends that samples be taken from 10–50 individuals per population as a reasonable middle ground between rigorous minima and overcollecting. The rationale for sampling at these levels for endangered species collections is as follows.

1. *The survival of propagules may be low in wild native species.* Unlike crops, which have been selected and bred for high germination rates, only a small portion of the seeds or cuttings of wild native species may be viable. Moreover, the horticultural and storage techniques for many endangered species are still unknown (R. Smith, Royal Botanical Gardens, U.K., pers. commun.). Although this problem is addressed largely by taking multiple propagules per individual, an additional safety margin may be established by sampling a larger number of individuals. If this is done, then even if all propagules of a particular individual prove to be nonviable (an entirely realistic possibility), the overall sample size will nonetheless remain adequate.

2. *"Populations" may in fact be fragmentary remains of original breeding gene-pools.* As noted above, natural biological populations of many endangered species have been anthropogenically reduced and fragmented. A conservation collector who treated each *site* as a population would in fact be enshrining an unnatural distribution of the species. A somewhat larger sample size in such cases can help avoid artificial intensification of founder and drift effects (Barrett and Kohn, Chapter 1; Lewin 1989).

3. *With endangered species, the consequences of underrepresenting allelic diversity in the collection are far greater than the consequences (and costs) of redundancy.* Especially where a reintroduction or restoration program is contemplated, ensuring the ample presence of all common alleles in the collection is well worth a somewhat larger sample size. Since the population genetic structure of most endangered species has never been characterized, there will be no second opportunity to establish a representative conservation collection if a species becomes extinct through a failure of conservation efforts. Thus a risk-avoidance strategy suggests higher levels of collecting within populations.

Life-history characteristics tabulated by Hamrick et al. (Chapter 5; Hamrick and Godt 1989) account for only 28% of observed differences among species for heterozygosity within populations. Thus sample size should be only partially influenced by these factors. Characteristics that restrict genetic mobility within populations are most influential, along with other factors that increase the degree of diversity among individuals in a single population. Other characteristics are significant largely because they influence diversity at the *species* level; for example, 50% of loci in outcrossing species are typically polymorphic, compared with only 20% in self-fertilizing species. This higher overall diversity at the species level is reflected within populations as well. Breeding system, seed dispersal mechanism, and geographical range appear particularly significant in determining the degree of variation among individuals within populations.

Within populations, genetic variation may be distributed unevenly or patchily. Microsite variation within populations, especially in endangered species, merits the additional effort to include in a conservation collection. Thus a *stratified random-sampling* approach should be taken, along the lines described by Brown and Briggs (Chapter 7). This will help to ensure the inclusion of ecologically significant genotypes within the population.

For taxa that occur in clonal clusters in the wild, including taxa that are known or suspected to reproduce by apomixis, the collector should try to determine the number of genets per site and collect one or more samples from each. This determination can be based on the number of apparent clones, phenotypic variation in the population, or the number of microhabitats represented at the site. Although less material from each "population" of clones is necessary than that required for nonclonal taxa, one should collect sufficient material to sample all suspected genotypes within the population. If in doubt about the reproductive strategy of the taxon, one should collect as if the taxon does not reproduce clonally.

Recommendation: Collect samples from 10–50 individuals per population, following a stratified sampling regime. In determining actual collection size, consideration should be given to relevant life-history characteristics, population history, and other factors that may influence the natural distribution of variation. The degree of endangerment, the consequences of introducing an artificial founder effect, and the prospect for use of the collection for reintroduction may also argue for larger collections.

Factors that correlate with high diversity within populations (and thus suggesting sample sizes at the upper end of the range) include

1. Observed microsite variation within populations
2. Mixed-mating or outcrossing
3. Fragmented historical population
4. Small breeding neighborhood size
5. Low survival of seeds
6. Extremely large populations
7. Boreal distribution
8. Gymnosperm or monocot
9. Long-lived woody perennial
10. Late-successional stage
11. Animal- or wind-dispersed seed

4. How Many Propagules Should Be Collected from Each Individual?

The primary objective in collecting multiple propagules per individual is to ensure the representation of a particular genotype in the collection. Thus the sampling regime at the level of the individual is driven overwhelmingly by the *predicted survivability* of propagules.

Consider two contrasting cases: a long-lived tree species with well-known vegetative propagation techniques, and a herbaceous annual in a genus unknown horticulturally. In the former case, an individual genotype can be represented with a high degree of certainty with a small number of cuttings, since the prospects for survival of each cutting to reproductive maturity are excellent. In the latter case, however, the degree of certainty is far lower. There may be unknown seed dormancy mechanisms, low germination rate, absence of key symbionts, and low survival of seedlings to reproductive age. If only 2% of the collected seeds eventually reached maturity, it would be necessary to collect 50 seeds in order to ensure the survival of at least one representative. Brown and Briggs (Chapter 7) discuss the survival factor and its impact on collecting strategies.

As a corollary to Question 4, **what rates of survival may be expected for propagules of wild native species?** Brown and Briggs found a survival ratio (s/N) of surviving propagules (s) to total sample size (N) of 0.74 for cuttings of some Australian endangered species, but this is probably not typical of most endangered species, particularly species propagated from seed. Experience among affiliated gardens in the Center for Plant Conservation suggests that s/N values of 0.10 are more typical, although there is still largely only anecdotal information for many rare wild species. Predicting 10% survival of rare species propagules is probably a conservative and realistic estimate in the absence of other information.

To determine total sample size within a population (C_p), the collector should first compute an attrition-correcting factor of $1/(s/N)$, where the term (s/N) is the proportion of propagules surviving to maturity; this gives the minimum number of propagules that need to be taken from each individual to ensure at least one survivor. This number is then multiplied by the optimal number of individuals (I) to be represented in the collection, as discussed in Question 3. The result is $C_p = I \times 1/(s/N)$. For example, suppose the target number of individuals in the collection from a given population is 25, for a species in which 20% of the seeds survive to maturity under maintenance conditions. In this case, $C_p = 25 \times 1/.20 = 125$ propagules from the population as a whole; the number of propagules taken per individual is C_p/I, or 5 per plant. In a different species with only 10% seed viability, if the goal were to represent 50 genetic individuals in a population, then 500 seeds would be collected from the population as a whole (ten per plant).

Recommendation: Collect a number of propagules from each individual sufficient to compensate for attrition in order to ensure the representation of each genotype. This number may be calculated for each individual as $1/(s/N)$ propagules per individual, where s/N is the survival ratio of propagules. Multiplying this quantity by the target number of individuals (Question 3) gives the total number of propagules collected C_p from the population as a whole, according to the formula $C_p = I \times 1/(s/N)$.

5. Under What Circumstances Is a Multiyear Collection Plan Indicated?

Some species are endangered because of poor net reproductive output; in such cases, it may be inadvisable to collect the target levels of propagules in a single year. When this is the case, the planned collection size should be obtained over a period of several years, depending on the fraction of the target level that can be collected each year. For example, if only one-third of the recommended number of propagules can be safely removed from the site, then collecting is spread over a three-year period. In *each* year of collecting activity, a stratified random-sampling procedure should be followed.

In *all* cases, collectors should take great care to avoid interfering with the reproductive ecology and demographics of rare plant populations. As Menges (Chapter 3), Barrett and Kohn (Chapter 1), and Gilpin and Soulé (1986) have observed, the probability of a catastrophic genetic or demographic event is inversely proportional to the size of the population. Special care should thus be taken with annuals and biennials, and in species that are monocarpic, for which the annual seed crop is the only link to the future of the population.

Table A.2. Center for Plant Conservation: Summary of Sampling Guidelines for Conservation Collections of Endangered Plants

	Decisions			
	Which species to collect?	**How many populations sampled per species?**	**How many individuals sampled per population?**	**How many propagules taken per individual?**
Recommended range	—	1–5	10–50	1–20
Target level of biological organization	Species	Ecotype and population	Individual	Allele
Key considerations	Probability of loss of unique genepool Potential for restoration or recovery	Degree of genetic difference among populations Population history	Log of population size Genetic mobility within population	Survivability of propagules Long-term use of collection
		Factors affecting sampling decisions		
Collect more	**High degree of endangerment**	**High diversity/limited gene flow among populations**	**High diversity among individuals within each population**	Low survivability of propagules
	Experiencing rapid decline	*Imminent destruction of populations*	*Observed microsite variation*	*Planned use for reintroduction or restoration program*
	Few protected sites	*High observed ecotypic or site variation*	*Mixed-mating or outcrossing*	
	Biological management required	*Isolated populations*	*Fragmented historical populations*	
			Small breeding neighborhood size	
	Recently or anthropogenically reduced	*Potential for biological management and recovery*	*Low survivability of propagules*	
	Feasibility of successful maintenance in cultivation or storage	*Recent or anthropogenic rarity*		
	Possibility of reintroduction or restoration	*Self-fertilization*	*Extremely large populations*	

Collect less/fewer →

Economic potential	*Herbaceous annual or short-lived perennial*	Boreal distribution	*Low annual reproductive output (indicates multiyear collecting strategy)*
	Early- to mid-successional stage	Gymnosperm or monocot	
	Gravity-, explosively- or animal-dispersed seed	Woody perennial	
	Dicot or monocot	Late-successional stage	
	Temperate–tropical distribution	Animal- or wind-dispersed seed	
	Wind-dispersed seed	Temperate–tropical distribution	
	Outcrossing wind-pollinated	Early- to mid-successional stage	
	Late-successional stage	Dicot	
High integrity of communities	*Observed similarity among populations*	Herbaceous annual or short-lived perennial	
Many protected sites	*Long-lived woody perennial* / *Gymnosperm*	Self-fertilizing	
Natural rarity	*Boreal–temperate distribution*	Explosively or gravity-dispersed seed	
	Protected populations or naturally rare	High survivability of propagules	
Stable condition	*Closely clustered populations*	Large breeding neighborhood size	
Low degree of endangerment	**Low diversity/extensive gene flow among populations**	**Low diversity among individuals within each population**	**High survivability of propagules**

Note: Four basic practical decisions are addressed: which species to conserve, and the number of populations, individuals, and propagules to be sampled for each species. For each decision, a recommended range is shown, along with an indication of the relevant level of biological organization and a summary of key considerations. Factors that influence the sampling decision are listed in detail, with the most significant factors shown in italics. Factors suggesting larger or more extensive collections are shown at the top of each column, while those suggesting smaller or less extensive collections are given at the bottom.

Assessment of the safe number of propagules to be removed from the site should ideally be made by an ecologist or a population biologist familiar with the species. Unfortunately, few species are monitored closely enough to permit these judgments to be based on actual data regarding annual reproductive output. As a practical matter, collectors should rely on guidance provided by agency botanists employed by state botanist's offices, Natural Heritage Programs, the U.S. Fish and Wildlife Service, or the stewards of particular preserves or public lands where the species is found. An effort should also be made to determine if university botanists have studied the ecology of the species in any detail; agency scientists are generally familiar with current research in their geographical area. Multiyear collecting is also indicated in cases where there is significant year-to-year variation in population size or structure due to factors such as climatic fluctuations, pathogenic effects, or population dynamics. The last may include species with alternate flower types in successive years, species with long blooming seasons or two or more blooming cycles in one year, annual species with variable reproductive output, and species such as *Isotria medeoloides* that have long population cycles (Mehrhoff 1989).

Recommendation: Collecting activities should be spread over more than one year if the target number of propagules cannot be safely removed from the site without interfering with the reproductive ecology or demographics of the population. This determination should be made by a trained ecologist, population biologist, or demographer and should be based on empirical data if possible. A stratified random sample should be taken for each year in which collecting activity occurs. In general, distribute the total planned collection over a number of years, corresponding to the fraction that may be taken in any given year.

CONCLUSIONS

Conservation collections are only as good as the diversity they contain. Thus the forethought and methods that go into sampling procedures play a critical role in determining the ultimate quality of the collection, as well as its usability for purposes such as reintroduction and restoration (Barrett and Kohn, Chapter 1; Lewin 1989; Falk 1990c). In the long run, the real significance of collections in biological conservation is their role in reinforcing the management and maintenance of natural populations. Collectors should view themselves not as "preserving" static entities, but as providing a stepping stone on the pathway to survival and evolution.

ACKNOWLEDGMENTS

The preparation of these guidelines was supported by grants from the National Science Foundation, the John D. and Catherine T. MacArthur Foundation, and the Pew Charitable Trusts. The Center acknowledges the important role of Jennifer A. Klein in preparing and editing the manuscript. Comments and suggestions were also made by Anthony H. D. Brown, William Brumback, Robert DeFilipps, Edward O. Guerrant, Jr., James L. Hamrick, Vernon E. Heywood, Kent E. Holsinger, Robert E. Jenkins, Richard W. Lighty, Larry E. Morse, Robert Ornduff, David M. Price, Peter H. Raven, and Norman Weeden.

Bibliography

Aarssen, L. W., and R. Turkington. 1985. Biotic specialization between neighboring genotypes in *Lolium perenne* and *Trifolium repens* from a permanent pasture. *J. Ecol.* 73:605–614.

Abrahamson, W. G. 1984. Species response to fire on the Florida Lake Wales ridge. *Am. J. Bot.* 71:35–43.

Aide, T. M. 1986. The influence of wind and animal pollination on variation in outcrossing rates. *Evolution* 40:434–435.

Aitken-Christie, J., and A. P. Singh. 1987. Cold storage of tissue cultures. In *Cell and Tissue Culture in Forestry,* Vol. 2, ed. J. M. Bonga and D. J. Durzan, pp. 285–304. Martinus Nijhoff, Dordrecht.

Akihama, T., and M. Omura. 1986. Preservation of fruit tree pollen. In *Biotechnology in Agriculture and Forestry,* Vol. 1: *Trees,* ed. Y. P. S. Bajaj, pp. 101–112. Springer-Verlag, Berlin.

Allard, R. W., S. K. Jain, and P. L. Workman. 1968. The gentics of inbreeding populations. *Adv. Genet.* 14:55–131.

Allard, R. W., R. D. Miller, and A. L. Kahler. 1978. The relationship between degree of environmental heterogeneity and genetic polymorphism. In *Structure and Functioning of Plant Populations,* ed. A. H. J. Freysen and J. W. Woldendorp, pp. 49–69. North-Holland, Amsterdam.

Allendorf, F. W., and R. F. Leary. 1986. Heterozygosity and fitness in natural populations of animals. In *Conservation Biology,* ed. M. E. Soulé, pp. 57–76. Sinauer, Sunderland, Mass.

Allendorf, F. W., and R. F. Leary. 1988. Conservation and distribution of genetic variation in a polytypic species, the cutthroat trout. *Conserv. Biol.* 2:170–184.

Allendorf, F. W., K. L. Knudsen, and G. M. Blake. 1982. Frequencies of null alleles at enzyme loci in natural populations of ponderosa and red pine. *Genetics* 100:497–504.

Allendorf, F. W., and C. Servheen. 1986. Genetics and the conservation of grizzly bears. *Trends Ecol. Evol.* 1:88–89.

Anderson, E. 1949. *Introgressive Hybridization.* Wiley, New York.

Anderson, E., and L. Hubricht. 1938. Hybridization in *Tradescantia* III. The evidence for introgressive hybridization. *Am. J. Bot.* 25:396–402.

Anderson, G. J. 1980. The status of the very rare *Prunus gravesii* Small. *Rhodora* 82:113–129.

Annest, L., and A. R. Templeton. 1978. Genetic recombination and clonal selection in *Drosophila mercatorum. Genetics* 89:193–210.

Antlfinger, A. E. 1986. Field germination and seedling growth of CH and CL progeny of *Impatiens capensis* (Balsaminaceae). *Am. J. Bot.* 73:1267–1273.

Antlfinger, A. E., W. F. Curtis, and O. T. Solbrig. 1985. Environmental and genetic determinants of plant size in *Viola sororia. Evolution* 39:1053–1064.

Antonovics, J. 1976. The nature of limits to natural selection. *Ann. Missouri Bot. Gard.* 63:224–247.

Antonovics, J. 1984. Genetic variation within populations. In *Perspectives on Plant Population Ecology,* ed. R. Dirzo and J. Sarukhan, pp. 229–241. Sinauer, Sunderland, Mass.

Ashmun, J. W., and L. F. Pitelka. 1984. Light-induced variation in the growth and dynamics of transplanted ramets of the understory herb, *Aster acuminatus. Oecologia* 64:255–262.

Ashton, P. S. 1984. Biosystematics of tropical forest plants: A problem of rare species. In *Plant Biosystematics,* ed. W. F. Grant, pp. 497–518. Academic Press, London.

Ashton, P. S. 1987. Biological considerations in *in situ* vs *ex situ* plant conservation. In *Botanic Gardens and the World Conservation Strategy,* ed. D. Bramwell, O. Hamann, V. Heywood, and H. Synge, pp. 117–130. Academic Press, London.

Ashton, P. S. (in prep). Comparative ecological studies in the mixed Dipterocarp forests of Northwest Borneo. III. Patterns of species richness. *J. Ecol.*

Ashwood-Smith, M., and E. Grant. 1977. Genetic stability in cellular systems stored in the frozen state. In *The Freezing of Mammalian Embryos,* ed. K. Elliot and J. Whelan, pp. 251–272. Elsevier, Amsterdam.

Asins, M. J., and E. A. Carbonell. 1987. Concepts involved in measuring genetic variability and its importance in conservation of plant genetic resources. *Evol. Trends Plants* 1:51–61.

Association of Official Seed Analysts. 1983. Seed vigor testing handbook. Contribution No. 32. Association of Official Seed Analysts, Boise, Idaho.

Averett, J. L. 1979. Biosystematics of the physaloid genera of the Solanaceae in North America. In *The Biology and Taxonomy of the Solanaceae,* ed. J. G. Hawkes, R. N. Lester, and A. D. Skelding, pp. 493–503. Academic Press, New York.

Avise, J., and R. A. Lansman. 1983. Polymorphism of mitochondrial DNA in populations of higher animals. In *Evolution of Genes and Proteins,* ed. M. Nei and R. K. Koehn, pp. 147–164. Sinauer, Sunderland, Mass.

Avise, J. C., and W. S. Nelson. 1989. Molecular genetic relationships of the extinct dusky seaside sparrow. *Science* 243:646–648.

Babbel, G. R., and R. K. Selander. 1974. Genetic variability in edaphically restricted and widespread plant species. *Evolution* 28:619–630.

Baccus, R., N. Ryman, M. M. Smith, C. Reuterwall, and D. Cameron. 1983. Genetic variability and differentiation of large grazing mammals. *J. Mammal.* 64:109–120.

Baker, A. J., and A. Moeed. 1987. Rapid genetic differentiation and founder effect in colonizing populations of common mynas (*Acridotheres tristis*). *Evolution* 41:525–538.

Baker, H. G. 1955. Self-compatibility and establishment after "long-distance" dispersal. *Evolution* 9:347–349.

Baker, H. G. 1967. Support for Baker's law—As a rule. *Evolution* 21:853–856.

Baker, H. G. 1974. The evolution of weeds. *Annu. Rev. Ecol. Syst.* 5:1–24.

Baker, J. K., and M. S. Allen. 1976. Hybrid *Hibiscadelphus* (Malvaceae) from Hawaii. *Phytologia* 33:276.

Banks, J. A., and C. W. Birky. 1985. Chloroplast DNA diversity in low in a wild plant, *Lupinus texensis. Proc. Natl. Acad. Sci. USA* 82:6950–6954.

Bannister, M. H., A. Williams, I. McDonald, and M. B. Forde. 1962. Variation in terpentine composition in five population samples of *Pinus radiata. N.Z. J. Sci.* 5:486–495.

Barbier, E. B. 1989. *Economics, Natural Resource Scarcity, and Development: Conventional and Alternative Views.* Earthscan, London.

Barker, J. E., and W. J. Libby. 1974. The use of selfing in the selection of forest trees. *J. Genet.* 61:152–168.

Barnabas, B., and E. Rajki. 1981. Fertility of deep-frozen maize (*Zea mays* L.) pollen. *Ann. Bot.* 48:861–864.

Barneby, R. C. 1964. Atlas of North American *Astragalus. Mem. N.Y. Bot. Gard.* 13:1–1188.

Barnes, B. V., and B. P. Dancik. 1988. Characteristics and origin of a new birch species, *Betula murryana*, from southeastern Michigan. *Can. J. Bot.* 63:223–226.

Barrett, S. C. H., and C. G. Eckert. 1990. Variation and evolution of mating systems in seed plants. In *Biological Approaches and Evolutionary Trends in Plants*, ed. S. Kawano, pp. 229–254. Academic Press, London.

Barrett, S. C. H., and B. C. Husband. 1989. The genetics of plant migration and colonization. In *Plant Population Genetics, Breeding, and Genetic Resources*, ed. A. H. D. Brown, M. T. Clegg, A. L. Kahler, and B. S. Weir, pp. 254–277. Sinauer, Sunderland, Mass.

Barrett, S. C. H., M. T. Morgan, and B. C. Husband. 1989. The dissolution of a complex genetic polymorphism: The evolution of self-fertilization in *Eichhornia paniculata* (Pontederiaceae). *Evolution* 43:1398–1416.

Barrett, S. C. H., and B. J. Richardson. 1986. Genetic attributes of invading species. In *Ecology of Biological Invasions*, ed. R. Groves and J. J. Burdon, pp. 21–33. Academy of Science, Canberra.

Barrett, S. C. H., and J. S. Shore. 1987. Variation and evolution of breeding systems in the *Turnera ulmifolia* complex (Turneraceae). *Evolution* 41:340–354.

Barrett, S. C. H., and J. S. Shore. 1989. Isozyme variation in colonizing plants. In *Isozymes in Plant Biology*, ed. D. Soltis and P. Soltis, pp. 106–126. Dioscorides Press, Washington, D.C.

Bartgis, R. L. 1985. Rediscovery of *Trifolium stoloniferum* Muhl. ex A. Eaton. *Rhodora* 87:425–429.

Barton, L. V. 1960. Storage of seeds of *Lobelia cardinalis* L. *Contrib. Boyce Thompson Inst.* 23:267–268.

Barton, L. V. 1961. *Seed Preservation and Longevity*. Wiley Interscience, New York.

Barton, N. H., and B. Charlesworth. 1984. Genetic revolutions, founder effects and speciation. *Annu. Rev. Ecol. Syst.* 15:133–164.

Barton, N. H., and G. M. Hewitt. 1985. Analysis of hybrid zones. *Annu. Rev. Ecol. Syst.* 15:113–148.

Baskin, J. M., and C. C. Baskin. 1986. Some considerations in evaluating and monitoring populations of rare plants in successional environments. *Nat. Areas J.* 6:26–30.

Bass, L. N. 1980. Seed viability during long-term storage. In *Horticultural Reviews*, ed. J. Janick, pp. 117–141. Avi, Westport, Conn.

Bass, L. N. 1981. Storage conditions for maintaining seed quality. In *Handbook of Transportation and Marketing in Agriculture*, ed. E. E. Finney, Jr., Vol. 2: *Field Crops*, ed. E. E. Finney, pp. 239–321. CRC Press, Boca Raton, Fla.

Bass, L. N. 1984. Germplasm preservation. In *Conservation of Crop Germplasm—An International Perspective*, pp. 55–67. Special Publ. No. 8. Crop Science Society of America, Madison, Wis.

Bawa, K. S., and D. M. O'Malley. 1987. Estudios genéticos y de sistemas de cruzamiento en algunas especies arbóreas de bosques tropicales. *Rev. Trop. Biol.* 35 (Suppl. 1):177–188.

Bawa, K. S., D. R. Perry, and J. H. Beach. 1985. Reproductive biology of tropical lowland rain forest trees. I. Sexual systems and incompatibility mechanisms. *Am. J. Bot.* 72:331–345.

Bayliss, M. W. 1980. Chromosomal variation in plant tissues in culture. *Int. Rev. Cytol.* [*Suppl.*] 11A:113–144.

Beardmore, J. A. 1983. Extinction, survival and genetic variation. In *Genetics and Conservation*, ed. C. M. Schoenwald-Cox, S. M. Chambers, B. MacBryde, and L. Thomas, pp. 125–151. Benjamin-Cummings, Menlo Park, Calif.

Belcher, E. W. 1980. Germplasm storage by *ex situ* method. In *Proceedings of Servicewide*

Workshop on Gene Resource Management. Timber Management Staff, U.S. Forest Service, Sacramento, Calif.

Bell, G. 1985. Two theories of sex and variation. *Experientia* 41:1235–1245.

Berlocher, S. H. 1984. Genetic changes coinciding with the colonization of California by the walnut huskfly, *Rhagoletis completa. Evolution* 38:906–918.

Berry, R. J. 1971. Conservation aspects of the genetical constitution of populations. In *The Scientific Management of Animal and Plant Communities for Conservation,* ed. E. Duffey and A. S. Watt, pp. 177–206. British Ecol. Soc. Symp. No. 11. Blackwell Scientific, Oxford.

Bewley, J. D., and M. Black. 1982. *Physiology and Biochemistry of Seeds in Relation to Germination.* Springer-Verlag, New York.

Biasutti O. E. 1956. *The Storage of Seeds for Maintenance of Viability.* Bull. No. 43, Commonwealth Bureau of Pastures and Field Crops. Commonwealth Agricultural Bureaux, Farnham Royal, England.

Bierzychudek, P. 1982. The demography of jack-in-the-pulpit, a forest perennial that changes sex. *Ecol. Monogr.* 52:335–351.

Billington, H. L., A. M. Mortimer, and T. McNelly. 1988. Divergence and genetic structure in adjacent grass populations. I. Quantitative genetics. *Evolution* 42:1267–1277.

Bonga, J. M. 1987. Clonal propagation of mature trees: Problems and possible solutions. In *Cell and Tissue Culture in Forestry,* Vol. 1: *General Principles and Biotechnology,* ed. J. M. Bonga and D. J. Durzan, pp. 249–271. Martinus Nijhoff, Dordrecht.

Bonnell, M. L., and R. K. Selander. 1974. Elephant seals: Genetic variation and near extinction. *Science* 184:908–909.

Bonner, F. T. 1988. Technologies to Maintain Tree Germplasm Diversity. Unpubl. paper prepared for the Office of Technology Assessment.

Bonner, F. T. 1989. Storage of seeds: Potential and limitations for germplasm conservation. *For. Ecol. Mgmt.* (in press).

Bowles, M. L., and S. I. Apfelbaum. 1989. Effects of land use and stochastic events on heart-leaved plantain (*Plantago cordata* Lam.) in an Illinois stream system. *Nat. Areas J.* 9:90–101.

Bowles, M. L., and M. M. DeMauro. 1990. Effects of anthropogenic disturbances on endangered and threatened plants at the Indiana Dunes National Lakeshore. *Nat. Areas J.* 10:187–200.

Bowles, M. L., W. J. Hess, M. M. DeMauro, and R. D. Hiebert. 1986. Endangered plant inventory and monitoring strategies of Indiana Dunes National Lakeshore. *Nat. Areas J.* 6:18–26.

Bradshaw, A. D. 1984. Ecological significance of genetic variation between populations. In *Perspectives on Plant Population Ecology,* ed. R. Dirzo and J. Sarukhan, pp. 213–228. Sinauer, Sunderland, Mass.

Bramwell, D. 1990. Conserving biodiversity in the Canary Islands. *Ann. Missouri Bot. Gard.* 77:28–37.

Bramwell, D., O. Hamann, V. Heywood, and H. Synge, eds. 1987. *Botanic Gardens and the World Conservation Strategy.* Academic Press, London.

Bremermann, H. J. 1980. Sex and polymorphism as strategies in host-pathogen interactions. *J. Theor. Biol.* 87:671–702.

Briggs, J. D., and J. H. Leigh. 1988. *Rare or Threatened Australian Plants,* rev. ed. Australian National Parks and Wildlife Service Special Publ. No. 14, Canberra.

Brown, A. H. D. 1975. Sample sizes required to detect linkage equilibrium between two and three loci. *Theor. Pop. Biol.* 8:184–201.

Brown, A. H. D. 1978. Isozymes, plant population genetic structure, and genetic conservation. *Theor. Appl. Genet.* 52:145–157.

Brown, A. H. D. 1979. Enzyme polymorphism in plant populations. *Theor. Pop. Biol.* 15:1–42.

Brown, A. H. D. 1989a. The case for core collections. In *The Use of Plant Genetic Resources*, ed. A. H. D. Brown, O. H. Frankel, D. R. Marshall, and J. T. Williams, pp. 136–156. Cambridge University Press, Cambridge.

Brown, A. H. D. 1989b. Core collections: A practical approach to genetic resources management. *Genome* (in press).

Brown, A. H. D. 1989c. Genetic characterization of plant mating systems. In *Plant Population Genetics, Breeding, and Genetic Resources*, ed. A. H. D. Brown, M. T. Clegg, A. L. Kahler, and B. S. Weir, pp. 145–162. Sinauer, Sunderland, Mass.

Brown, A. H. D. and J. J. Burdon. 1987. Mating systems and colonizing success in plants. In *Colonization, Succession and Stability*, ed. A. J. Gray, M. J. Crawley, and P. J. Edwards, pp. 115–131. Blackwell Scientific, Oxford.

Brown, A. H. D., J. E. Grant, and R. Pullen. 1986. Outcrossing and paternity in *Glycine argyrea* by paired fruit analysis. *Biol. J. Linn. Soc.* 29:283–294.

Brown, A. H. D., and G. F. Moran. 1980. Isozymes and the genetic resources of forest trees. In *Isozymes of North American Forest Trees and Forest Insects*, ed. M. T. Conkle, pp. 1–10. U.S.D.A., Washington, D.C.

Brown, A. H. D., and G. T. Moran. 1981. Isozymes and the genetic resources of forest trees. In *Proc. Symp. Isozymes of North American Forest Trees and Forest Insects*, ed. M. T. Conkle, pp. 1–10. Gen. Tech. Rep. PSW-48. U.S.D.A. Forest Service, P.S.W. Forest and Range Experiment Station, Berkeley, Calif.

Brown, A. H. D., and B. S. Weir. 1983. Measuring genetic variability in plant populations. In *Isozymes in Plant Genetics and Breeding*, Part A, ed. S. D. Tanksley and T. J. Orton, pp. 219–239. Elsevier, Amsterdam.

Bryant, E. H., L. M. Combs, and S. A. McCommas. 1986a. Morphometric differentiation among experimental lines of the housefly in relation to a bottleneck. *Genetics* 114:1213–1223.

Bryant, E. H., S. A. McCommas, and L. M. Combs. 1986b. The effect of an experimental bottleneck upon quantitative genetic variation in the housefly. *Genetics* 114:1191–1211.

Bryant, E. H., H. van Dijk, and W. van Delden. 1981. Genetic variability of the face fly, *Musca autumnalis* De Gear in relation to a population bottleneck. *Evolution* 35:872–881.

Buckley, D. P., D. M. O'Malley, V. Apsit, G. T. Prance, and K. S. Bawa. 1988. Genetics of Brazil nut (*Bertholletia excelsa* Humb. & Bonpl.; Lecythidaceae). I. Genetic variation in natural populations. *Theor. Appl. Genet.* 76:923–928.

Burdon, J. J. 1980a. Intra-specific diversity in a natural population of *Trifolium repens*. *J. Ecol.* 68:717–736.

Burdon, J. J. 1980b. Variation in disease-resistance within a population of *Trifolium repens*. *J. Ecol.* 68:737–744.

Burdon, J. J. 1985. Pathogens and the genetic structure of plant populations. In *Studies on Plant Demography: A Festschrift for John L. Harper*, ed. J. White, pp. 313–325. Academic Press, London.

Burdon, J. J. 1987. *Diseases and Plant Population Biology*. Cambridge University Press, Cambridge.

Burdon, J. J., and D. R. Marshall. 1981. Biological control and the reproductive mode of weeds. *J. Appl. Ecol.* 18:649–658.

Burdon, J. J., D. R. Marshall, and A. H. D. Brown. 1983. Demographic and genetic changes in populations of *Echium plantagineum*. *J. Ecol.* 71:667–680.

Burdon, J. J., D. R. Marshall, and R. H. Groves. 1980. Isozyme variation in *Chondrilla juncea* L. in Australia. *Aust. J. Bot.* 28:193–198.

Burgman, M. A., H. R. Akcakaya, and S. S. Loew. 1988. The use of extinction models for species conservation. *Biol. Conserv.* 43:9–25.

Butlin, J. A., ed. 1981. *Economics and Resource Policy.* Longman, London.

Byron, P. A. 1980. On the ecology and systematics of Coloradan bumblebees. Ph.D. diss., University of Colorado, Boulder.

Cade, T. J. 1983. Hybridization and gene exchange among birds in relation to conservation. In *Genetics and Conservation: A Reference for Managing Wild Animal and Plant Populations,* ed. C. M. Schonewald-Cox, S. M. Chambers, B. MacBryde, and W. L. Thomas, pp. 288–310. Benjamin-Cummings, Menlo Park, Calif.

Cain, S. A. 1944. *Foundations of Plant Geography.* Harper, New York.

Callaham, R. Z., and A. R. Liddicoet. 1961. Altitudinal variation at 20 years in ponderosa and Jeffrey pines. *J. Forestry* 59:814–820.

Campbell, J. J. N., M. Evans, M. E. Medley, and N. L. Taylor. 1988. Buffalo clovers in Kentucky (*Trifolium stoloniferum* and *T. reflexum*): Historical records, presettlement environment, rediscovery, endangered status, cultivation, and chromosome number. *Rhodora* 90:399–418.

Campbell, R. B. 1986. The interdependence of mating structure and inbreeding depression. *Theor. Pop. Biol.* 30:232–244.

Carr, G., and D. W. Kyhos. 1981. Adaptive radiation in the Hawaiian silversword alliance (Compositae–Madiinae). I. Cytogenetics of spontaneous hybrids. *Evolution* 35:543–556.

Carr, K. M. 1988. The BCD System: The next generation of Heritage computer systems. *Biodiversity Network News,* Fall 1988. Nature Conservancy, Arlington, Va.

Carson, H. L. 1968. The population flush and its genetic consequences. In *Population Biology and Evolution,* ed. R. C. Lewontin, pp. 123–127. Syracuse University Press, Syracuse, N.Y.

Carson, H. L. 1983. The genetics of the founder effect. In *Genetics and Conservation,* ed. C. M. Schonewald-Cox, S. M. Chambers, B. MacBryde and W. L. Thomas, pp. 189–200. Benjamin-Cummings, Menlo Park, Calif.

Carson, H. L. 1987. Colonization and speciation. In *Colonization, Succession and Stability,* The 26th Symp. of Br. Ecol. Soc., ed. A. J. Gray, M. J. Crawley, and P. J. Edwards, pp. 187–206. Blackwell Scientific, Oxford.

Carson, H. L., and A. R. Templeton. 1984. Genetic revolutions in relation to speciation phenomena: The founding of new populations. *Annu. Rev. Ecol. Syst.* 15:97–131.

Carulli, J. P., and D. E. Fairbrothers. 1988. Allozyme variation in three eastern United States species of *Aeschynomene* (Fabaceae), including the rare *A. virginica. Syst. Bot.* 13:559–566.

Case, T. J., and M. L. Taper. 1986. On the coexistence and coevolution of asexual and sexual competitors. *Evolution* 40:366–387.

Caswell, H. 1982a. Optimal life histories and the maximization of reproductive value: A general theorem for complex life cycles. *Ecology* 63:1218–1222.

Caswell, H. 1982b. Stable population structure and reproductive value for populations with complex life cycles. *Ecology* 63:1223–1231.

Caswell, H. 1983. Phenotypic plasticity in life history traits: Demographic effects and evolutionary consequences. *Am. Zool.* 23:35–46.

Caswell, K. L., N. J. Tyler, and C. Stushnoff. 1986. Cold hardening of in vitro apple and Saskatoon shoot cultures. *HortScience* 21:1207–1209.

Center for Plant Conservation. 1986. Recommendations for the collection and *ex situ* management of germplasm resources from rare wild plants. Jamaica Plain, Mass.

Center for Plant Conservation. 1987. Criteria for selecting highest priority taxa for accession to the national collection of endangered plants. Jamaica Plain, Mass.

Center for Plant Conservation. 1988a. Endangerment survey. Jamaica Plain, Mass.

Center for Plant Conservation. 1988b. Database ranking program for highest priority species. Jamaica Plain, Mass.

Chandler, J. M., C. Jan, and B. H. Beard. 1986. Chromosomal differentiation among the annual *Helianthus* species. *Syst. Bot.* 11:363–371.

Chapin, F. S., III, and M. C. Chapin. 1981. Ecotypic differentiation of growth processes in *Carex aquatilis* along latitudinal and local gradients. *Ecology* 62:1000–1009.

Chapman, C. G. O. 1984. On the size of a genebank and the genetic variation it contains. In *Crop Genetic Resources, Conservation and Evaluation*, ed. J. H. W. Holden and J. T. Williams, pp. 102–118. Allen and Unwin, London.

Chapman, C. G. O. 1989. Collection strategies for the wild relatives of field crops. In *The Use of Plant Genetic Resources*, ed. A. H. D. Brown, O. H. Frankel, D. R. Marshall, and J. T. Williams, pp. 263–278. Cambridge University Press, Cambridge.

Charlesworth, D. 1988. A method of estimating outcrossing rates in natural populations of plants. *Heredity* 61:469–471.

Charlesworth, D., and B. Charlesworth. 1987. Inbreeding depression and its evolutionary consequences. *Annu. Rev. Ecol. Syst.* 18:237–268.

Charlesworth, D., and B. Charlesworth. 1990. Inbreeding depression with heterozygote advantage and its effect on selection for modifiers changing the outcrossing rate. *Evolution* 44:870–888.

Chen, T., and K. K. Kartha. 1987. Cryopreservation of woody species. In *Cell and Tissue Culture in Forestry*, Vol. 2, ed. J. M. Bonga and D. J. Durzan, pp. 305–319. Martinus Nijhoff, Dordrecht.

Cheplick, G. P. 1987. The ecology of amphicarpic plants. *Trends Ecol. Evol.* 2:97–101.

Chesser, R. K., M. H. Smith, and I. L. Brisbin. 1980. Management and maintenance of genetic variability in endangered species. *Int. Zool. Ybk.* 20:146–154.

Chin, H. F., and E. H. Roberts. 1980. *Recalcitrant Crop Seeds*. Tropical Press, Kuala Lumpur, Malaysia.

Cicmanec, J. C., and A. K. Campbell. 1977. Breeding the owl monkey (*Aotus trivirgattus*) in a laboratory environment. *Lab. Animal Sci.* 27:517.

Clampitt, C. A. 1987. Reproductive biology of *Aster curtus* (Asteraceae), a Pacific Northwest endemic. *Am. J. Bot.* 74:941–946.

Clausen, J., D. D. Keck, and W. M. Hiesey. 1940. *Experimental Studies on the Nature of Species. I. The Effect of Varied Environments on Western North American Plants*. Carnegie Institute of Washington, Publ. No. 520.

Clausen, J., D. D. Keck, and W. M. Hiesey. 1947. Heredity of geographically and ecologically isolated races. *Am. Nat.* 81:114–133.

Clausen, J., D. D. Keck, and W. M. Hiesey. 1948. *Experimental Studies on the Nature of Species. III. Environmental Responses of Climatic Races of Achillea*. Carnegie Institute of Washington, Publ. No. 581.

Clay, K., and J. Antonovics. 1985a. Demographic genetics of the grass *Danthonia spicata:* Success of progeny from chasmogamous and cleistogamous flowers. *Evolution* 39:205–210.

Clay, K., and J. Antonovics. 1985b. Quantitative variation of progeny from chasmogamous and cleistogamous flowers in the grass *Danthonia spicata. Evolution* 39:335–348.

Clegg, M. T. 1980. Measuring plant mating systems. *BioScience* 30:814–818.

Clegg, M. T., and R. W. Allard. 1972. Patterns of genetic differentiation in the slender wild oat species *Avena barbata. Proc. Nat. Acad. Sci USA* 69:1820–1824.

Clegg, M. T., and R. W. Allard. 1973. Viability versus fecundity selection in the slender wild oat, *Avena barbata. Science* 181:667–668.

Clegg, M. T., and A. H. D. Brown. 1983. The founding of plant populations. In *Genetics and*

Conservation, ed. C. M. Schonewald-Cox, S. M. Chambers, B. MacBryde, and L. Thomas, pp. 216–228. Benjamin-Cummings, Menlo Park, Calif.

Clegg, M. T., A. L. Kahler, and R. W. Allard. 1978a. Estimation of the life cycle components of selection in an experimental plant population. *Genetics* 89:765–792.

Clegg, M. T., A. L. Kahler, and R. W. Allard. 1978b. Genetic demography of plant populations. In *Ecological Genetics: The Interface,* ed. P. F. Brussard, pp. 173–188. Springer-Verlag, New York.

Cockerham, C. C., 1969. Variance of gene frequencies. *Evolution* 23:72–84.

Cockerham, C. C. 1973. Analyses of gene frequencies of mates. *Genetics* 74:701–712.

Cohen, D. 1966. Optimizing reproduction in a randomly varying environment. *J. Theor. Biol.* 12:119–129.

Cohen, J. E. 1979. Comparative statics and stochastic dynamics of age-structured populations. *Theor. Pop. Biol.* 16:159–171.

Cole, W. W., and S. C. H. Barrett. 1989. Genetic variation and patterns of sterility in the rare, disjunct, aquatic *Eichhornia paradoxa* (Pontederiaceae). *Am. J. Bot.* 76 (Suppl.):145.

Conkle, M. T. 1973. Growth data for 29 years from the California elevation transect study of ponderosa pine. *For. Sci.* 19:31–39.

Conkle, M. T. 1981. Isozyme variation and linkage in six conifer species. In *Proceed. Symp. Isozymes of North American Forest Trees and Forest Insects,* ed. M. T. Conkle, pp. 11–17. Gen. Tech. Rep. PSW-48. U.S.D.A. Forest Service, P.S.W. Forest and Range Experiment Station, Berkeley, Calif.

Conkle, M. T. 1987. Electrophoretic analysis of variation in native Monterey cypress (*Cupressus macrocarpa* Hartw.). In *Conservation and Management of Rare and Endangered Plants,* Proc. California Conference on Conservation and Management of Rare and Endangered Plants, ed. T. S. Elias, pp. 249–256. California Native Plant Society, Sacramento.

Conrad, J. M., and C. W. Clark. 1987. *Natural Resource Economics.* Cambridge University Press, Cambridge.

Cook, R. E., and E. E. Lyons. 1983. The biology of *Viola fimbriatula* in a natural disturbance. *Ecology* 64:654–660.

Copes, D. L. 1981. Isozyme uniformity in western red cedar seedlings from Oregon and Washington. *Can. J. For. Res.* 11:451–453.

Crawford, T. J. 1984. What is a population? In *Evolutionary Ecology,* ed. B. Shorrocks, pp. 135–174. British Ecol. Soc. Symp. No. 23. Blackwell Scientific, Oxford.

Critchfield, W. B. 1984. Impact of the Pleistocene on the genetic structure of North American conifers. In *Proc. 8th North American Forest Biology Workshop,* ed. R. M. Lanner, pp. 70–118. Logan, Utah.

Croat, R. B. 1978. *Flora of Barro Colorado Island.* Stanford University Press, Stanford, Calif.

Crocker, W. 1938. Life-span of seeds. *Bot. Rev.* 4:235–274.

Cromarty, A. S., R. H. Ellis, and E. H. Roberts. 1982. *The Design of Seed Storage Facilities for Genetic Conservation.* International Board for Plant Genetic Resources. IBPGR Secretariat, Rome.

Crossa, J. 1989. Methodologies for estimating the sample size required for genetic conservation of outbreeding crops. *Theor. Appl. Genet.* 77:153–161.

Crow, J. F., and C. Denniston. 1988. Inbreeding and variance effective population numbers. *Evolution* 42:482–495.

Crow, J. F., and M. Kimura. 1970. *An Introduction to Population Genetics Theory.* Harper & Row, New York.

Crumpacker, D. W. 1967. Genetic loads in maize (*Zea mays* L.) and other cross-fertilized plants and animals. *Evol. Biol.* 1:306–423.

Cwynar, L. C., and G. M. MacDonald. 1987. Geographical variation of lodgepole pine in relation to population history. *Am. Naturalist* 123:463–469.

Dallas, J. F. 1988. Detection of DNA "fingerprints" of cultivated rice by hybridization with a human minisatellite probe. *Proc. Natl. Acad. Sci. USA* 85:6831–6835.

D'Amato, F. 1985. Cytogenetics of plant cell and tissue cultures and their regenerates. *CRC Crit. Rev. Plant Sci.* 3:73–112.

Darwin, C. 1859. *On the Origin of Species.* John Murray, London.

Darwin, C. R. 1876. *The Effects of Cross and Self-Fertilization in the Vegetable Kingdom.* John Murray, London.

Davis, M. B. 1976. Pleistocene biogeography of temperate deciduous forests. *Geoscience and Man* 13:13–26.

Davis, S. D., S. J. M. Droop, P. Gregerson, L. Henson, C. J. Leon, J. L. Villa-Lobos, H. Synge, and J. Zantovska. 1986. *Plant in Danger: What Do We Know?* IUCN, Conservation Monitoring Center, Gland, Switzerland.

de Boer, L. E. M. 1982. Karyological problems in breeding owl monkeys, *Aotus trivirgatus. Int. Zoo Yrbk.* 22:119–124.

de Langhe, E. A. L. 1984. The role of in vitro techniques in germplasm conservation. In *Crop Genetic Resources: Conservation and Evaluation,* ed. J. H. W. Holden and J. T. Williams, pp. 131–137. Allen and Unwin, London.

Delouche, J. C. 1965. An accelerated aging technique for predicting relative storability of crimson clover and tall fescue seed lots. *Agron. Abstr.,* p. 40.

Delouche, J. C., and C. C. Baskin. 1973. Accelerated aging techniques for predicting the relative storability of seed lots. *Seed Sci. Technol.* 1:427–452.

De Mauro, M. 1989. Aspects of the reproductive biology of the endangered *Hymenoxys acaulis* var. *glabra:* Implications for conservation. M.Sc. Thesis, University of Illinois, Chicago.

Dennis, B. 1981. Extinction and waiting times in birth–death processes: Applications to endangered species and insect pest control. In *Statistical Distributions in Scientific Work,* Vol. 6, ed. C. Jaillie, G. P. Patil, and B. A. Baldessari, pp. 289–301. Reidel, Dordrecht.

Dennis, B., P. L. Munholland, and J. M. Scott. 1991. Estimation of growth and extinction parameters for endangered species. *Ecol. Monogr.* 61:115–143.

Denniston, C. 1978. Small population size and genetic diversity: Implications for endangered species. In *Endangered Birds: Management Techniques for Preserving Threatened Species,* ed. S. A. Temple. University of Wisconsin Press, Madison.

Diamond, J. M. 1975. The island dilemma: Lessons of modern biogeographic studies for the design of natural reserves. *Biol. Conserv.* 7:129–146.

Diamond, J. M., K. D. Bishop, and S. Van Balen. 1987. Bird survival in isolated Javan woodlands: Island or mirror? *Conserv. Biol.* 1:132–142.

Dirzo, R., and J. L. Harper. 1982. Experimental studies on slug–plant interactions. III. Differences in the acceptability of individual plants of *Trifolium repens* to slugs and snails. *J. Ecol.* 70:101–118.

Dobzhansky, T. 1948. The genetics of natural populations. XVIII. Experiments on chromosomes of *Drosophila pseudoobscura* from different geographical regions. *Genetics* 33:588–602.

Dobzhansky, T. 1970. *Genetics of the Evolutionary Process.* Columbia University Press, New York.

Dobzhansky, T., and B. Spassky. 1960. Release of genetic variability through recombination: V. Breakup of synthetic lethals by crossing-over in *Drosophila pseudoobscura. Zool. Jahrb. Abt. Syst.* 88:57–66.

Doyle, J. J., D. E. Soltis, and P. S. Soltis. 1985. An intergeneric hybrid in the Saxifragaceae: Evidence from ribosomal RNA genes. *Am. J. Bot.* 72:1388–1391.

Drury, W. H. 1980. Rare species of plants. *Rhodora* 82:3–48.

Dudash, M. R. 1990. Relative fitness of selfed and outcrossed progeny in a self-compatible, protandrous species, *Sabatia angularis* L. (Gentianaceae): A comparison in three environments. *Evolution* 44:1129–1139.

Echelle, A. A., and P. J. Conner. 1989. Rapid geographically extensive genetic introgression after secondary contact between two pupfish species (*Cyprinidon*, Cyprinodontidae). *Evolution* 43:717–727.

Echelle, A. F., A. E. Echelle, and D. R. Edds. 1989. Conservation genetics of a spring-dwelling desert fish, the Pecos gambusia (*Gambusia nobilis*, Poeciliidae). *Conserv. Biol.* 3:159–169.

Eckenwalder, J. E. 1984. Natural intersectional hybridization between North American species of *Populus* (Salicaceae) in sections *Aigeiros* and *Tacamahaca*. III. Paleobotany and evolution. *Cand. J. Bot.* 62:336–342.

Edwards, S. W. 1989. The role of botanical gardens in a deteriorating global environment. *Fremontia* 18(1):3–8.

Ehrlich, P. R., and D. D. Murphy. 1987. Conservation lessons from long-term studies of checkerspot butterflies. *Conserv. Biol.* 1:122–131.

Ehrlich, P. R., and P. H. Raven. 1969. Differentiation of populations. *Science* 165:1228–1232.

Ellis, R. H., T. D. Hong, and E. H. Roberts. 1985a. *Handbook of Seed Technology for Genebanks*, Vol. 1: *Principle and Methodology*. Handbooks for Genebanks No. 2. International Board for Plant Genetic Resources, IBPGR Secretariat, Rome.

Ellis, R. H., T. D. Hong, and E. H. Roberts. 1985b. *Handbook of Seed Technology for Genebanks*, Vol. 2: *Compendium of Specific Germination Information and Test Recommendations*. Handbooks for Genebanks No. 3. International Board for Plant Genetic Resources, IBPGR Secretariat, Rome.

Ellis, R. H., and E. H. Roberts. 1981. The quantification of ageing and survival in orthodox seeds. *Seed Sci. Technol.* 9:373–409.

Ellstrand, N. C., and D. A. Levin. 1980. Recombination system and population structure in *Oenothera*. *Evolution* 34:923–933.

Ellstrand, N. C., and M. L. Roose. 1987. Patterns of genotypic diversity in clonal plant species. *Am. J. Bot.* 74:132–135.

Ellstrand, N. C., A. M. Torres, and D. A. Levin. 1978. Density and the rate of apparent outcrossing in *Helianthus annuus* (Asteraceae). *Syst. Bot.* 3:403–407.

Elton, C. S. 1958. *The Ecology of Invasions by Animals and Plants.* Methuen, London.

Emerson, S. 1939. A preliminary survey of the *Oenothera organensis* population. *Genetics* 24:524–537.

Emerson, S. 1940. Growth of incompatible pollen tubes in *Oenothera organensis*. *Bot. Gaz.* 101:890–911.

Endler, J. A. 1986. *Natural Selection in the Wild.* Princeton University Press, Princeton, N.J.

Ennos, R. A. 1985a. The mating system and genetic structure in a perennial grass *Cynosurus cristatus* L. *Heredity* 55:121–126.

Ennos, R. A. 1985b. The significance of genetic variation for root growth within a natural population of white clover (*Trifolium repens*). *J. Ecol.* 73:615–624.

Enright, N. J. 1982. Recognition of successional pathways in forest communities using size–class ordination. *Vegetation* 48:133–140.

Erickson, R. O. 1945. The *Clematis fremontii* var. *riehlii* population in the Ozarks. *Ann. Missouri Bot. Gard.* 32:413–460.

Ewart, A. J. 1908. On the longevity of seeds. *Proc. R. Soc. Victoria* 21:1–210.

Ewens, W. J. 1972. The sampling theory of selectively neutral alleles. *Theor. Pop. Biol.* 3:87–112.

Ewens, W. J., P. J. Brockwell, J. M. Gani, and S. I. Resnick. 1987. Minimum viable population size in the presence of catastrophes. In *Viable Populations for Conservation*, ed. M. E. Soulé, pp. 59–86. Cambridge University Press, Cambridge.

Fahrig, L., and G. Merriam. 1985. Habitat patch connectivity and population survival. *Ecology* 66:1762–1768.

Falconer, D. S. 1981. *Introduction to Quantitative Genetics*, 2nd ed. Longman, London.

Falk, D. A. 1987a. Integrated conservation strategies for endangered plants. *Natural Areas J.* 7(3):118–123.

Falk, D. A. 1987b. Strategies for endangered plant conservation: Integrating on-site and off-site approaches. In Proc. 1st Annu. Meetg. Soc. Conserv. Biologists, Bozeman, Montana, June 1987.

Falk, D. A. 1990a. The theory of integrated conservation strategies for biological diversity. In *Ecosystem Management: Rare Species and Significant Habitats*, Proc. 15th Annual Natural Areas Conference, ed. R. S. Mitchell, C. J. Sheviak, and D. J. Leopold, pp. 5–10.

Falk, D. A. 1990b. From conservation biology to conservation practice: Strategies for protecting plant diversity. In *Conservation Biology: The Theory and Practice of Nature Conservation, Preservation, and Management*, ed. P. L. Fiedler and S. Jain. Chapman and Hall, New York (in press).

Falk, D. A. 1990c. Integrated strategies for conserving plant genetic diversity. *Ann. Missouri Bot. Gard.* 77:38–47.

Fenster, C. B. 1991a. Gene flow in *Chamaecrista fasciculata* (Leguminosae). I. Gene dispersal. *Evolution* 45:398–409.

Fenster, C. B. 1991b. Gene flow in *Chamaecrista fasciculata* (Leguminosae). II. Gene establishment. *Evolution* 45:410–422.

Fenster, C. B., and V. L. Sork. 1988. Effect of crossing distance and male parent on in vivo pollen tube growth in *Chamaecrista fasciculata*. *Am. J. Bot.* 75:1898–1903.

Fernald, M. L. 1926. The antiquity and dispersal of vascular plants. *Q. Rev. Biol.* 1:212–245.

Ferriera, V. 1980. Introgressive hybridization between *Helianthus annuus* L. and *Helianthus petiolaris* Nutt. *Mendeliana* 4:81–93.

Fiedler, P. L. 1985. Heavy metal accumulation and the nature of edaphic endemism in the genus *Calochortus* (Liliaceae). *Am. J. Bot.* 72:1712–1718.

Fiedler, P. L. 1986. Concepts of rarity in vascular plant species, with special reference to the genus *Calochortus* Pursh (Liliaceae). *Taxon* 35:502–518.

Fiedler, P. L. 1987. Life history and population dynamics of rare and common mariposa lilies (*Calochortus* Pursh; Liliaceae). *J. Ecol.* 75:977–995.

Finkle, B. J., M. E. Zavala, and J. M. Ulrich. 1985. Cryoprotective compounds in the viable freezing of plant tissues. In *Cryopreservation of Plant Cells and Organs*, ed. K. K. Kartha, pp. 75–113. CRC Press, Boca Raton, Fla.

Fins, L., and W. J. Libby. 1982. Population variation in *Sequoiadendron:* Seed and seedling studies, vegetative propagation, and isozyme variation. *Silvae Genetica* 31:101–148.

Fisher, A. C. 1981. *Resource and Environmental Economics*. Cambridge University Press, Cambridge.

Fisher, R. A. 1930. *The Genetical Theory of Natural Selection*. Oxford University Press, Oxford.

Fisher, R. A. 1941. Average excess and average effect of gene substitution. *Ann. Eugenics* 11:53–63.

Fisher, R. A. 1949. *The Theory of Inbreeding*. Oliver and Boyd, Edinburgh.

Flora of North America Editorial Committee. 1992. *Flora of North America*. Oxford University Press, New York.

Forde, M. B. 1964. Variation in natural populations of *Pinus radiata* in California. Parts 1–4. *N.Z. J. Bot.* 2:213–259; 459–501.

Fowler, D. P., and R. W. Morris. 1977. Genetic diversity in red pine: Evidence for low genetic heterogeneity. *Can. J. For. Res.* 7:343–347.

Frankel, O. H. 1970. Variation, the essence of life. Sir William Macleay Memorial Lecture. *Proc. Linn. Soc.* 95:158–169.

Frankel, O. H. 1983. The place of management in conservation. In *Genetics and Conservation,* ed. C. M. Schonewald-Cox, S. M. Chambers, B. MacBryde, and L. Thomas, pp. 1–14. Benjamin-Cummings, Menlo Park, Calif.

Frankel, O. H., and E. Bennett, eds. 1970. *Genetic Resources in Plants: Their Exploration and Conservation.* Blackwell Scientific, Oxford.

Frankel, O. H., and A. H. D. Brown. 1984. Current plant genetic resources—A critical appraisal. In *Genetics: New Frontiers,* Vol. 4, pp. 1–11. India Book House, New Delhi.

Frankel, O. H., and J. G. Hawkes, eds. 1975. *Crop Genetic Resources for Today and Tomorrow.* Cambridge University Press, Cambridge.

Frankel, O. H., and M. E. Soulé. 1981. *Conservation and Evolution.* Cambridge University Press, Cambridge.

Franklin, I. R. 1980. Evolutionary change in small populations. In *Conservation Biology: An Evolutionary–Ecological Perspective,* ed. M. E. Soulé and B. A. Wilcox, pp. 135–150. Sinauer, Sunderland, Mass.

Franz, G., and W. Kunz. 1981. Intervening sequences in ribosomal RNA genes and bobbed phenotype in *Drosophila hydei. Nature* 292:683–640.

Fritz, R. S. 1979. Consequences of insular population structure: Distribution and extinction of spruce grouse populations. *Oecologia* 42:57–65.

Fuerst, P. A., and T. Maruyama. 1986. Considerations on the conservation of alleles and of genic heterozygosity in small managed populations. *Zoo Biol.* 5:171–180.

Furnier, G. R., and W. T. Adams. 1986. Geographic patterns of allozyme variation in Jeffrey pine. *Am. J. Bot.* 73:1009–1015.

Futuyma, D. J. 1983. Interspecific interactions and the maintenance of genetic diversity. In *Genetics and Conservation,* ed. C. M. Schonewald-Cox, S. M. Chambers, B. MacBryde, and L. Thomas, pp. 364–373. Benjamin-Cummings, Menlo Park, Calif.

Fyfe, J. L., and N. T. J. Bailey. 1951. Plant breeding studies in leguminous forage crops. I. Natural cross-breeding in winter beans. *J. Agric. Sci.* 41:371–378.

Gallez, G. P., and L. D. Gottlieb. 1982. Genetic evidence for the hybrid origin of the diploid plant *Stephanomeria diegensis. Evolution* 36:1158–1167.

Gan, Y. Y., F. W. Robertson, P. S. Ashton, E. Soepadmo, and D. W. Lee. 1977. Genetic variation in wild populations of rain forest trees. *Nature* 269:323–325.

Ganders, F. R., and K. M. Nagata. 1984. The role of hybridization in the evolution of *Bidens* on the Hawaiian Islands. In *Plant Biosystematics,* ed. W. F. Grant, pp. 179–194. Academic Press, Toronto.

Garbutt, K., and F. A. Bazzaz. 1987. Population niche structure: Differential response of *Abutilon theophrasti* progeny to resource gradients. *Oecologia* 72:291–296.

Garbutt, K., F. A. Bazzaz, and D. A. Levin. 1985. Population and genotype niche width in clonal *Phlox paniculata. Am. J. Bot.* 72:640–648.

Gentry, A. H. 1988. Tree species richness of upper Amazonian forests. *Proc. Natl. Acad. Sci. U.S.A.* 85:156–159.

Gentry, A. H., and C. Dodson. 1987. Diversity and phytogeography of neotropical vascular epiphytes. *Ann. Missouri Bot. Gard.* 74:205–233.

Georgiadis, N., P. Dunham, B. Read, and A. R. Templeton. 1988. HK/LK polymorphism and its genetic determination in Speke's gazelle. *J. Hered.* 79:325–331.

Gilbert, N., and S. B. Lee. 1980. Two perils of plant population dynamics. *Oecologia* 46:283–284.

Giles, B. E. 1983. A comparison of quantitative and biochemical variation in the wild barley *Hordeum murinum*. *Evolution* 38:34–41.

Gillett, G. W. 1966. Hybridization and its taxonomic implications in the *Scaveola quadichandiana* complex in the Hawaiian Islands. *Evolution* 20:206–216.

Gillett, G. W. 1972. The role of hybridization in the evolution of the Hawaiian flora. In *Taxonomy, Phytogeography, and Evolution*, ed. D. H. Valentine, p. 205. Academic Press, London.

Gillett, G. W., and E. K. S. Lim. 1970. An experimental study of the genus *Bidens* (Asteraceae) in the Hawaiian Islands. *Univ. Calif. Publ. Bot.* 56:1–63.

Gilpin, M. E. 1987. Spatial structure and population vulnerability. In *Viable Populations for Conservation*, ed. M. E. Soulé, pp. 125–139. Cambridge University Press, Cambridge.

Gilpin, M. E., and M. E. Soulé. 1986. Minimum viable populations: Processes of species extinction. In *Conservation Biology*, ed. M. E. Soulé, pp. 19–34. Sinauer, Sunderland, Mass.

Given, D. R. 1987. What the conservationist requires of *ex situ* collections. In *Botanic Gardens and the World Conservation Strategy*, ed. D. Branwell, O. Hamann, V. Heywood, and H. Synge, pp. 103–116. Academic Press, London.

Given, D. R. 1990. Conserving botanical diversity on a global scale. *Ann. Missouri Bot. Gard.* 77:48–62.

Glover, D. E., and S. C. H. Barrett. 1987. Genetic variation in continental and island populations of *Eichhornia paniculata* (Pontederiaceae). *Heredity* 59:7–17.

Goff, F. G., and D. West. 1975. Canopy–understory interaction effects on forest population structure. *For. Sci.* 22:98–108.

Golenberg, E. M. 1987. Estimation of gene flow and genetic neighborhood size by indirect methods in a selfing annual. *Evolution* 41:1326–1334.

Good, R. B., and P. S. Lavarack. 1981. The status of Australian plants at risk. In *The Biological Aspects of Rare Plant Conservation*, ed. H. Synge, pp. 81–91. Wiley, New York.

Goodman, D. 1984. Risk spreading as an adaptive strategy in iteroparous life histories. *Theor. Pop. Biol.* 25:1–20.

Goodman, D. 1987a. How do any species persist? Lessons for conservation biology. *Conserv. Biol.* 1:59–62.

Goodman, D. 1987b. The demography of chance extinction. In *Viable Populations for Conservation*, ed. M. E. Soulé, pp. 11–34. Cambridge University Press, Cambridge.

Goodnight, C. J. 1987. On the effect of founder events on epistatic genetic variance. *Evolution* 41:80–91.

Goodnight, C. J. 1988. Epistasis and the effect of founder events on epistatic genetic variance. *Evolution* 42:441–454.

Gottlieb, L. D. 1973a. Enzyme differentiation and phylogeny in *Clarkia franciscana, Clarkia rubicunda*, and *Clarkia amoena*. *Evolution* 27:205–214.

Gottlieb, L. D. 1973b. Genetic differentiation, sympatric speciation, and the origin of a diploid species of *Stephanomeria*. *Am. J. Bot.* 60:545–553.

Gottlieb, L. D. 1974. Genetic confirmation of the origin of *Clarkia lingulata*. *Evolution* 28:244–250.

Gottlieb, L. D. 1977a. Genotypic similarity of large and small individuals in a natural population of the annual plant *Stephanomeria exigua* ssp. *coronaria*. *J. Ecol.* 65:127–134.

Gottlieb, L. D. 1977b. Electrophoretic evidence and plant systematics. *Ann. Missouri Bot. Gard.* 64:161–180.

Gottlieb, L. D. 1979. The origin of phenotype in a recently evolved species. In *Topics in Plant Population Biology*, ed. O. T. Solbrig, S. Jain, G. B. Johnson, and P. H. Raven, pp. 264–286. Columbia University Press, New York.

Gottlieb, L. D. 1981. Electrophoretic evidence and plant populations. *Prog. Phytochem.* 7:1–46.

Gottlieb, L. D., and G. Pilz. 1976. Genetic similarity between *Gaura longiflora* and its obligately outcrossing derivative *G. demareei. Syst. Bot.* 1:181–187.

Gottlieb, L. D., S. I. Warwick, and V. S. Ford. 1985. Morphological and electrophoretic divergence between *Layia discoidea* and *L. glandulosa. Syst. Bot.* 10:484–495.

Grant, V. 1981. *Plant Speciation,* 2nd ed. Columbia University Press, New York.

Griffiths, R. C. 1979. Exact sampling distributions from the infinite neutral alleles model. *Adv. Appl. Prob.* 11:326–354.

Griggs, F. T., and S. K. Jain. 1983. Conservation of vernal pool plants in California, II. Population biology of a rare and unique grass genus *Orcuttia. Biol. Conserv.* 27:171–193.

Gross, K. L. 1981. Predictions of fate from rosette size in four "biennial" plant species: *Verbascum thapsus, Oenothera biennis, Daucus carota,* and *Tragopogon dubious. Oecologia* 48:209–213.

Grout, B. W. W., K. Shelton, and H. W. Pritchard. 1984. Orthodox behavior of oil palm seed and cryopreservation of the excised embryo for genetic conservation. *Ann. Bot.* 52:381–384.

Grubb, P. J. 1977. The maintenance of species richness in plant communities: The importance of the regeneration niche. *Biol. Rev.* 52:107–145.

Guries, R. P., and F. T. Ledig. 1982. Genetic diversity and population structure in pitch pine (*Pinus rigida*). *Evolution* 36:387–402.

Haldane, J. B. S. 1930. Theoretical genetics of autopolyploids. *J. Genet.* 22:359–373.

Hamilton, N. R., B. Schmid, and J. L. Harper. 1987. Life-history concepts and the population biology of clonal organisms. *Proc. R. Soc. Biol. Lond.* 232:35–57.

Hamilton, W. D. 1982. Pathogens as causes of genetic diversity in their host populations. In *Population Biology of Infectious Diseases,* ed. R. M. Anderson and R. M. May, pp. 269–296. Springer-Verlag, New York.

Hamrick, J. L. 1976. Variation and selection in western montane species. II. Variation within and between populations of white fir on an environmental transect. *Theor. Appl. Genet.* 47:27–34.

Hamrick, J. L. 1983. The distribution of genetic variation within and among natural plant populations. In *Genetics and Conservation,* ed. C. M. Schonewald-Cox, S. M. Chambers, B. MacBryde and W. L. Thomas, pp. 335–348. Benjamin-Cummings, Menlo Park, Calif.

Hamrick, J. L. 1989. Isozymes and analysis of genetic structure of plant populations. In *Isozymes in Plant Biology,* ed. D. Soltis and P. Soltis, pp. 87–105. Dioscorides Press, Washington, D.C.

Hamrick, J. L., and M. J. W. Godt. 1989. Allozyme diversity in plant species. In *Plant Population Genetics, Breeding, and Genetic Resources,* ed. A. H. D. Brown, M. T. Clegg, A. L. Kahler, and B. S. Weir, pp. 43–63. Sinauer, Sunderland, Mass.

Hamrick, J. L., and L. R. Holden. 1979. Influence of microhabitat heterogeneity on gene frequency distribution and gametic phase disequilibrium in *Avena barbata. Evolution* 33:521–533.

Hamrick, J. L., and W. J. Libby. 1972. Variation and selection in western U.S. montane species. I. White fir. *Silvae genetica* 21(1–2):29–35.

Hamrick, J. L., Y. B. Linhart, and J. B. Mitton. 1979. Relationships between life history characteristics and electrophoretically detectable genetic variation in plants. *Annu. Rev. Ecol. Syst.* 10:173–200.

Hamrick, J. L., and M. D. Loveless. 1986. Allozyme variation in tropical trees: Techniques and preliminary results. *Brotropica* 18:201–207.

Hamrick, J. L., and M. D. Loveless. 1989. Associations between the breeding system and the genetic structure of tropical tree populations. In *Evolutionary Ecology of Plants*, ed. J. Bock and Y. B. Linhart, pp. 129–146. Westview Press, Boulder, Colo.

Hamrick, J. L., J. B. Mitton, and Y. B. Linhart. 1981. Levels of genetic variation in trees: Influence of life history characteristics. In *Proc. Symp. Isozymes of North American Forest Trees and Forest Insects*, ed. M. T. Conkle, pp. 35–41. Gen. Tech. Rep. PSW-48. U.S.D.A. Forest Service, P.S.W. Forest and Range Experiment Station, Berkeley, Calif.

Hanna, W. W., G. W. Burton, and W. G. Monson. 1986. Long-term storage of pearl millet pollen. *J. Hered.* 77:361–362.

Harding, J., and C. B. Mankinen. 1972. Genetics of *Lupinus*. IV. Colonization and genetic variability in *Lupinus succulentus*. *Theor. Appl. Genet.* 42:267–271.

Harlan, J. R. 1983. Some merging of plant populations. In *Genetics and Conservation: A reference for Managing Wild Animal and Plant Populations*, ed. C. M. Schonewald-Cox, S. M. Chambers, B. MacBryde, and W. L. Thomas, pp. 267–276. Benjamin-Cummings, Menlo Park, Calif.

Harlan, J. R., and J. M. J. de Wet. 1963. The compilospecies concept. *Evolution* 17:497–501.

Harper, J. L. 1977. *Population Biology of Plants*. Academic Press, London.

Harper, J. L. 1981. The meanings of rarity. In *The Biological Aspects of Rare Plant Conservation*, ed. H. Synge, pp. 189–203. Wiley, Chichester.

Harrington, J. F. 1972. Seed storage and longevity. In *Seed Biology*, Vol. 3, ed. T. T. Kozlowski, pp. 145–245. Academic Press, New York.

Harris, L. D. 1984. *The Fragmented Forest: Island Biogeographic Theory and the Preservation of Biotic Diversity*. University of Chicago Press, Chicago.

Harris, L. D., M. E. McGlothlen, and M. N. Manlowe. 1984. Genetic resources and biotic diversity. In *The Fragmented Forest: Island Biogeographic Theory and the Preservation of Biotic Diversity*, ed. L. D. Harris, pp. 93–107. University of Chicago Press, Chicago.

Harrison, S., D. D. Murphy, and P. R. Ehrlich. 1988. Distribution of the bay checkerspot butterfly, *Euphydras bayensis:* Evidence for a metapopulation model. *Am. Naturalist* 132:360–382.

Harry, D. E. 1984. Genetic structure of incense-cedar populations. Ph.D. diss., University of California, Berkeley.

Harry, D. E. 1987. Shoot elongation and growth plasticity in incense-cedar. *Can. J. For. Res.* 17:484–489.

Haskins, R. H., and K. K. Kartha. 1980. Freeze preservation of pea meristems: Cell survival. *Can. J. Bot.* 58:833–840.

Hawkes, J. G. 1976. Sampling gene pools. In *Conservation of Threatened Plants*, ed. J. B. Simmons, R. I. Benzer, P. C. Brandham, G. L. Lucas, and V. T. H. Parry, pp. 145–154. Plenum, New York.

Hawkes, J. G. 1986. Suggested sampling and conservation system for coffee in Ethiopia. PGRCE-ILCA Germplasm Newsletter No. 11, pp. 25–28. Addis Ababa, Ethiopia.

Hawkes, J. G. 1987a. A strategy for seed banking in botanic gardens. In *Botanic Gardens and the World Conservation Strategy*, ed. D. Bramwell, O. Hamman, V. Heywood, and H. Synge, pp. 131–149. Academic Press, London.

Hawkes, J. G. 1987b. World strategies for collecting, preserving, and using genetic resources. In *Improving Vegetatively Propagated Crops*, pp. 285–301.

Hecker, R. J., P. C. Stanwood, and C. A. Soulis. 1986. Storage of sugarbeet pollen. *Euphytica* 35:777–783.

Hedrick, P. W. 1983. *Genetics of Populations*. Science Books International, Boston.

Hedrick, P. W. 1987. Genetic load and the mating system in homosporous ferns. *Evolution* 41:1282–1289.

Heiser, C. B. 1947. Hybridization between the sunflower species *Helianthus annuus* and *H. petiolaris. Evolution* 1:249–262.

Heiser, C. B. 1949. Study in the evolution of the sunflower species *Helianthus annuus* and *H. bolanderi. Univ. Calif. Publ. Bot.* 23:157–196.

Heiser, C. B. 1958. Three new annual sunflowers (*Helianthus*) from the southwestern United States. *Rhodora* 60:272–283.

Heiser, C. B. 1965. Sunflowers, weeds and cultivated plants. In *The Genetics of Colonizing Species,* ed. H. G. Baker and G. L. Stebbins, pp. 147–172. Academic Press, New York.

Heiser, C. B. 1973. Introgression re-examined. *Bot. Rev.* 39:347–366.

Heiser, C. B. 1978. Taxonomy of *Helianthus* and origin of domesticated sunflower. In *Sunflower Science and Technology,* ed. J. F. Carter, pp. 31–53. American Society of Agronomy, Madison, Wis.

Heiser, C. B., D. M. Smith, S. B. Clevenger, and W. C. Martin. 1969. The North American sunflowers (*Helianthus*). *Mem. Torrey Bot. Club* 22:1–218.

Helenurm, K., and F. R. Ganders. 1985. Adaptive radiation and genetic differentiation in the Hawaiian *Bidens. Evolution* 39:753–765.

Heuch, I. 1980. Loss of incompatibility types in finite populations of the heterostylous plant *Lythrum salicaria. Hereditas* 92:53–57.

Heusmann, H. W. 1974. Mallard–black duck relationships in the Northeast. *Wildlife Soc. Bull.* 2:171–177.

Heywood, J. S. 1986. The effect of plant size variation on genetic drift in populations of annuals. *Am. Naturalist* 127:851–861.

Heywood, J. S., and D. A. Levin. 1985. Associations between allozyme frequencies and soil characteristics in *Gaillardia pulchella. Evolution* 39:1076–1086.

Hiebert, R. D., and J. L. Hamrick. 1983. Patterns and levels of genetic variation in Great Basin bristlecone pine, *Pinus longaeva. Evolution* 37:302–310.

Hirsh, A. G. 1987. Vitrification in plants as a natural form of cryoprotection. *Cryobiology* 24:214–228.

Hirsh, A. G., R. J. Williams, and H. T. Meryman. 1985. A novel method of natural cryoprotection. *Plant Physiol.* 79:41–56.

Hoekstra, F. A. 1986. Water content in relation to stress in pollen. In *Membranes, Metabolism and Dry Organisms,* ed. A. C. Leopold, pp. 102–122. Cornell University Press, Ithaca, N.Y.

Holsinger, K. E. 1985. A phenetic study of *Clarkia unguiculata* Lindley (Onagraceae) and its relatives. *Syst. Bot.* 10:155–165.

Holsinger, K. E. 1986. Dispersal and plant mating systems: The evolution of self-fertilization in sub-divided populations. *Evolution* 40:405–413.

Holsinger, K. E. 1988a. The evolution of self-fertilization in plants: Lessons from population genetics. *Oecologia Plantarum* 9:95–102.

Holsinger, K. E. 1988b. Inbreeding depression doesn't matter: The genetic basis of mating system evolution. *Evolution* 42:1235–1244.

Holsinger, K. E., and L. D. Gottlieb. 1989. The conservation of rare and endangered plants. *Trends Ecol. Evol.* 4:193–194.

Holsinger, K. E., and J. Roughgarden. 1985. A model for the dynamics of an annual plant population. *Theor. Pop. Biol.* 28:288–313.

Holtsford, T. 1989. Genetic causes and consequences of variation in outcrossing rate in *Clarkia tembloriensis* (Onagraceae). Ph.D. diss., University of California, Riverside.

Homoya, M. L., J. R. Aldrich, and E. M. Jacquart. 1989. The rediscovery of the globally endangered clover, *Trifolium stoloniferum,* in Indiana. *Rhodora* 91:207–212.

Hopper, S. D., and G. F. Moran. 1981. Bird pollination and the mating system of *Eucalyptus stoatei. Aust. J. Bot.* 29:625–638.

Hu, C. Y., and P. J. Wang. 1983. Meristem, shoot tip and bud culture. In *Handbook of Plant Cell Culture,* Vol. 1: *Techniques for Propagation and Breeding,* ed. D. A. Evans, W. R. Sharp, P. V. Ammirato, and Y. Yamada, pp. 177–227. Macmillan, New York.

Hubbell, S. P., and R. B. Foster. 1983. Diversity of canopy trees in a neotropical forest and implications for conservation. In *Tropical Rain Forest: Ecology and Management,* ed. T. C. Whitmore and S. Chadwick, pp. 25–41. Blackwell Scientific, Oxford.

Hubbell, S. P., and R. B. Foster. 1986a. Biology, chance, and history and the structure of tropical rain forest tree communities. In *Community Ecology,* ed. T. J. Case and J. Diamond, pp. 314–329. Harper & Row, New York.

Hubbell, S. P., and R. B Foster. 1986b. Commonness and rarity in a neotropical forest. In *Conservation Biology,* ed. M. Soulé, pp. 205–231. Sinauer, Sunderland, Mass.

Huenneke, L. F., K. E. Holsinger, and M. E. Palmer. 1986. Plant population biology and the management of viable plant populations. In *The Management of Viable Populations: Theory, Applications, and Case Studies,* ed. B. A. Wilcox, P. F. Brussard, and B. G. Marcot, pp. 169–183. Center for Conservation Biology, Stanford, Calif.

Huettel, M., P. A. Fuerst, T. Maruyama, and R. Chakraborty. 1980. Genetic effects of multiple bottlenecks in the Mediterranean fruit fly (*Ceratitis capitata*). *Genetics* 94:547–548.

Hume, L., and P. B. Cavers. 1981. A methodological problem in genecology: Seeds versus clones as source material for uniform gardens. *Can. J. Bot.* 59:763–768.

Hutchinson, J. F., and R. H. Zimmerman. 1987. Tissue culture of temperate fruit and nut trees. *Hort. Rev.* 9:273–349.

Hyman, E. L., and B. Stiftel. 1988. *Combining Facts and Values in Environmental Impact Assessment.* Social Impact Assessment Series No. 16. Westview Press, Boulder, Colo.

Imrie, B. C., C. J. Kirkman, and D. R. Ross. 1972. Computer simulation of a sporophytic self-incompatibility breeding system. *Aust. J. Biol. Sci.* 25:343–349.

Jain, S. K. 1976. The evolution of inbreeding in plants. *Annu. Rev. Ecol. Syst.* 7:469–495.

Jain, S. 1979. Adaptive strategies, polymorphism, plasticity, and homeostasis. In *Topics in Plant Population Biology,* ed. O. T. Solbrig, S. Jain, G. B. Johnson, and P. H. Raven, pp. 160–187. Columbia University Press, New York.

Jain, S. K., and P. S. Martins, 1979. Ecological genetics of the colonizing ability of rose clover (*Trifolium hirtum* All.). *Am. J. Bot.* 66:361–366.

Jain, S. K., and K. N. Rai. 1974. Population biology of *Avena.* IV. Polymorphism in small populations of *Avena fatua. Theor. Appl. Genet.* 44:7–11.

Janzen, D. H. 1983. No park is an island: Increase in interference from outside as park size decreases. *Oikos* 41:402–410.

Janzen, D. H. 1987. Insect diversity of Costa Rican dry forest: Why keep it and how? *Biol. J. Linn. Soc.* 30:343–356.

Janzen, D. H., and P. S. Martin. 1982. Neotropical anachronisms: The fruits the gomphotheres ate. *Science* 215:19–27.

Jarvinen, O. 1982. Conservation of endangered plant populations: Single large or several small reserves? *Oikos* 38:301–307.

Jefferies, R. L., and L. D. Gottlieb. 1982. Genetic differentiation of the microspecies *Salicornia europaea* L. (*Sensu stricto*) and *S. ramosissima* J. Woods. *New Phytol.* 92:123–129.

Jeffreys, A. J., V. Wilson, and S. L. Thein. 1985. Individual-specific "fingerprints" of human DNA. *Nature* 316:76–79.

Jenkins, R. E. 1989. Long-term conservation and preserve complexes. *Nature Conservancy Mag.* 39:4–7.

Johansen, C. A. 1977. Pesticides and pollinators. *Annu. Rev. Entomol.* 22:177–192.

Johnston, R. F., D. Siegel-Causey, and S. G. Johnson. 1988. European populations of the rock dove *Columba livia* and genotypic extinction. *Am. Midland Nat.* 120:1–10.

Jolly, D. 1989. Letter on USFS Southwestern Region Policy on protection of hybrid plants. Memo 2670, U.S.F.S. Region 3, Albuquerque, N.M.

Justice, O. L., and L. N. Bass. 1978. *Principles and Practices of Seed Storage.* U.S.D.A. Handbk. 506. Government Printing Office, Washington, D.C.

Karron, J. D. 1987a. A comparison of levels of genetic polymorphisms and self-compatibility in geographically restricted and widespread plant congeners. *Evol. Ecol.* 1:47–58.

Karron, J. D. 1987b. The pollination ecology of co-occurring geographically restricted and widespread species of *Astragalus* (Fabaceae). *Biol. Conserv.* 39:179–193.

Karron, J. D. 1989. Breeding systems and levels of inbreeding depression in geographically restricted and widespread species of *Astragalus* (Fabaceae). *Am. J. Bot.* 76:331–340.

Karron, J. D., Y. B. Linhart, C. A. Chaulk, and C. A. Robertson. 1988. Genetic structure of populations of geographically restricted and wide-spread species of *Astragalus* (Fabaceae). *Am. J. Bot.* 75:1114–1119.

Kartha, K. K. ed. 1985a. *Cryopreservation of Plant Cells and Organs.* CRC Press, Boca Raton, Fla.

Kartha, K. K. 1985b. Meristem culture and germplasm preservation. In *Cryopreservation of Plant Cells and Organs,* ed. K. K. Kartha, pp. 115–134. CRC Press, Boca Raon, Fla.

Katano, M., A. Ishihara, and A. Sakai. 1983. Survival of dormant apple shoot tips after immersion in liquid nitrogen. *HortScience* 18:707–708.

Keiding, N. 1975. Extinction and exponential growth in random environments. *Theor. Pop. Biol.* 8:49–63.

Kelley, S. E., and K. Clay. 1987. Interspecific competitive interactions and the maintenance of genotypic variation within two perennial grasses. *Evolution* 41:92–103.

Kerhoas, C., G. Gay, and C. Dumas. 1987. A multidisciplinary approach to the study of the plasma membrane of *Zea mays* pollen during controlled dehydration. *Planta* 171:1–10.

Kesseli, R. V., and S. K. Jain. 1984. New variation and biosystematic patterns detected by allozyme and morphological comparisons in *Limnanthes* sect. Reflexae (Limnanthaceae). *Plant Syst. Evol.* 147:133–165.

Kesseli, R. V., and S. K. Jain. 1986. Breeding systems and population structure in *Limnanthes. Theor. Appl. Genet.* 71:292–299.

Kevan, P. G. 1975. Forest application of the insecticide fenitrothion and its effect on wild bee pollinators (Hymenoptera: Apoidea) of lowbush blueberries (*Vaccinium* spp.) in southern New Brunswick, Canada. *Biol. Conserv.* 7:301–309.

Kevers, C., M. Coumans, M.-F. Coumans-Gilles, and Th. Gaspar. 1984. Physiological and biochemical events leading to vitrification of plants cultured in vitro. *Physiol. Plantarum* 61:69–74.

Kimura, M., and J. F. Crow. 1963. The measurement of effective population numbers. *Evolution* 17:279–288.

Kimura, M., and J. F. Crow. 1964. The number of alleles that can be maintained in a finite population. *Genetics* 49:725–738.

King, L. A. M. 1989. Genetic variation in *Taraxacum:* An analysis of ribosomal and chloroplast DNA. Ph.D. diss., Washington University, St. Louis.

King, L. A. M., and B. A. Schaal. 1989. Ribosomal-DNA variation and distribution in *Rudbeckia missouriensis. Evolution* 43:1117–1119.

King, M. W., and E. H. Roberts. 1979. *The Storage of Recalcitrant Seeds—Achievements and Possible Approaches.* International Board for Plant Genetic Resources Report, IBPGR Secretariat, Rome.

King, M. W., and E. H. Roberts. 1980. Maintenance of recalcitrant seeds in storage. In *Recalcitrant Crop Seeds,* ed. H. F. Chin and E. H. Roberts, pp. 53–89. Tropical Press, Kuala Lumpur, Malaysia.

Kingsolver, R. W. 1986. Vegetative reproduction as a stabilizing feature of the population dynamics of *Yucca glauca*. *Oecologia* 69:380–387.

Kitzmiller, J. H. 1976. Tree improvement master plan for the California region. U.S.D.A. Forest Service, San Francisco.

Klemow, K. M., and D. J. Raynal. 1981. Population ecology of *Melilotus alba* in a limestone quarry. *J. Ecol.* 69:33–44.

Klinkhamer, P. G. L., and T. J. deJong. 1983. Is it profitable for biennials to live longer than two years? *Ecol. Model.* 20:223–232.

Knapp, A. K., and L. C. Hurlbert. 1986. Production, density, and height of flower stalks of three grasses in annually burned and unburned eastern Kansas tallgrass prairie: A four year record. *Southwest. Naturalist* 31:235–241.

Knight, R. R., and L. L. Eberhardt. 1985. Population dynamics of Yellowstone grizzly bears. *Ecology* 66:323–334.

Knowles, P., G. P. Furnier, M. A. Aleksnik, and D. J. Perry. 1987. Significant levels of self-fertilization in natural populations of tamarack. *Can. J. Bot.* 1087–1091.

Kohn, J. R. 1988. Why be female? *Nature* 355:431–433.

Koski, V. 1970. A study of pollen dispersal as a mechanism of gene flow in conifers. *Commun. Inst. For. Fenniae* 3:1–78.

Kreitman, M. 1983. Nucleotide polymorphism at the alcohol dehydrogenase locus of *Drosophila melanogaster*. *Nature* 304:412–417.

Kress, W. J., and J. H. Beach. 1990. Reproductive systems of La Selva plants. In *Ecology and Natural History of a Neotropical Rain Forest,* ed. L. McDade, K. Bawa, H. Hispenheide, and G. Hartshord. University of Chicago Press, Chicago (in review).

Kruckeberg, A. R. 1984. *California Serpentines*. University of California Press, Berkeley.

Kruckeberg, A. R., and D. Rabinowitz. 1985. Biological aspects of endemism in higher plants. *Annu. Rev. Ecol. Syst.* 16:447–479.

Krugman, S. L. 1984. Policies, strategies, and means for genetic conservation in forestry. In *Plant Genetic Resources: A Conservation Imperative,* ed. C. W. Yeatman, D. Kafton, and G. Wilkes, AAAS Selected Symp. No. 87. Westview Press, Boulder, Colo.

Lacey, E. P. 1986. Onset of reproduction in plants: Size- versus age-dependency. *Trends Ecol. Evol.* 1:72–75.

Lacy, R. C. 1987. Loss of genetic diversity from managed populations: Interacting effects of drift, mutation, immigration, selection and population subdivision. *Conserv. Biol.* 1:143–158.

Lande, R. 1976. The maintenance of genetic variability by mutation in a polygenic character with linked loci. *Genet. Res.* 26:221–235.

Lande, R. 1977. The influence of the mating system on the maintenance of genetic variability in polygenic characters. *Genetics* 86:485–498.

Lande, R. 1980. Genetic variation and phenotypic evolution during allopatric speciation. *Am. Naturalist* 116:463–479.

Lande, R. 1987. Extinction thresholds in demographic models of territorial populations. *Am. Naturalist* 130:624–635.

Lande, R. 1988a. Demographic models of the northern spotted owl (*Strix occidentalis caurina*). *Oecologia* 75:601–607.

Lande, R. 1988b. Genetics and demography in biological conservation. *Science* 241:1455–1460.

Lande, R., and G. F. Barrowclough. 1987. Effective population size, genetic variation, and their use in population management. In *Viable Populations for Management,* ed. M. E. Soulé, pp. 87–124. Cambridge University Press, Cambridge.

Lande, R., and D. W. Schemske. 1985. The evolution of self-fertilization and inbreeding depression in plants. I. Genetic models. *Evolution* 39:24–40.

Lanner, R. M. 1966. Needed: A new approach to the study of pollen dispersal. *Silvae Genetica* 15:50–52.

Lazell, J. D., Jr. 1986. The search for rare animals: Statistics and probability. *Cryptozoology* 5:135–138.

Learn, G. H., Jr., and B. A. Schaal. 1987. Population subdivision for ribosomal DNA repeat variants in *Clematis fremontii. Evolution* 41:433–438.

Ledig, F. T. 1986a. Conservation strategies for forest gene resources. *For. Ecol. Mgmt.* 14:77–90.

Ledig, F. T. 1986b. Heterozygosity, heterosis, and fitness in outbreeding plants. In *Conservation Biology: The Science of Scarcity and Diversity,* ed. M. E. Soulé, pp. 77–104. Sinauer, Sunderland, Mass.

Ledig, F. T. 1987. Genetic structure and the conservation of California's endemic and near-endemic conifers. In *Conservation and Management of Rare and Endangered Plants,* Proc. California Conference on Conservation and Management of Rare and Endangered Plants, ed. T. S. Elias, pp. 587–594. California Native Plant Society, Sacramento.

Ledig, F. T. 1988. The conservation of diversity in forest trees. *BioScience* 38:431–439.

Ledig, F. T., and M. T. Conkle. 1983. Gene diversity and genetic structure in a narrow endemic, Torrey pine (*Pinus torreyana* Parry ex carr.). *Evolution* 37:79–86.

Lefkovitch, L. P. 1965. The study of population growth in organisms grouped by stages. *Biometrics* 21:1–18.

Lehmkuhl, J. F. 1984. Determining size and dispersion of minimum viable populations for land management planning and species conservation. *Environ. Mgmt.* 8:167–176.

Leigh, E. G., Jr. 1981. The average lifetime of a population in a varying environment. *J. Theor. Biol.* 90:213–239.

Leigh, J. H., R. Boden, and J. D. Briggs. 1984. *Extinct and Endangered Plants of Australia.* Macmillan, Melbourne.

Leigh, J. H., J. D. Briggs, and W. Hartley. 1982a. The conservation status of Australian plants. In *Species at Risk: Research in Australia,* ed. R. H. Groves and W. D. L. Ride, pp. 13–25. Australian Academy of Science, Canberra.

Leigh, E. G., A. S. Rand, and D. M. Windsor. 1982b. *The Ecology of a Tropical Forest.* Smithsonian Institution Press, Washington, D.C.

Leps, J., and P. Kindlman. 1987. Models of the development of spatial pattern of an even-aged plant population over time. *Ecol. Modelling* 39:45–57.

Lesica, P., R. F. Leary, F. W. Allendorf, and D. E. Bilderback. 1988. Lack of genic diversity within and among populations of an endangered plant, *Howellia aquatilis. Conserv. Biol.* 2:275–282.

Leslie, P. H. 1945. On the use of matrices in certain population mathematics. *Biometrika* 33:183–212.

Leveille, H. 1911. Plantae novae sandwichenses. II. *Repert. Spec. Nov. Regni Veg.* 10:149–157.

Levin, D. A. 1978. Some genetic consequences of being a plant. In *Ecological Genetics: The Interface,* ed. P. Brussard, pp. 189–912. Springer-Verlag, New York.

Levin, D. A. 1981. Dispersal versus gene flow in plants. *Ann. Missouri Bot. Gard.* 68:233–253.

Levin, D. A. 1983. Immigration in plants: An exercise in the subjunctive. In *Genetics and Conservation,* ed. C. M. Schonewald-Cox, S. M. Chambers, B. MacBryde, and L. Thomas, pp. 242–260. Benjamin-Cummings, Menlo Park, Calif.

Levin, D. A. 1984. Inbreeding depression and proximity-dependent crossing success in *Phlox drummondii. Evolution* 38:116–127.

Levin, D. A., and H. W. Kerster. 1974. Gene flow in seed plants. *Evol. Biol.* 7:139–220.

Levin, D. A., K. Ritter, and N. C. Ellstrand. 1979. Protein polymorphism in the narrow endemic *Oenothera organensis*. *Evolution* 33:534–542.

Levy, M., and D. A. Levin. 1975. Genic heterozygosity and variation in permanent translocation heterozygotes of the *Oenothera biennis* complex. *Genetics* 79:493–512.

Lewin, R. 1986a. Damage to tropical forests, or why were there so many kinds of animals? *Science* 234:149–150.

Lewin, R. 1986b. A mass extinction without asteroids. *Science* 234:14–15.

Lewin, R. 1989. How to get plants into the conservationists' ark. *Science* 244:32–33.

Lewontin, R. C. 1965. Comment. In *The Genetics of Colonizing Species,* ed. H. G. Baker and G. L. Stebbins, pp. 481–485. Academic Press, New York.

Lewontin, R. C. 1974. *The Genetic Basis of Evolutionary Change.* Columbia University Press, New York.

Lewontin, R. C., and L. C. Birch. 1966. Hybridization as a source of variation for adaptation to new environments. *Evolution* 20:315–336.

Libby, W. J. 1976. Closing remarks and summary. In *Proc. IUFRO Meeting on Advanced Generation Breeding,* pp. 181–188. Bordeaux, France, June 14–18, 1976.

Libby, W. J. 1990. Genetic conservation of Monterey pine and coast redwood. *Fremontia* 18:15–21.

Libby, W. J., and W. B. Critchfield. 1987. Patterns of genetic architecture. In Proc. 18th IUFRO Congress, September 20–21, 1986, Dubrovnik, Yugoslavia. *Annales Forestales* 13 (1–2):77–92.

Libby, W. J., K. Isik, and J. P. King. 1980. Variation in flushing time among white fir samples. *Annales Forestales* 8(6):123–138.

Libby, W. J., B. G. McCutchan, and C. I. Millar. 1981. Inbreeding depression in selfs of redwood. *Silvae Genetica* 29:15–25.

Linhart, Y. B. 1974. Intra-population differentiation in annual plants. I. *Veronica peregrina* raised under non-competitive conditions. *Evolution* 28:232–243.

Linhart, Y. B., and J. B. Mitton. 1985. Relationships among reproduction, growth rates, and protein heterozygosity in ponderosa pine. *Am. J. Bot.* 72:181–184.

Lloyd, D. G. 1979. Some reproductive factors affecting the selection of self-fertilization in plants. *Am. Naturalist* 113:67–79.

Long, E. O., and I. B. Dawid. 1980. Repeated genes in eukaryotes. *Annu. Rev. Biochem.* 49:727–764.

Lovejoy, T. E. 1976. We must decide which species will go forever. *Smithsonian* 7:52–59.

Loveless, M. P., and J. L. Hamrick. 1984. Ecological determinants of genetic structure in plant populations. *Annu. Rev. Ecol. Syst.* 15:65–95.

Loveless, M. D., and J. L. Hamrick. 1987. Distribución de la variación genética en especies arbóreas tropicales. *Revista de Biología Tropical* 35:165–176.

Lucas, G., and H. Synge. 1978. *The IUCN Plant Red Data Book.* International Union for Conservation of Nature and Natural Resources, Morges, Switzerland.

MacArthur, R. H., and E. O. Wilson. 1967. *The Theory of Island Biogeography.* Princeton University Press, Princeton, N.J.

MacCleur, J. W., J. L. Vandeberg, B. Read, and O. A. Ryder. 1986. Pedigree analysis by computer simulation. *Zoo Biol.* 5:147–160.

Mack, R. N., and J. L. Harper. 1977. Interference in dune annuals: Spatial pattern and neighborhood effects. *J. Ecol.* 65:345–364.

MacKay, I. J. 1981. Population genetics of *Papaver dubium*. Ph.D. diss., University of Birmingham.

Malécot, G. 1975. Heterozygosity and relationship in regularly subdivided populations. *Theor. Pop. Biol.* 8:212–214.

Margules, C., A. J. Higgs, and R. W. Rafe. 1982. Modern biogeographic theory: Are there lessons for nature reserve design? *Biol. Conserv.* 24:115–128.

Marshall, D. R. 1989. Crop genetic resources—Current and emerging issues. In *Plant Population Genetics, Breeding, and Genetic Resources,* ed. A. H. D. Brown, M. T. Clegg, A. L. Kahler, and B. S. Weir, pp. 370–391. Sinauer, Sunderland, Mass.

Marshall, D. R., and A. H. D. Brown. 1975. Optimum sampling strategies in genetic conservation. In *Crop Genetic Resources for Today and Tomorrow,* ed. O. H. Frankel and J. G. Hawkes, pp. 53–80. Cambridge University Press, Cambridge.

Marshall, D. R., and A. H. D. Brown. 1983. Theory of forage plant collection. In *Genetic Resources of Forage Plants,* ed. J. G. McIver and R. A. Bray, pp. 135–148. CSIRO, Melbourne.

Martin, B., J. Nienhuis, G. King, and A. Schaefer. 1989. Restriction fragmentation length polymorphisms associated with water use efficiency in tomato. *Science* 243:1725–1728.

Martin, T. 1984. Status report for *Cercocarpus traskiae.* California Natural Diversity Data Base. Element Code SPLC.EB3.

Martins, P. S., and S. K. Jain. 1979. Role of genetic variation in the colonizing ability of rose clover (*Trifolium hirtum* All.). *Am. Naturalist* 114:591–595.

Maruyama, T. 1970. Rate of decrease of genetic variability in a subdivided population. *Biometrika* 57:299–312.

Maruyama, T., and P. A. Fuerst. 1984. Population bottlenecks and non-equilibrium models in population genetics. I. Allele numbers when populations evolve from zero variability. *Genetics* 108:745–763.

Maruyama, T., and P. A. Fuerst. 1985a. Population bottlenecks and non-equilibrium models in population genetics. II. Number of alleles in a small population derived from a large steady-state population by means of a bottleneck. *Genetics* 111:675–689.

Maruyama, T., and P. A. Fuerst. 1985b. Population bottlenecks and non-equilibrium models in population genetics. III. Genic homozygosity in populations which experience periodic bottlenecks. *Genetics* 111:691–703.

Mashburn, S. J., R. R. Sharitz, and M. H. Smith. 1978. Genetic variation among *Typha* populations of the southeastern United States. *Evolution* 32:681–685.

Mason, H. L. 1946a. The edaphic factor in narrow endemism. I. The nature of environmental influences. *Madroño* 8:209–257.

Mason, H. L. 1946b. The edaphic factor in narrow endemism. II. The geographic occurrence of plants of highly restricted patterns of distribution. *Madroño* 8:241–257.

Mason, H. L., and J. H. Langenheim. 1957. Language analysis and the concept environment. *Ecology* 38:325–340.

Mayo, O. 1971. Rates of change in gene frequency in tetrasomic organisms. *Genetica* 42:329–337.

Mayr, E. 1963. *Animal Species and Evolution.* Harvard University Press, Cambridge, Mass.

Mazur, P. 1966. Physical and chemical basis of drying in single celled micro-organisms subjected to freezing and thawing. In *Cryobiology,* ed. H. T. Meryman, pp. 214–315. Academic Press, New York.

McClenaghan, L. R., Jr., and A. C. Beauchamp. 1986. Low genic differentiation among isolated populations of the California fan palm (*Washingtonia filifera*). *Evolution* 40:315–322.

McGraw, J. B., and J. Antonovics. 1983. Experimental ecology of *Dryas octopetala* ecotypes. I. Ecotypic differentiation and life-cycle stages of selection. *J. Ecol.* 71:879–897.

McInerny, J. 1981. Natural resource economics: The basic analytical principals. In *Economics and Resource Policy,* ed. J. A. Butlin, pp. 30–58. Longman, London.

McLeod, M. J., S. I. Guttman, W. H. Eshbaugh, and R. E. Rayle. 1983. An electrophoretic study of evolution in *Capsicum* (Solanaceae). *Evolution* 37:562–574.

McMahan, L. R. 1989. Rarest U.S. plants are literally one of a kind. *Plant Conserv.* 4:6.

McNeilly, T., and A. D. Bradshaw. 1968. Evolutionary processes in populations of copper tolerant *Agrostis tenuis. Evolution* 22:108–118.

Meagher, T. R. 1986. Analysis of paternity within a natural population of *Chamaelirium luteum.* I. Identification of most-likely male parents. *Am. Naturalist* 128:199–215.

Meagher, T. R., J. Antonovics, and R. Primack. 1978. Experimental ecological genetics in *Plantago.* III. Genetic variation and demography in relation to survival of *Plantago cordata,* a rare species. *Biol. Conserv.* 14:243–257.

Mech, L. D. 1970. *The Wolf: The Ecology and Behavior of an Endangered Species.* Natural History Press, Garden City, N.Y.

Mehrhoff, L. A., III. 1983. Pollination in the genus *Isotria* (Orchidaceae). *Am. J. Bot.* 70:1444–1453.

Mehrhoff, L. A. 1989. The dynamics of declining populations of an endangered orchid, *Isotria medeoloides. Ecology* 70:783–786.

Menges, E. S. 1986. Predicting the future of rare plant populations: Demographic monitoring and modelling. *Nat. Areas J.* 6:13–25.

Menges, E. S. 1990. Population viability analysis for a rare plant. *Conserv. Biol.* 4:52–62.

Menges, E. S. 1991a. Seed germination percentage increases with population size in a fragmented prairie species. *Cons. Biol.* 5:158–164.

Menges, E. S. 1991b. Stochastic modeling of extinction in plant populations. In *Conservation Biology: The Theory and Practice of Nature Conservation, Preservation, and Management,* ed. P. L. Fiedler and S. Jain. Chapman and Hall, New York.

Menges, E. S., and S. C. Gawler. 1986. Four-year changes in population size of the endemic Furbish's lousewort: Implications for endangerment and management. *Nat. Areas J.* 6:6–17.

Menges, E. S., S. C. Gawler, and D. M. Waller. 1985. Population biology of the endemic plant, Furbish's lousewort (*Pedicularis furbishiae*). A report to the U.S. Fish and Wildlife Service.

Menges, E. S., and O. L. Loucks. 1984. Modeling a disease-caused patch disturbance: Oak wilt in the Midwestern United States. *Ecology* 65:487–498.

Menken, S. B. J. 1987. Is the extremely low heterozygosity level in *Yponomeuta rorellus* caused by bottlenecks? *Evolution* 41:630–637.

Meredith, L. D., and M. M. R. Richardson. 1987. *Rare and Threatened Australian Plants in Cultivation in Australia.* Australian National Botanic Gardens, Canberra (in press).

Milkman, R., and I. P. Crawford. 1983. Clustered third-base substitutions among wild strains of *Escherichia coli. Science* 221:378–379.

Millar, C. I. 1983. A steep cline in *Pinus muricata. Evolution* 37:311–319.

Millar, C. I. 1986. The California closed-cone pines (subsection Oocarpae Little and Critchfield): A taxonomic history and review. *Taxon* 35:657–670.

Millar, C. I. 1987. The California forest germplasm conservation project: A case for genetic conservation of temperate tree species. *Conserv. Biol.* 1:191–193.

Millar, C. I. 1989. Allozyme variation in bishop pine associated with pygmy forest soils in northern California. *Can. J. For. Res.* 19:870–879.

Millar, C. I. 1991. Conservation of germplasm in forest trees. In *Clonal Forestry: Genetics, Biotechnology, and Applications,* ed. W. J. Libby and M. R. Ahuja. Springer-Verlag, New York.

Millar, C. I., and W. J. Libby. 1989. Restoration: Disneyland or native ecosystem? *Rest. Mgmt. Notes* 7:18–24.

Millar, C. I., S. H. Strauss, M. T. Conkle, and R. D. Westfall. 1988. Allozyme differentiation and biosystematics of the Californian closed-cone pines (*Pinus* subsect. Oocarpae). *Syst. Bot.* 13:351–370.

Mitchell-Olds, T. 1986. Quantitative genetics of survival and growth in *Impatiens capensis*. *Evolution* 40:107–117.

Mitchell-Olds, T., and J. J. Rutledge. 1986. Quantitative genetics in natural plant populations: A review of the theory. *Am. Naturalist* 127:379–402.

Mitchell-Olds, T., and D. M. Waller. 1985. Relative performance of selfed and outcrossed progeny in *Impatiens capensis*. *Evolution* 39:533–544.

Mitton, J. B., and M. C. Grant. 1984. Associations among protein heterozygosity, growth rate, and developmental homeostasis. *Annu. Rev. Ecol. Syst.* 15:479–499.

Mlot, C. 1989. Blueprint for conserving plant diversity. *BioScience* 39:364–368.

Moldenke, H. N. 1959. *A Resume of the Verbenaceae, Avicenniaceae, Stilbaceae, Symphoremaceae, and Eriocaulaceae of the World as to Valid Taxa, Geographic Distribution and Synonymy,* Suppl. 1. Publ. privately. Yonkers, N.Y.

Moore, F. D., III, A. E. McSay, and E. E. Roos. 1983. Probit analysis: A computer program for evaluation of seed germination and viability loss rate. Colo. State Univ. Exp. Stn. Tech. Bull. No. 147.

Moore, F. D., III, and E. E. Roos. 1982. Determining differences in viability loss rates during seed storage. *Seed Sci. Technol.* 10:283–300.

Moran, G. F., and J. C. Bell. 1983. *Eucalyptus.* In *Isozymes in Plant Genetics and Breeding,* Part B, ed. D. S. Tanksley and T. J. Orton, pp. 423–441. Elsevier, Amsterdam.

Moran, G. F., J. C. Bell, and K. C. Eldridge. 1988. The genetic structure and conservation of the five natural populations of *Pinus radiata. Can. J. For. Res.* 18:506–514.

Moran, G. F., J. C. Bell, and A. C. Matheson. 1980. The genetic structure and levels of inbreeding in a *Pinus radiata* D. Don seed orchard. *Silvae Genetica* 29:190–193.

Moran, G. F., and S. D. Hopper. 1983. Genetic diversity and the insular population structure of *Eucalyptus caesia* Beuth. *Aust. J. Bot.* 31:161–172.

Moran, G. F., and S. D. Hopper. 1987. Conservation of the genetic resources of rare and widespread eucalypts in remnant vegetation. In *Nature Conservation: The Role of Remnants of Native Vegetation,* ed. D. A. Saunders, G. W. Arnold, A. A. Burbidge, and A. J. M. Hopkins, pp. 151–162. Surrey Beatty and Sons, Chipping Norton, Australia.

Moran, G. F., and D. R. Marshall. 1978. Allozyme uniformity within and variation between races of the colonizing species *Xanthium strumarium* L. (Noogoora Burr). *Aust. J. Biol. Sci.* 31:282–291.

Moran, G. F., D. R. Marshall, and W. J. Muller. 1981. Phenotypic variation and plasticity in the colonizing species *Xanthium strumarium* L. (Noogoora Burr). *Aust. J. Biol. Sci.* 34:639–638.

Moran, G. F., O. Muona, and J. C. Bell. 1989. *Acacia mangium:* A tropical forest tree of the coastal lowlands with low genetic diversity. *Evolution* 43:231–235.

Morgan, M. T., and S. C. H. Barrett. 1988. Historical factors and anisoplethic population structure in tristylous *Pontederia cordata:* A reassessment. *Evolution* 42:496–504.

Moriguchi, T., T. Akihama, and I. Kozaki. 1985. Freeze-preservation of dormant pear shoot apices. *Jpn. J. Breed.* 35:196–199.

Murata, M., E. E. Roos, and T. Tsuchiya. 1980. Mitotic delay in root tips of peas induced by artificial seed aging. *Bot. Gaz.* 141:19–23.

Murata, M., E. E. Roos, and T. Tsuchiya. 1981. Chromosome damage induced by artificial seed aging in barley. I. Germinability and frequency of aberrant anaphases at first mitosis. *Can. J. Genet. Cytol.* 23:267–280.

Murawski, D. A., J. L. Hamrick, S. P. Hubbell, and R. B. Foster. 1990. Mating systems of two Bombacaceous trees of a neotropical moist forest. *Oecologia* 82:501–506.

Myers, N. 1979. *The Sinking Ark.* Pergamon Press, Oxford.

Myers, N. 1988. Threatened biotas: "Hotspots" in tropical forests. *Environmentalist* 8:1–20.

Namkoong, G. 1988. Sampling for germplasm collections. *HortScience* 23:79–81.

Nature Conservancy, The. 1988. Annual Report. Arlington, Va.

Nature Conservancy, The. 1990. Central Scientific Databases. Arlington, Va.

Neale, D. B., M. A. Saghai-Maroof, R. W. Allard, Q. Zhang, and R. A. Jorgensen. 1988. Chloroplast DNA diversity in populations of wild and cultivated barley. *Genetics* 120:1105–1110.

Nei, M. 1973. Analysis of gene diversity in subdivided populations. *Proc. Natl. Acad. Sci. USA* 70:3321–3323.

Nei, M. 1975. *Molecular Population Genetics and Evolution*. Elsevier, New York.

Nei, M. 1987. *Molecular Evolutionary Genetics*. Columbia University Press, New York.

Nei, M., T. Maruyama, and R. Chakraborty. 1975. The bottleneck effect and genetic variability in populations. *Evolution* 29:1–10.

Nevo, E. 1978. Genetic variation in natural populations: Patterns and theory. *Theor. Pop. Biol.* 13:121–177.

Nevo, E., A. Beiles, and R. Ben-Shlomi. 1984. The evolutionary significance of genetic diversity: Ecological, demographic and life history correlates. In *Evolutionary Dynamics of Genetic Diversity*, ed. G. S. Mani, pp. 13–213. Lecture Notes in Biomathematics, Vol. 53. Springer-Verlag, New York.

Nevo, E., A. Beiles, D. Kaplan, E. M. Golenberg, L. Olsvig-Whittaker, and Z. Naveh. 1986. Natural selection of allozyme polymorphisms: A microsite test revealing ecological genetic differentiation in wild barley. *Evolution* 40:13–20.

Newmark, W. D. 1985. Legal and biotic boundaries of western North American National Parks: A problem of congruence. *Biol. Conserv.* 33:197–208.

Newport, M. E. A. 1989. A test for proximity-dependent outcrossing in the alpine skypilot, *Polemonium viscosum*. *Evolution* 43:1110–1113.

Nickrent, D. L., and D. Wiens. 1979. Genetic diversity in the rare California shrub *Dedeckera eurekensis* (Polygonacene). *Syst. Bot.* 14:245–253.

Niebling, C. R., and M. T. Conkle. 1990. Washoe pine allozyme diversity and genetic comparisons with races of ponderosa pine. *Can. J. For. Res.* (submitted).

Nijkamp, P. 1977. *Theory and Application of Environmental Economics*, Vol. 1 of *Studies in Regional Science and Urban Economics*. North-Holland, Amsterdam.

Norton, B. G. 1987. *Why Preserve Natural Variety?* Princeton University Press, Princeton, NJ.

Noss, R. F. 1987. Protecting natural areas in fragmented landscapes. *Nat. Areas J.* 7:2–13.

Noss, R. F., and L. D. Harris. 1986. Nodes, networks, and MUMS: Preserving diversity at all scales. *Environ. Mgmt.* 10:299–309.

Nybom, H., and B. A. Schaal. 1990a. DNA "fingerprints" reveal genotypic distributions in natural populations of blackberries and raspberries (*Rubus* L., Rosaceae). *Am. J. Bot.* 77:883–887.

Nybom, H., and B. A. Schaal. 1990b. DNA "fingerprints" applied to paternity analysis in apples (*Malus* × *domestica*). *Theor. Appl. Genet.* 79:763–768.

O'Brien, S. J., and J. F. Evermann. 1988. Interactive influence of infectious disease and genetic diversity in natural populations. *Trends Ecol. Evol.* 3:254–259.

O'Brien, S. J., M. E. Roelke, L. Marker, A. Newman, C. A. Winkler, D. Meltzer, L. Colly, J. F. Evermann, M. Bush, and D. E. Wildt. 1985. Genetic basis for species vulnerability in the cheetah. *Science* 227:1428–1434.

Odum, E. P. 1971. *Fundamentals of Ecology*. Saunders, Philadelphia.

Oldfield, M. L. 1984. *The Value of Conserving Genetic Resources*. U.S.D.I. National Park Service, Washington, D.C.

Oliveri, A. M., and S. K. Jain. 1977. Variation in the *Helianthus exilis-bolanderi* complex. *Madroño* 24:177–189.

O'Malley, D. M., F. W. Allendorf, and G. M. Blake. 1979. Inheritance of isozyme variation and heterozygosity in *Pinus ponderosa*. *Biochem. Genet.* 17:233–250.

O'Malley, D. M., and K. S. Bawa. 1987. Mating system of a tropical rain forest tree species. *Am. J. Bot.* 74:1143–1149.

O'Malley, D. M., D. P. Buckley, G. T. Prance, and K. S. Bawa. 1988. Genetics of Brazil nut (*Bertholletia excelsa* Humb. & Bonpl.: Lecythidaceae) 2. Mating system. *Theor. Appl. Genet.* 76:929–932.

Palmer, J. D. 1987. Chloroplast DNA evolution and biosystematic uses of chloroplast DNA variation. *Am. Naturalist* 130:S6–S29.

Palmer, J. D., C. R. Sheilds, D. B. Cohen, and T. J. Orton. 1983. Chloroplast DNA evolution and the origin of amphidiploid *Brassica* species. *Theor. Appl. Genet.* 65:181–189.

Palmer, J. D., and D. Zamir. 1982. Chloroplast DNA evolution and phylogenetic relationships in *Lycopersicon. Proc. Natl. Acad. Sci. USA* 79:5006–5010.

Paques, M., and Ph. Boxus. 1987. "Vitrification": Review of the literature. *Acta Hort.* 212:155–166.

Parker, M. A. 1985. Local population differentiation for compatibility in an annual legume and its host-specific fungal pathogen. *Evolution* 39:713–723.

Parkin, D. T. 1979. *An Introduction to Evolutionary Genetics.* Edward Arnold, London.

Parsons, P. A. 1983. *The Evolutionary Biology of Colonizing Species.* Cambridge University Press, Cambridge.

Pavlik, B. M., and M. J. Barbour. 1988. Demographic monitoring of endemic sand dune plants, Eureka Valley, California. *Biol. Conserv.* 46:217–242.

Philip, J. R. 1957. Sociality and sparse populations. *Ecology* 38:107–111.

Phillips, B. G., A. M. Phillips, and D. J. Howard. 1988. Final report, starch gel electrophoresis of *Purshia subintegra* and *Purshia stansburiana.* Contract from U.S. Bureau of Reclamation to Museum of Northern Arizona, Flagstaff.

Pickersgill, B. P., S. C. H. Barrett, and D. Andrade-Lima. 1975. Wild cotton in N.E. Brazil. *Biotropica* 7:42–54.

Pickett, S. T. A., and P. S. White. 1985. *The Ecology of Natural Disturbance and Patch Dynamics.* Academic Press, Orlando, Fla.

Pielou, E. C. 1960. A single mechanism to account for regular, random and aggregated populations. *J. Ecol.* 48:575–584.

Pimm, S. L., H. L. Jones, and J. Diamond. 1988. On the risk of extinction. *Am. Naturalist* 132:757–785.

Pindyck, R. S. 1978. The optimal exploration and production of nonrenewable resources. *J. Polit. Econ.* 86:841–861.

Pinero, D., M. Martinez-Ramos, and J. Sarukhan. 1984. A population model of *Astrocaryum mexicanum* and a sensitivity analysis of its finite rate of increase. *J. Ecol.* 72:977–991.

Polans, N. O., and R. W. Allard. 1989. An experimental evaluation of the recovery potential of ryegrass populations from genetic stress resulting from restriction of population size. *Evolution* 43:1320–1323.

Pollard, J. H. 1966. On the use of the direct matrix product in analyzing certain stochastic population models. *Biometrika* 53:397–415.

Poore, M. E. D. 1968. Studies in Malaysian rain forest. I. The forest on the Triassic sediments in Jengka Forest Reserv. *J. Ecol.* 56:143–196.

Powell, E. 1985. The Mauna Kea silversword: A species on the brink of extinction. *Newslett. Hawai. Bot. Soc.* 24:44–57.

Prance, G. T. 1984. Completing the inventory. In *Current Concepts in Plant Taxonomy,* ed. V. H. Heywood and D. N. Moore, pp. 365–396. Academic Press, London.

Prance, G. T., and T. S. Elias, eds. 1977. *Extinction Is Forever: Threatened and Endangered Species of Plants in the Americas and Their Significance in Ecosystems Today and in the*

Future. Proceedings of a Symposium, May 11–13, 1976, New York Botanical Garden, Bronx, N.Y.

Prescott-Allen, C., and R. Prescott-Allen. 1986. *The First Resource: Wild Species in the North American Economy.* Yale University Press, New Haven, Conn.

Price, S. C., K. N. Schumaker, A. L. Kahler, R. W. Allard, and J. E. Hill. 1984. Estimates of population differentiation obtained from enzyme polymorphisms and quantitative characters. *J. Hered.* 75:141–142.

Price, M. V., and N. M. Waser. 1979. Pollen dispersal and optimal outbreeding in *Delphinium nelsonii. Nature* 277:294–298.

Priestley, D. A. 1986. *Seed Aging: Implications for Seed Storage and Persistence in the Soil.* Cornell University Press, Ithaca, N.Y.

Priestley, D. A., V. I. Cullinan, and J. Wolfe. 1985. Differences in seed longevity at the species level. *Plant Cell Environ.* 8:577–562.

Primack, R. B. 1980. Phenotypic variation of rare and widespread species of *Plantago. Rhodora* 82:87–95.

Pulliam, H. R. 1988. Sources, sinks and population regulation. *Am. Naturalist* 132:652–665.

Quinn, J. A. 1978. Plant ecotypes: ecological or evolutionary units? *Bull. Torrey Bot. Club* 105:58–64.

Quinn, J. F., and A. Hastings. 1987. Extinction in subdivided habitats. *Conserv. Biol.* 1:198–208.

Rabinowitz, D. 1981. Seven forms of rarity. In *The Biological Aspects of Rare Plant Conservation,* ed. H. Synge, pp. 205–218. Wiley, New York.

Rabinowitz, D., J. K. Rapp, and P. M. Dixon. 1984. Competitive abilities of sparse grass species: Means of persistence or cause of abundance. *Ecology* 65:1144–1154.

Ralls, K., J. D. Ballou, and A. Templeton. 1988. Estimates of lethal equivalents and the cost of inbreeding in mammals. *Conserv. Biol.* 2:185–193.

Ralls, K., P. H. Harvey, and A. M. Lyles. 1986. Inbreeding in natural populations of birds and mammals. In *Conservation Biology: The Science of Scarcity and Diversity,* ed. M. E. Soulé, pp. 35–56. Sinauer, Sunderland, Mass.

Raven, P. H. 1976. Ethics and attitudes. In *Conservation of Threatened Plants,* ed. J. B. Simmons, R. I. Benzer, P. C. Brandham, G. L. Lucas, and V. T. H. Parry, pp. 155–179. Plenum, New York.

Raven, P. H. 1987. The scope of the plant conservation problem world-wide. In *Botanic Gardens and the World Conservation Strategy,* ed. D. Bramwell, O. Hamann, V. Heywood, and H. Synge, pp. 19–29. Academic Press, London.

Raven, P. H., and D. I. Axelrod. 1978. Origin and relationships of the California flora. *Univ. Calif. Publ. Bot.* 72:1–134.

Read, R. A. 1980. Genetic variation in seedling progeny of ponderosa pine provenances. *For. Sci. Monogr.* 23:1–59.

Reed, B. M. 1988. Cold acclimation as a method to improve survival of cryopreserved *Rubus* meristems. *Cryoletters* 9:166–171.

Renner, S. 1986. The neotropical epiphytic Melastomataceae: Phytogeographic patterns, fruit types and floral biology. *Selbyana* 9:104–111.

Richards, A. J. 1986. *Plant Breeding Systems.* Allen and Unwin, London.

Richardson, B. J., P. M. Rogers, and G. M. Hewitt. 1980. Ecological genetics of the wild rabbit in Australia. II. Protein variation in British, French, and Australian rabbits and the geographical distribution of the variation in Australia. *Aust. J. Biol. Sci.* 33:371–383.

Rieseberg, L. H. 1987. A re-examination of introgression in *Helianthus.* Ph.D. diss., Washington State University, Pullman.

Rieseberg, L. H., R. Carter, and S. Zona. 1990. Molecular tests of the hypothesized hybrid origin of two diploid *Helianthus* species (Asteraceae). *Evolution* (in press).

Rieseberg, L. H., D. E. Soltis, and J. D. Palmer. 1988. A molecular reexamination of introgression between *Helianthus annuus* and *H. bolanderi* (Compositae). *Evolution* 42:227–238.

Rieseberg, L. H., S. Zona, L. Aberbom, and T. D. Martin. 1989. Hybridization in the island endemic, Catalina mahogany. *Conserv. Biol.* 3:52–58.

Riggs, L. R. 1982. Douglas-fir genetic resources: An assessment and plan for California. California gene resource program, Sacramento.

Rincker, C. M. 1980. Effect of long-term subfreezing storage of seed on legume forage production. *Crop Sci.* 20:574–577.

Rincker, C. M. 1981. Long-term subfreezing storage of forage crop seeds. *Crop Sci.* 21:424–427.

Rincker, C. M. 1983. Germination of forage crop seeds after 20 years of subfreezing storage. *Crop Sci.* 23:229–231.

Rincker, C. M., and J. D. Maguire. 1979. Effect of seed storage on germination and forage production of seven grass cultivars. *Crop Sci.* 19:857–860.

Ritland, K. 1983. Estimation of mating systems. In *Isozymes in Plant Genetics and Breeding,* Part A, ed. S. D. Tanksley and T. J. Orton, pp. 289–302. Elsevier, Amsterdam.

Ritland, K. 1984. The effective proportion of self-fertilization with consanguineous matings in inbred populations. *Genetics* 106:139–152.

Ritland, K. 1989. Gene identity and the genetic demography of plant populations. In *Plant Population Genetics, Breeding, and Genetic Resources,* ed. A. H. D. Brown, M. T. Clegg, A. L. Kahler, and B. S. Weir, pp. 181–199. Sinauer, Sunderland, Mass.

Ritland, K., 1990. Inferences about inbreeding depression based on changes of the inbreeding coefficient. *Evolution* 44:1230–1241.

Ritland, K., and F. R. Ganders. 1987a. Covariation of selfing rates with parental gene fixation indices within populations of *Mimulus guttatus. Evolution* 41:760–771.

Ritland, K., and F. R. Ganders. 1987b. Crossability of *Mimulus guttatus* in relation to components of gene fixation. *Evolution* 41:772–786.

Ritland, K., and S. K. Jain. 1981. A model for the estimation of outcrossing rate and gene frequencies using *N* independent loci. *Heredity* 47:35–52.

Roberts, E. H. 1972. *Viability of Seeds.* Chapman and Hall, London.

Roberts, E. H. 1973. Predicting the storage life of seeds. *Seed Sci. Technol.* 1:499–514.

Roberts, E. H. 1975. Problems of long-term storage of seed and pollen for genetic resources conservation. In *Crop Genetic Resources for Today and Tomorrow,* ed. O. H. Frankel and J. G. Hawkes, pp. 269–296. Cambridge University Press, Cambridge.

Roberts, E. H. 1983. Loss of seed viability during storage. In *Advances in Research and Technology of Seeds,* Part 8, ed. J. R. Thompson, pp. 9–34. Pudoc, Wageningen.

Roberts, E. H. 1986. Quantifying seed deterioration. In *Physiology of Seed Deterioration.* Spec. Publ. No. 11, ed. M. B. McDonald, Jr., and C. J. Nelson, pp. 101–123. Crop Science Society of America, Madison, Wis.

Roberts, E. H., and R. H. Ellis. 1984. The implications of the deterioration of orthodox seeds during storage for genetic resources conservation. In *Crop Genetic Resources: Conservation and Evaluation,1 ed. J. H. W. Holden and J. T. Williams, pp. 18–37. Allen and Unwin, London.*

Roberts, E. H., and M. W. King. 1981. Storage of recalcitrant seeds. In *Crop Genetic Resources, the Conservation of Difficult Material,* ed. L. A. Withers and J. T. Williams, pp. 39–48. IUBS Series B42.

Roberts, E. H., M. W. King, and R. H. Ellis. 1984. Recalcitrant seeds: Their recognition and storage. In *Crop Genetic Resources: Conservation and Evaluation,* ed. J. H. W. Holden, pp. 38–52. Allen and Unwin, London.

Rock, J. F. 1913. *The Indigenous Trees of the Hawaiian Islands.* Publ. privately, Honolulu.

Rogers, C. E., T. E. Thompson, and G. J. Seiler. 1982. *Sunflower Species of the United States.* National Sunflower Association, Bismarck, N.D.

Rogers, S. O., S. Honda, and A. J. Bendich. 1986. Variation in the ribosomal RNA genes among individuals of *Vicia faba. Plant Mol. Biol.* 6:339–345.

Rogstad, S. H., J. C. Patton, II, and B. A. Schaal. 1988a. M13 repeat probe detects DNA minisatellite-like sequences in gymnosperms and angiosperms. *Proc. Natl. Acad. Sci. USA* 85:9176–9178.

Rogstad, S. H., J. C. Patton, II, and B. A. Schaal. 1988b. A human minisatellite probe reveals RFLPs among individuals of two angiosperms. *Nucl. Acids Res.* 16:11378.

Rogstad, S. H., J. C. Patton, II, and B. A. Schaal. 1989. A human minisatellite probe reveals RFLPs among individuals of two angiosperms. *Nucl. Acids Res.* 16:11378.

Roos, E. E. 1982. Induced genetic changes in seed germplasm during storage. In *The Physiology and Biochemistry of Seed Development, Dormancy and Germination,* ed. A. A. Khan, pp. 409–434. Elsevier, New York.

Roos, E. E. 1986. Precepts of successful seed storage. In *Physiology of Seed Deterioration,* Spec. Publ. No. 11, ed. M. B. McDonald, Jr., and C. J. Nelson, pp. 1–25. Crop Science Society of America, Madison, Wis.

Roos, E. E. 1989. Long-term seed storage. *Plant Breeding Rev.* 7:129–158.

Roos, E. E., and C. M. Rincker. 1982. Genetic stability in 'Pennlate' orchardgrass seed following artificial aging. *Crop Sci.* 22:611–613.

Roos, E. E., P. C. Stanwood, and C. M. Rincker. 1986. Long-term subfreezing storage of forage seeds—A second look. *Agron. Abstr.* pp. 128–129.

Roos, E. E., P. C. Stanwood, and C. M. Rincker. 1987. Effect of year and location of production on long-term storage of forage grass and legume seeds. *Agron. Abstr.,* pp. 130–131.

Rozema, J., E. Rozema-Dijst, A. H. J. Freijsen, and J. J. L. Huber. 1978. Population differentiation within *Festuca rubra* with regard to soil salinity and soil water. *Oecologia* 34:329–341.

Russell, J., and D. A. Levin. 1988. Competitive relationships of *Oenothera* species with different recombination systems. *Am. J. Bot.* 75:1175–1180.

Russell, R. R., and M. Wilkinson. 1979. *Microeconomics: A Synthesis of Modern and Neoclassical Theory.* Wiley, New York.

Ryder, E. J. 1988. Efficient sampling from a collection. *HortScience* 23:82–84.

Ryskov, A. P., A. G. Jincharadze, M. I. Prosnyak, P. L. Ivanov, and S. A. Limborska. 1988. M13 phage DNA as a universal marker for DNA fingerprinting of animals, plants, and microorganisms. *FEBS Lett.* 233:388–392.

Sachs, M. M., E. S. Dennis, W. L. Gerlach, and W. J. Peacock. 1986. Two alleles of maize alcohol dehydrogenase 1 have 3′ structural and poly(A) addition polymorphisms. *Genetics* 113:449–467.

Sage, R. D., D. Heyneman, K. C. Lim, and A. C. Wilson. 1986. Wormy mice in a hybrid zone. *Nature* 324:60–62.

Saghai-Maroof, M. A., K. M. Soliman, R. A. Jorgensen, and R. W. Allard. 1984. Ribosomal DNA spacer-length polymorphism in barley: Mendelian inheritance, chromosomal location, and population dynamics. *Proc. Natl. Acad. Sci. USA* 81:8014–8018.

Sakai, A. 1984. Cryopreservation of apical meristems. *Hort. Rev.* 6:357–372.

Sakai, A. 1986. Cryopreservation of germplasm of woody plants. In *Biotechnology in Agriculture and Forestry,* Vol. 1: *Trees,* ed. Y. P. S. Bajaj, pp. 113–129. Springer-Verlag, Berlin.

Sakai, A., and Y. Nishiyama. 1978. Cryopreservation of winter vegetative buds of hardy fruit trees in liquid nitrogen. *HortScience* 13:225–227.

Salwasser, H. 1987. Editorial. *Conserv. Biol.* 1:275–276.

Salzman, A. G., and M. A. Parker. 1985. Neighbors ameliorate local salinity stress for a rhizomatous plant in a heterogeneous environment. *Oecologia* 65:273–277.

Sampson, J. F., S. D. Hopper, and S. H. James. 1988. Genetic diversity and the conservation of *Eucalyptus crucis. Aust. J. Bot.* 36:447–460.

Samson, F. B., F. Perez-Trejo, H. Salwaser, L. F. Ruggiero, and M. L. Shaffer. 1985. On determining and managing minimum population size. *Wild. Soc. Bull.* 13:425–433.

Samuelson, P. A. 1989. *Economics,* 12th ed. McGraw-Hill, New York.

Sarukhan, J., M. Martinez-Ramos, and D. Piñero. 1984. The analysis of demographic variability at the individual level and its population consequences. In *Perspectives on Plant Population Ecology,* ed. R. Dirzo and J. Sarukhan, pp. 83–106. Sinauer, Sunderland, Mass.

Sarvas, R. 1967. Pollen dispersal within and between subpopulations. In *XIV IUFRO Cong.,* Sect. 22, pp. 332–345, Munich.

Sawyer, S. A., D. E. Dykhuizen, and D. L. Hartl. 1987. Confidence interval for the number of selectively neutral amino acid polymorphisms. *Proc. Natl. Acad. Sci. USA* 84:6225–6228.

Schaal, B. A. 1975. Population structure and local differentiation in *Liatris cylindraceae. Am. Naturalist* 109:511–528.

Schaal, B. A., and G. H. Learn. 1988. Ribosomal DNA variation within and among plant populations. *Ann. Missouri Bot. Gard.* 75:1207–1216.

Schaal, B. A., W. J. Leverich, and J. Nieto-Sotelo. 1987. Ribosomal DNA variation in the native plant, *Phlox divaricata. Mol. Biol. Evol.* 4:611–621.

Schaal, B. A., and D. A. Levin. 1976. The demographic genetics of *Liatris cylindracea. Am. Naturalist* 110:191–206.

Schemske, D. W., and R. Lande. 1985. The evolution of self-fertilization and inbreeding depression in plants. II. Empirical observations. *Evolution* 39:41–52.

Schilling, E. E., and C. B. Heiser. 1981. An infrageneric classification of *Helianthus* (Compositae). *Taxon* 30:393–403.

Schlichting, C. D. 1986. The evolution of phenotypic plasticity in plants. *Annu. Rev. Ecol. Syst.* 17:667–693.

Schmid, B. 1985. Clonal growth in grassland perennials. III. Genetic variation and plasticity between and within populations of *Bellis perennis* and *Prunella vulgaris. J. Ecol.* 73:819–830.

Schmidt, K. P., and D. A. Levin. 1985. The comparative demography of reciprocally sown populations of *Phlox drummondii.* I. Survivorships, fecundities, and finite rates of increase. *Evolution* 39:396–404.

Schoen, D. J. 1982. The breeding system of *Gilia achilleifolia:* Variation in floral characteristics and outcrossing rate. *Evolution* 36:352–370.

Schoen, D. J. 1983. Relative fitnesses of selfed and outcrossed progeny in *Gilia achilleifolia* (Polemoniaceae). *Evolution* 37:292–301.

Schoen, D. J., and S. C. Stewart. 1987. Variation in male fertilities and pairwise mating probabilities in *Picea glauca. Genetics* 116:141–152.

Schoener, T. W. 1987. The geographical distribution of rarity. *Oecologia* 74:161–173.

Schoenike, R. E., and C. F. Bey. 1981. Conserving genes through pollen storage. In *Pollen Management Handbook,* ed. E. D. Franklin, pp. 72–73. U.S.D.A. Handbook 587. Government Printing Office, Washington, D.C.

Schwaegerle, K. E., and F. A. Bazzaz. 1987. Differentiation among nine populations of *Phlox:* Response to environmental gradients. *Ecology* 68:54–64.

Schwaegerle, K. E., and B. A. Schaal. 1979. Genetic variability and founder effect in the pitcher plant *Sarracenia purpurea* L. *Evolution* 33:1210–1218.

Schwartz, O. A. 1985. Lack of protein polymorphism in the endemic relict *Chrysosplenium iowense* (Saxifraga). *Can. J. Bot.* 63:2031–2034.

Scott, A. W., and R. D. B. Whalley. 1984. The influence of intensive sheep grazing on genotypic differentiation in *Danthonia linkii, D. richardsonii* and *D. racemosa* on the New England Tablelands. *Aust. J. Ecol.* 9:419–429.

Scottsberg, C. 1939. A hybrid violet from the Hawaiian Islands. *Bot. Not.* 1939:805–812.

Sears, B. B. 1983. Genetics and evolution of the chloroplast. *Stadler Symp.* 15:119–139.

Shaffer, M. L. 1978. Determining minimum viable population sizes: A case study of the grizzly bear (*Ursus arctos* L.). Ph.D. diss., Duke University, Durham, N.C.

Shaffer, M. L. 1981. Minimum population sizes for species conservation. *BioScience* 31:131–134.

Shaffer, M. L. 1987. Minimum viable populations: Coping with uncertainty. In *Viable Populations for Conservation,* ed. M. E. Soulé, pp. 69–86. Cambridge University Press, Cambridge.

Shaffer, M. L., and F. B. Samson. 1985. Population size and extinction: A note on determining critical population sizes. *Am. Naturalist* 125:144–152.

Sharitz, R. R., S. A. Wineriter, M. H. Smith, and E. H. Liu. 1980. Comparisons of isozymes among *Typha* species in the eastern United States. *Am. J. Bot.* 67:1297–1303.

Shaw, D. V., A. L. Kahler, and R. W. Allard. 1981. A multilocus estimator of mating system parameters in plant populations. *Proc. Natl. Acad. Sci. USA* 78:1298–1302.

Shaw, R. G. 1987. Maximum-likihood approaches applied to quantitative genetics of natural populations. *Evolution* 41:812–826.

Sherf, E. E. 1933a. New or noteworthy Compositae. IX. *Bot. Gaz. (Crawfordsville)* 95:78–103.

Sherf, E. E. 1933b. Some new or otherwise important Compositae from the Hawaiian Islands. *Am. J. Bot.* 20:616–619.

Sherf, E. E. 1934. Some new or otherwise noteworthy members of the families Labiatae and Compositae. *Bot. Gaz. (Crawfordsville)* 96:136–153.

Sherf, E. E. 1944. Some additions to our knowledge of the flora of the Hawaiian Islands. *Am. J. Bot.* 31:161–161.

Sherf, E. E. 1951. Miscellaneous notes on new or otherwise noteworthy dicotyledonous plants. *Am. J. Bot.* 38:54–73.

Sherf, E. E. 1952. Some new or otherwise noteworthy Compositae from the Hawaiian Islands. *Bot. Leafl.* 7:2–6.

Sherf, E. E. 1956. Some recently collected dicotyledonous Hawaiian Island and Peruvian plants. *Am. J. Bot.* 43:475–478.

Shumaker, K. M., and G. R. Babbel. 1980. Patterns of allozymic similarity in ecologically central and marginal populations of *Hordeum jubatum* in Utah. *Evolution* 34:110–116.

Silander, J. A. 1979. Microevolution and clone structure in *Spartina patens*. *Science* 203:658–660.

Silander, J. A. 1985a. Microevolution in clonal plants. In *Population Biology and Evolution of Clonal Organisms,* ed. J. B. C. Jackson, L. W. Buss, and R. E. Cook, pp. 107–152. Yale University Press, New Haven, Conn.

Silander, J. A. 1985b. The genetic basis of the ecological amplitude of *Spartina patens*. II. Variance and correlation analysis. *Evolution* 39:1034–1052.

Silvertown, J. 1985. Survival, fecundity and growth of wild cucumber, *Echinocystis lobata*. *J. Ecol.* 73:841–849.

Simberloff, D. 1986. Are we on the verge of a mass extinction in tropical rain forests? In *Dynamics of Extinction,* ed. D. K. Elliott, pp. 165–180. Wiley, New York.

Simberloff, D. 1988. The contribution of population and community biology to conservation science. *Annu. Rev. Ecol. Syst.* 19:473–512.

Simberloff, P. S., and L. G. Abele. 1982. Refuge design and island biogeographic theory. Effects of fragmentation. *Am. Naturalist* 120:41–50.

Simmons, J. B., R. I. Beyer, P. E. Brandham, G. Lucas, and V. T. H. Parry, eds. 1976. *Conservation of Threatened Plants*. Plenum, New York.

Sirkkomaa, S. 1983. Calculations on the decrease of genetic variation due to the founder effect. *Hereditas* 99:11–20.

Slade, A. J., and M. J. Hutchings. 1987. The effects of nutrient availability on foraging in the clonal herb *Glechoma hederacea*. *J. Ecol.* 75:95–112.

Slatkin, M., and N. Barton. 1989. A comparison of three indirect methods for estimating average levels of gene flow. *Evolution* 43:1349–1368.

Smith, J. P., Jr., and K. Berg. 1988. *Inventory of Rare and Endangered Vascular Plants of California*, 4th ed. California Native Plant Society, Berkeley.

Smith, R. A., R. L. Peloquin, and P. C. Passof. 1969. Local and regional variation in the monoterpenes of ponderosa pine wood oleoresin. U.S.D.A. Forest Service Research Paper PSW-56. Berkeley, California.

Smith, R. D. 1985. Maintaining viability during collecting expeditions. Appendix III. In *IBPGR Advisory Committee on Seed Storage: Report of the Third Meeting*, pp. 13–22. IBPGR Secretariat, Rome.

Snaydon, R. W. 1962. The microdistribution of *Trifolium repens* and its relation to soil factors. *J. Ecol.* 50:133–143.

Snaydon, R. W., and M. S. Davies. 1976. Rapid population differentiation in a mosaic environment. IV. Populations of *Anthoxanthum odoratum* at sharp boundaries. *Heredity* 37:9–25.

Snaydon, R. W., and T. M. Davies. 1982. Rapid divergence of plant populations in response to recent changes in soil conditions. *Evolution* 36:289–297.

Sobrevilla, C. 1988. Effects of distance between pollen donor and pollen recipient in fitness components in *Espeletia schultzii*. *Am. J. Bot.* 75:701–724.

Soltz, D. L., and R. J. Naiman. 1978. *The Natural History of Native Fishes in the Death Valley System*. Serial Publ. of the Natural History Museum of Los Angeles County, Science Series No. 30.

Sorensen, F. C. 1969. Embryonic genetic load in Douglas-fir, *Pseudotsuga menziesii* var. *menziesii*. *Am. Naturalist* 103:389–398.

Sorensen, F. C. 1979. Provenance variation in *Pseudotsuga menziesii* seedlings from the *glauca-menziesii* transition zone. *Silvae Genetica* 28:96–103.

Sorensen, F. C., and R. S. Miles. 1982. Inbreeding depression in height, height growth, and survival of Douglas-fir, ponderosa pine, and noble fir to 10 years of age. *For. Sci.* 28:283–292.

Soulé, M. E. 1985. What is conservation biology? *BioScience* 335:727–734.

Soulé, M. E., ed. 1986. *Conservation Biology: The Science of Scarcity and Diversity*. Sinauer, Sunderland, Mass.

Soulé, M. E., and D. Simberloff. 1986. What do genetics and ecology tell us about the design of nature preserves? *Biol. Conserv.* 35:19–40.

Soulé, M. E., and B. A. Wilcox. 1980. Conservation biology: Its scope and challenge. In *Conservation Biology: An Evolutionary–Ecological Perspective*, ed. M. E. Soulé and B. A. Wilcox, pp. 1–8. Sinauer, Sunderland, Mass.

Spring, O., and E. E. Schilling. 1989. A chemosystematic investigation of the annual species of *Helianthus* (Asteraceae). *Biochem. Syst. Ecol.* 17:535–538.

Stanton, M. L. 1985. Seed size and emergence time within a stand of wild radish (*Raphanus raphanistrum* L.): The establishment of a fitness hierarchy. *Oecologia* 67:524–531.

Stanwood, P. C. 1985. Cryopreservation of seed germplasm for genetic conservation. In *Cry-*

opreservation of Plant Cells and Organs, ed. K. K. Kartha, pp. 199–226. CRC Press, Boca Raton, Fla.

Stanwood, P. C., and L. N. Bass. 1981. Seed germplasm preservation using liquid nitrogen. *Seed Sci. Technol.* 9:423–437.

Start, A. N., and A. G. Marshall. 1976. Nectarivorous bats as pollinators of trees in West Malaysia. In *Tropical Trees: Variation, Breeding and Conservation,* ed., J. Burley and B. T. Styles, pp. 141–150. Academic Press, New York.

Stebbins, G. L. 1942. The genetic approach to problems of rare and endemic species. *Madroño* 6:241–272.

Stebbins, G. L. 1950. *Variation and Evolution in Plants.* Columbia University Press, New York.

Stebbins, G. L. 1980. Rarity of plant species: A synthetic viewpoint. *Rhodora* 82:77–86.

Stebbins, G. L., and K. Daly. 1961. Changes in the variation of a hybrid population of *Helianthus* over an eight-year period. *Evolution* 15:60–71.

Stebbins, G. L., and J. Major. 1965. Endemism and speciation in the California flora. *Ecol. Monogr.* 35:1–35.

Steinhoff, R. J., D. G. Joyce, and L. Fins. 1983. Isozyme variation in *Pinus monticola. Can. J. For. Res.* 13:1122–1131.

Stohlgren, T. J., and P. W. Rundel. 1986. A population model for a long-lived, resprouting chaparral shrub: *Adenostoma fasciculatum. Ecol. Modelling* 34:245–257.

Strauss, S. H. 1987. Heterozygosity and developmental stability under inbreeding and crossbreeding in *Pinus attenuata. Evolution* 41:331–339.

Stushnoff, C. 1987. Cryopreservation of apple genetic resources. *Can. J. Plant Sci.* 67:1151–1154.

Stushnoff, C., and C. Fear. 1985. The potential use of in vitro storage for temperate fruit germplasm: A status report. IBPGR, Rome.

Styer, D. J., and C. K. Chin. 1983. Meristem and shoot-tip culture for propagation, pathogen elimination and germplasm preservation. *Hort. Rev.* 5:221–227.

Sugden, E. A. 1985. Pollinators of *Astragalus monoensis* Barneby (Fabaceae): New host records; potential impact of grazing. *Great Basin Nat.* 45:299–312.

Sun, M., and F. R. Ganders. 1988. Mixed mating systems in Hawaiian *Bidens* (Asteraceae). *Evolution* 42:516–527.

Sutter, R. 1988. Status file determining priorities of rare species for *ex situ* protection. North Carolina Botanical Garden, Chapel Hill.

Svennson, L. 1988. Inbreeding, crossing and variation in stamen number in *Scleranthus annuus* (Caryophyllaceae), a selfing annual. *Evol. Trends Plants* 2:31–37.

Sykes, Z. M. 1969. Some stochastic versions of the matrix model for population dynamics. *Am. Stat. Assoc. J.* 64:111–130.

Synge, J., ed. 1981. *The Biological Aspects of Rare Plant Conservation.* Wiley, New York.

Sytsma, K. J., and L. D. Gottlieb. 1986. Chloroplast DNA evolution and phylogenetic relationships in *Clarkia* Sect. *Peripetasma* (Onagraceae). *Evolution* 40:1248–1261.

Sytsma, K. J., and B. A. Schaal. 1985a. Genetic variation, differentiation, and evolution in a species complex of tropical shrubs based on isozymic data. *Evolution* 39:582–598.

Sytsma, K. J., and B. A. Schaal. 1985b. Phylogenetics of the *Lisianthius skinneri* (Gentianaceae) species complex in Panama utilizing DNA restriction fragment analysis. *Evolution* 39:594–608.

Tansley, S. A. 1988. The status of threatened Proteaceae in the Cape flora, South Africa, and the implications for their conservation. *Biol. Conserv.* 43:227–239.

Taylor, C. E., and G. C. Gorman. 1975. Population genetics of a "colonizing" lizard: Natural selection for allozyme morphs in *Anolis grahami. Heredity* 35:241–247.

Taylor, R. H. 1975. Some ideas on speciation in New Zealand parakeets. *Nortornis* 22:110–121.

Templeton, A. R. 1979. The unit of selection in *Drosophila mercatorum*. II. Genetic revolutions and the origin of coadapted genomes in parthenogenetic strains. *Genetics* 92:1265–1282.

Templeton, A. R. 1980. The theory of speciation via the founder principle. *Genetics* 94:1011–1038.

Templeton, A. R. 1982. The crisis of partial extinction. *Natural Areas J.* 2(3):25–38.

Templeton, A. R. 1983. Phylogenetic inference from restriction endonuclease cleavage site maps with particular reference to the evolution of humans and apes. *Evolution* 37:221–244.

Templeton, A. R. 1986. Coadaptation and outbreeding depression. In *Conservation Biology: The Science of Scarcity and Diversity,* ed. M. E. Soulé, pp. 105–116. Sinauer, Sunderland, Mass.

Templeton, A. R. 1987. Inferences on natural population structure from genetic studies on captive mammalian populations. In *Mammalian Dispersal Patterns,* ed. B. D. Chepko-Sade and Z. T. Halpin, pp. 257–272. University of Chicago Press, Chicago.

Templeton, A. R. 1989. The meaning of species and speciation: A genetic perspective. In *Speciation and Its Consequences,* ed. D. Otte and J. A. Endler, pp. 3–27. Sinauer, Sunderland, Mass.

Templeton, A. R., S. K. Davis, and B. Read. 1987. Genetic variability in a captive herd of Speke's gazelle (*Gazella spekei*). *Zoo Biol.* 6:305–313.

Templeton, A. R., H. Hemmer, G. Mace, U. S. Seal, W. M. Shields, and D. S. Woodruff. 1986. Local adaptation, coadaptation and population boundaries. *Zool. Biol.* 5:115–125.

Templeton, A. R., H. Hollocher, S. Lawler, and J. S. Johnston. 1989. Natural selection and ribosomal DNA in *Drosophila. Genome* 31:296–303.

Templeton, A. R., and B. Read. 1983. The elimination of inbreeding depression in a captive herd of Speke's gazelle. In *Genetics and Conservation: A Reference for Managing Wild Animal and Plant Populations,* ed. C. M. Schonewald-Cox, S. M. Chambers, B. MacBryde, and L. Thomas, pp. 241–261. Benjamin-Cummings, Menlo Park, Calif.

Templeton, A. R., and B. Read. 1984. Factors eliminating inbreeding depression in a captive herd of Speke's Gazelle. *Zoo Biol.* 3:177–199.

Templeton, A. R., K. Shaw, E. Routman, and S. K. Davis. 1990. The genetic consequences of habitat fragmentation. *Ann. Missouri Bot. Gard.* 77:13–27.

Templeton, A. R., C. F. Sing, and B. Brokaw. 1977. The unit of selection in *Drosophila mercatorum*. I. The interaction of selection and meiosis in parthenogenetic strains. *Genetics* 82:349–376.

Tepedino, V. J. 1979. The importance of bees and other insect pollinators in maintaining floral species composition. In *The Endangered Species: A Symposium,* pp. 139–150. Great Basin Nat. Mem. No. 3.

Terborgh, J. 1986. Keystone plant resources in the tropical forest. In *Conservation Biology: The Science of Scarcity and Diversity,* ed. M. E. Soulé, pp. 330–344. Sinauer, Sunderland, Mass.

Terborgh, J. 1988. The big things that run the world—A sequel to E. O. Wilson. *Conserv. Biol.* 3:403–403.

Thibodeau, F. R. 1983. Endangered species: Deciding which species to save. *Envir. Mgmt.* 7:102–107.

Thomson, J. D., and R. C. Plowright. 1985. Matacil insecticide spraying, pollinator mortality, and plant fecundity in New Brunswick forests. *Can. J. Bot.* 63:2056–2061.

Thorne, R. F. 1967. A flora of Santa Catalina Island, California. *Aliso* 6:1–77.

Toole, V. K. 1986. Ancient seeds: Seed longevity. *J. Seed Technol.* 10:1–10.

Toppings, P. 1989. The significance of inbreeding depression to the evolution of self-fertilization in *Eichhornia paniculata* (Spreng.) Solms (Pontederiaceae). M.Sc. thesis, University of Toronto.

Towill, L. E. 1983. Improved survival after cryogenic exposure of shoot-tips derived from in vitro plant cultures of potato. *Cryobiology* 20:567–573.

Towill, L. E. 1985. Low temperature and freeze-/vacuum-drying preservation of pollen. In *Cryopreservation of Plant Cells and Organs,* ed. K. K. Kartha, pp. 171–198. CRC Press, Boca Raton, Fla.

Towill, L. E. 1988. Genetic considerations for germplasm preservation of clonal materials. *HortScience* 23:91–97.

Tracey, J. G. 1981. Australia's rain forests: Where are the rare plants and how do we keep them? In *The Biological Aspects of Rare Plant Conservation,* ed. H. Synge, pp. 165–178. Wiley, New York.

Travis, J., and R. Sutter. 1986. Experimental designs and statistical methods for demographic studies of rare plants. *Nat. Areas J.* 6:3–12.

Tuljapurkar, S. D. 1985. Population dynamics in variable environments. VI. Cyclic environments. *Theor. Pop. Biol.* 28:1–17.

Tuljapurkar, S. D., and S. H. Orzack. 1980. Population dynamics in variable environments. I. Long-run growth rates and extinction. *Theor. Pop. Biol.* 18:314–342.

Turesson, G. 1922. The genotypical response of the plant species to the habitat. *Hereditas* 3:211–350.

Turkington, R., and J. L. Harper. 1979. The growth, distribution and neighbour relationships of *Trifolium repens* in permanent pastures. IV. Fine-scale biotic differentiation. *J. Ecol.* 67:245–254.

Turner, B. L. 1981. Letter to G. Seiler, University of Texas Herbarium, Austin.

Tyler, N., C. Stushnoff, and L. V. Gusta. 1988. Freezing of water in dormant vegetative apple buds in relation to cryopreservation. *Plant Physiol.* 87:201–205.

U.S. Fish and Wildlife Service. 1985. Endangered and threatened wildlife and plants: review of plant taxa for listing as endangered or threatened species; notice of review. *Federal Register* 50(17).

U.S. Fish and Wildlife Service. 1987. Draft recovery plan for the large-flowered fiddleneck (*Amsinckia grandiflora*). U.S.F.W.S., Portland, Ore.

U.S. Fish and Wildlife Service. 1990. Endangered and threatened wildlife and plants: Review of plant taxa for listing as endangered or threatened species; notice of review. 50 CFR Part 17. *Federal Register.*

Uyenoyama, M. K. 1986. Inbreeding and the cost of meiosis: The evolution of selfing in populations practising biparental inbreeding. *Evolution* 40:388–404.

Vandermeer, J. 1982. To be rare is to be chaotic. *Ecology* 63:1167–1168.

Van Dijk, H. 1989. Genetic variability in *Plantago* species in relation to their ecology. 4. Ecotypic differentiation in *P. major. Theor. Appl. Genet.* 77:749–759.

van Sloten, D. H., and M. Holle. 1988. Temperate fruit genetic resources and the IBPGR. *HortScience* 23:73–74.

Van Valen, L. 1965. Morphological variation and width of ecological niche. *Am. Naturalist* 99:377–390.

Vaquero, F. J., P. Garcia, L. Ramirez, and M. Perez de la Vega. 1989. Mating system in rye: Variability in relation to the population and plant density. *Heredity* 62:17–26.

Vasek, F. C. 1964. Two new species related to *Clarkia unguiculata. Madroño* 17:219–221.

Venable, D. L. 1984. Using intra-specific variation to study the ecological significance and evolution of plant life-histories. In *Perspectives on Plant Population Ecology,* ed. R. Dirzo and J. Sarukhan, pp. 166–187. Sinauer, Sunderland, Mass.

Verkleij, J. A. C., A. M. de Boer, and T. F. Lugtenborg. 1980. On the ecogenetics of *Stellaria media* and *Stellaria pallida* from abandoned arable field. *Oecologia* 46:354–359.

Wagner, W. L., D. R. Herbst, and S. H. Sohmer. 1990. *Manual of the Flowering Plants of Hawaii*. 2 vols. University of Hawaii Press/Bishop Museum Press, Honolulu.

Wagner, D. B., G. R. Furnier, M. A. Saghi-Maroof, S. M. Williams, B. P. Dancik, and R. W. Allard. 1987. Chloroplast DNA polymorphisms in lodgepole and jack pines and their hybrids. *Proc. Natl. Acad. Sci. USA* 84:2097–2100.

Waller, D. M. 1981. Neighborhood competition in several violet populations. *Oecologia* 51:116–122.

Waller, D. M. 1986. Is there disruptive selection for self-fertilization? *Am. Naturalist* 128:421–426.

Waller, D. M., D. M. O'Malley, and S. C. Gawler. 1988. Genetic variation in the extreme endemic *Pedicularis furbishiae* (Scrophulariaceae). *Conserv. Biol.* 1:335–340.

Wang, C. H., T. O. Perry, and A. G. Johnson. 1969. Pollen dispersion of slash pine (*Pinus elliottii*) with special reference to seed orchard management. *Silvae Genetica* 9:78–86.

Warwick, S. I., and L. D. Black. 1986. Genecological variation in Canadian and European populations of the colonizing weed species *Setaria faberi*. *Can. J. Bot.* 65:1396–1402.

Warwick, S. I., and L. D. Gottlieb. 1985. Genetic divergence and geographic speciation in *Layia* (Compositae). *Evolution* 39:1236–1241.

Warwick, S. I., B. K. Thompson, and L. D. Black. 1987. Life history and allozyme variation in populations of the weed species *Setaria faberi*. *Can. J. Bot.* 65:1396–1402.

Waser, N. M., and M. V. Price. 1983. Optimal and actual outcrossing, and the nature of plant-pollinator interaction. In *Handbook of Experimental Pollination Biology*, ed. C. E. Jones and R. J. Little, pp. 341–359. Van Nostrand, New York.

Waser, N. M., and M. V. Price. 1985. Reciprocal transplant experiments with *Delphinium nelsonii* (Ranunculaceae): Evidence for local adaptation. *Am. J. Bot.* 72:1726–1732.

Waser, N. M., and M. V. Price. 1989. Optimal outcrossing in *Ipomopsis aggregata:* Seed set and offspring fitness. *Evolution* 43:1097–1109.

Waser, N. M., M. V. Price, A. M. Montalvo, and R. N. Gray. 1987. Female mate choice in a perennial herbaceous wildflower, *Delphinium nelsonii*. *Evol. Trends Plants* 1:29–33.

Watterson, G. A. 1984. Allele frequencies after a bottleneck. *Theor. Pop. Biol.* 26:387–407.

Weir, B. S., and C. C. Cockerham. 1984. Estimating F-statistics for the analysis of population structure. *Evolution* 38:1358–1370.

Welch, S., and R. Miewald, eds. 1983. *Scarce Natural Resources: The Challenge to Public Policymaking*, Vol. 11 of *Sage Yearbooks in Politics and Public Policy*. Sage, Beverly Hills, Calif.

Wells, O. O. 1964. Ecotypes of ponderosa pine. 1. The ecotypes and their distribution. *Silvae genetica* 13:89–103.

Werner, P. A. 1975. Predictions of fate from rosette size in teasel, *Dipsacus follonum* L. *Oecologia* 20:197–201.

Werner, P. A., and H. Caswell. 1977. Population growth rates and age vs. size-distribution models for teasel (*Dipsacus silvestris* Huds.). *Ecology* 58:1103–1111.

Wheeler, N. C., and R. P. Guries. 1982. Population structure, genic diversity, and morphological variation in *Pinus contorta* Doug. *Can. J. For. Res.* 12:595–606.

Whelan, E. D. P. 1978. Cytology and interspecific hybridization. In *Sunflower Science and Technology*, ed. J. F. Carter, pp. 339–370. Society of Agronomy, Madison, Wis.

White, J. 1979. The plant as metapopulation. *Annu. Rev. Ecol. Syst.* 10:109–145.

White, P. S. 1979. Pattern, process, and natural disturbance in vegetation. *Bot. Rev.* 45:229–299.

Wiens, D., D. L. Nickrent, C. I. Davern, C. L. Calvin, and N. J. Vivrette. 1989. Developmental failure and loss of reproductive capacity in the rare palaeoendemic shrub *Dedeckera eurekensis. Nature* 338:65–67.

Wilcove, D. S. 1988. *National Forests: Policies for the Future,* Vol. 2: *Protecting Biological Diversity.* Wilderness Society, Washington, D.C.

Wilcox, M. D. 1983. Inbreeding depression and genetic variances estimated from self- and cross-pollinated families of *Pinus radiata. Silvae Genetica* 32:89–96.

Wildt, D. E., M. Bush, K. L. Goodrowe, C. Packer, A. E. Pusey, J. L. Brown, P. Joslin, and S. J. O'Brien. 1987. Reproductive and genetic consequences of founding isolated lion populations. *Nature* 329:328–330.

Wildt, D. E., and U. S. Seal. 1988. *Research Priorities for Single Species Biology.* National Science Foundation, Washington, D.C.

Wilson, E. O. ed. 1988. *Biodiversity.* National Academy Press, Washington, D.C.

Withers, L. A. 1984. Germplasm conservation in vitro: Present state of research and its application. In *Crop Genetic Resources: Conservation and Evaluation,* ed. J. H. W. Holden and J. T. Williams, pp. 138–157. Allen and Unwin, London.

Withers, L. A. 1987. The low temperature preservation of plant cell, tissue and organ cultures and seed for genetic conservation and improved agricultural practices. In *The Effects of Low Temperatures on Biological Systems,* ed. B. W. W. Grout and G. J. Morris, pp. 389–409. Edward Arnold, London.

Withers, L. A., and J. T. Williams. 1986. *In Vitro Conservation.* IBPGR, Rome.

Wolf, P. G., C. H. Haufler, and E. Sheffield. 1988. Electrophoretic variation and mating system of the clonal weed *Pteridium aquilinum* (bracken). *Evolution* 42:1350–1354.

Wolff, S. L., and R. L. Jefferies. 1987a. Morphological and isozyme variation in *Salicornia europaea* (S. L.) (Chenopodiaceae) in northeastern North America. *Can. J. Bot.* 65:1410–1419.

Wolff, S. L., and R. L. Jefferies. 1987b. Taxonomic status of diploid *Salicornia europaea* (S. L.) (Chenopodiaceae) in northeastern North America. *Can. J. Bot.* 65: 1420–1426.

Woodstock, L. W. 1975. Freeze-drying as an alternative method for lowering seed moisture. *Proc. Assoc. Offic. Seed Anal.* 65:159–163.

Woodstock, L. W., S. Maxon, K. Faul, and L. Bass. 1983. Use of freeze-drying and acetone impregnation with natural and synthetic anti-oxidants to improve storability of onion, pepper, and parsley seeds. *J. Am. Soc. Hort. Sci.* 108:692–696.

Woodstock, L. W., J. Simkin, and E. Schroeder. 1976. Freeze-drying to improve seed storability. *Seed Sci. Technol.* 4:301–311.

Wright, S. 1922. Coefficients of inbreeding and relationship. *Am. Naturalist* 56:330–338.

Wright, S. 1931. Evolution in Mendelian populations. *Genetics* 16:97–159.

Wright, S. 1932. The roles of mutation, inbreeding, crossbreeding, and selection in evolution. *Proc. 6th Int. Cong. Genet.* 1:356–366.

Wright, S. 1933. Inbreeding and homozygosis. *Proc. Natl. Acad. Sci. USA* 19:411–420.

Wright, S. 1938. Size of population and breeding structure in relation to evolution. *Science* 87:430–431.

Wright, S. 1939. The distribution of self-sterility alleles in populations. *Genetics* 24:538–552.

Wright, S. 1946. Isolation by distance under diverse systems of mating. *Genetics* 31:39–59.

Wright, S. 1960. On the number of self-incompatibility alleles maintained by a given mutation rate in a population of a given size: A reexamination. *Biometrics* 16:61–85.

Wright, S. 1965. The distribution of self-incompatibility alleles in populations. *Evolution* 18:609–619.

Wright, S. 1977. *Evolution and the Genetics of Populations,* Vol. 3: *Experimental Results and Evolutionary Deductions.* University of Chicago Press, Chicago.

Wright, S. J., and S. P. Hubbell. 1983. Stochastic extinction in reserve size: A focal species approach. *Oikos* 41:466–477.

Yazdani, R., O. Muona, D. Rudin, and A. E. Szmidt. 1985. Genetic structure of a *Pinus sylvestris* seed-tree stand and naturally regenerated understory. *For. Sci.* 31:430–436.

Yazdi-Samadi, B., K. K. Wu, and S. K. Jain. 1978. Population biology of *Avena*. VI. The role of genetic polymorphisms in the outcome of interspecific competition. *Genetica* 48:151–159.

Yeh, F. C., and Y. A. El-Kassaby. 1980. Enzyme variation in natural populations of Sitka spruce (*Picea sitchensis*). 1. Genetic variation patterns among trees from 10 IUFRO provenances. *Can. J. For. Res.* 10:415–422.

Yeh, F. C., and C. Layton. 1979. The organization of genetic variability in central and marginal populations of lodgepole pine *Pinus contorta* ssp. *latifolia. Can. J. Genet. Cytol.* 21:487–503.

Yeh, F. C., and D. M. O'Malley. 1980. Enzyme variation in natural populations of Douglas-fir (*Pseudotsuga menziesii* var. *menziesii* (Mirb) Franco) from British Columbia. 1. Genetic patterns in coastal populations. *Silvae Genetica* 29:83–92.

Yonezawa, K. 1985. A definition of the optimal allocation of effort in conservation of plant genetic resources with application to sample size determination for field collections. *Euphytica* 34:345–354.

Zangerl, A. R., and F. A. Bazzaz. 1984. Niche partitioning between two phosphoglucoisomerase genotypes in *Amaranthus retroflexus. Ecology* 65:218–222.

Zavarin, E., W. Hathaway, T. Reichert, and Y. B. Linhart. 1967. Chemotaxonomic study of *Pinus torreyana* Parry turpentine. *Phytochemistry* 6:1019–1023.

Ziv, M. 1986. In vitro hardening and acclimatization of tissue culture plants. In *Plant Tissue Culture and Its Agricultural Applications,* ed. L. A. Withers and P. G. Alderson, pp. 187–196. Butterworth, London.

Zurawski, G., and M. T. Clegg. 1987. Evolution of higher-plant chloroplast DNA encoded genes: Implications for structural function and phylogenetic studies. *Annu. Rev. Plant. Physiol.* 38:391–418.

Index

Abies
  bracteata, 151
  concolor, 153
Acacia mangium, 12
Accelerated aging test, 140–41
Achillea lanulosa, 124
Adaptation to captive environment, 187
Allee effect, 52
Allelic richness, 102
Allozyme variation, 12, 75–76, 88, 126–28,
    172–73, 176, 179, 184, 185. *See also*
    Genetic diversity
  correspondence of, with morphological
    variation, 154–55
  correspondence of, with other measures,
    127–28
  limitations of, 126–27
Alseis blackiana, 81
Amsinckia grandiflora, 41, 204–5
Ancestry coefficient, 192. *See also* Hybridity
    coefficient; Inbreeding coefficient
Aphanisma blitoides, 198–99
Araucaria cunninghamii, 69
Arctostaphylos hookeri ssp. *ravenii*, 215
Argyroxiphium sandwicense, 41
Arisaema triphyllum, 55
Artificial aging, 141
Asexual sports, 199
Astragalus, 40–42, 84, 87
  lentiginosus, 203
  linifolius, 88, 89
  lonchocarpus, 88
  osterhouti, 88, 89
  pattersoni, 88, 89
  pectinatus, 88, 89
Atriplex vallicola, 199
Australia, rare and endangered species in,
    116–18

Avena, 32, 39
  barbata, 32, 84
Avoidance of inbreeding, 190

Barro Colorado Island, 81
Benefit–cost analysis, 212–13, 216
Bertholletia excelsa, 66
Betula murryana, 199
Biased sample, 113
Bidens, 128
Biparental inbreeding, 18, 19
Bombax, 64
Botanical gardens, 118, 216, 221
Brassica, 130
Breeding systems. *See* Mating systems
Bromus inermis, 137

California Endangered Species Act,
    amendments to, 196
California Native Plant Protection Act, 196
Calochortus, 200
  albus, 202
Captive breeding, 182, 189, 206, 207
  number of generations, used for, 190
Captive environment, adaptation to, 187
Captive management, goals of, 182
Castilleja uliginosa, 215
Ceiba, 64
Center for Plant Conservation, 195, 206,
    211, 215, 221
Cercocarpus
  betuloides var. *blancheae*, 172–73
  traskiae, 172–75
Chamaecrista fasciculata, 26
Checkerspot butterflies, 51, 56
Chloroplast DNA, 130–31, 176, 179
Clarkia, 131
  tembloriensis, 23, 24

Clarkia (continued)
  tembloriensis ssp. *calientensis,* 205
  williamsonii, 85
Clematis
  fremontii, 131
  fremonti ssp. *riehlii,* 124
Clinal variation, 152–55. *See also* Ecotypes
Clonal growth, 54–55
Clonal plants, population viability of, 55
Clonal preservation, 142–44
Clonal reproduction, 35
Coadaptation, 25, 192
Coelostegia, 64
Collared lizard, 184
Collection priorities, 99
Collection records, 116
Collection strategy, 220–21. *See also*
  Sampling; Sampling design
Colonizing species, 12
Common alleles, 103
Competitor exclusion, 205
Composite mixtures, 29, 192
Conservation
  of biological hierarchy, 214–15
  collection for, 221, 225
  cost vs. benefit of, 209
  definition of, 211
  of diversity, objectives, 158
  as economic activity, 210
  priorities in, 157–58, 197–99, 215–16,
    227
  strategy in, 210
Core collection, 112
Core variability, 162
Critical-zone biological entities, 223
Critical-zone resource, 213–14
Cryogenic storage, 138, 206
Cryopreservation, 143–44
Cryoprotectants, 140
Cucurbita foetidissima, 23, 24
Cupressus
  macnabiana, 156–57
  sempervirens, 156

Decision-making model, 220–21
Delphinium nelsonii, 26
Demographic models, 49, 57
Demographic stochasticity, 37, 47, 61, 188
Depletion function, 217
Depletion models, 213, 217–18

Disjunct populations, 220–21
Disturbance, 52–53
Diversity, in tropical rain forests, 62
DNA fingerprinting, 133, 184, 193
DNA variation, 128–33
Domestication, 30
Drosophila
  hydei, 187
  mercatorum, 187
Durio, 64

Echinochloa, 39
  microstachya, 13
Echium plantagineum, 38
Economic activity, 210
Economic-resource allocation, 213–15
Economic scarcity and depletion models,
    213
Economic theory of production, 216–17
Ecosystem conservation, 214
Ecotypes, 32, 111, 152, 206, 214, 221,
    229–30. *See also* Clinal variation;
    Habitat specialization
Edaphic specialization, 32, 53, 204. *See also*
    Habitat specialization
Effective population size, 7
Eichhornia
  paniculata, 5, 23
  paradoxa, 27
Endangered species, 211–12
  definition of, 196
  resource analysis of, 213–14
Endangered Species Act, 177, 215
Endemism, 32, 42, 53, 204
Endonuclease restriction site variation,
    128–30
Environmental heterogeneity, 26
Environmental stochasticity, 36, 46, 50,
    53–55, 60
Eonycteris spelaea, 68
Espeletia, 27
Eucalyptus, 40, 67
  pendens, 107
  stoatei, 7
Ex situ collections, 44, 69–70, 71, 158–59,
    166, 204, 221
  founder effect in, 39
  management of, 61, 168–69
  types of, 100
Excised embryos, 140

Extinction, 50–51
  rates of, in tropical areas, 62
  of U.S. plants, 211
Extinction probability, 45

Festuca, 33
Ficus, 82
Founder effect, 8–10, 39–40, 104
  with *ex situ* propagation, 39
Founder event, 185, 188
Founder number, 185
Founder population, 183–84
Founder representation, 185–86, 191

Gaillardia pulchella, 32
Gene banks, 166
Gene flow, 56, 65–66, 80, 164, 186, 188,
  193
  and genetic diversity, 186
Gene resource management unit, 160–61,
  206. *See also* Genetic management
  as a buffer, 164
  geographic placement of, 162–63
  size of, 163–64
Genet diversity, 102, 107
Genetic architecture, 11, 150, 155, 161, 218
Genetic assimilation, 171, 181
Genetic bottleneck, 8–17, 59, 88, 104–6,
  151
  consequences of, 9–10
  effects of, on polygenic traits, 17
  experimental, 13
  loss of alleles with, 9–10
  loss of variation with, 9–10
Genetic covariance, 125
Genetic differentiation, 89, 98
Genetic diversity, 12, 67, 76–80. *See also*
  Allelic richness; Allozyme variation;
  Heterozygosity
  among populations, 77, 79–80
  breeding system and, 83
  clonal reproduction and, 35
  in common species, 81–82
  conservation goals and, 101–2, 209
  geographic patterns and, 40–42
  geographic range and, 31–33, 77, 88
  geographic structure and, 184
  intraspecific, 150
  long-term viability and, 203
  loss of, in small populations, 184–85

management goals for preservation of, 85
mating system and, 78–79, 83
measures of, 77, 102, 134
niche width and, 32
partitioning of, 218
patterns of, 76. *See also* Genetic
  architecture
pollination mechanisms and, 78–79
in rare species, 97–98
sample size and, 207
sensitivity to loss of, 123
short-term viability and, 200–201
at species level, 77
in uncommon species, 82–83
within populations, 78–79
Genetic drift, 4, 8, 163, 186
Genetic environment, 188–89
Genetic gap analysis, 160
Genetic load, 21, 89, 191
Genetic management, 190, 205–6
Genetic resources, 198, 213, 214
  manipulation of, 28
  sampling of, 99
Genetic stochasticity, 47–48, 60–61
Genetic surveys, 184, 189, 206
Genetic transilience, 188
Germination potential, 137
Germination tests, 140
Germplasm conservation, 166
Ginkgo biloba, 158
Glycine
  argyrea, 106
  latrobeana, 111
Gossypium mustelinum, 7
Grizzly bear, 50, 51

Habitat area, 51
Habitat fragmentation, 202
Habitat preservation, 49, 182
Habitat specialization, 42–43
Helianthus, 131
  annuus, 175, 178
  bolanderi, 178–80
  exilis, 178
  paradoxus, 175–78, 199
  petiolaris, 175
Herbivory, 37, 201
Heritability, 125
Hermaphroditism, 11
Heterozygosity, 36, 38, 106

Hibiscus cannabinus, 137
High-grading, 156
Holcus lanatus, 126
Hordeum
  jubatum, 84
  vulgare, 85, 131
Howellia aquatilis, 42
Hybrid derivatives, 199
Hybrid swarm, 175. *See also* Introgression;
  Syngameon
Hybridity coefficient, 191. *See also* Ancestry
  coefficient; Inbreeding coefficient
Hybridization, 171, 181, 193–94. *See also*
  Introgression
  management of area, 174
Hybrids
  protection status of, 177–78
  reintroduction of, 178
Hymenoxys acaulis var. *glabra,* 29
Hypervariable sequences, 133

Immigration, 163
Impatiens, 38
  pallida, 27
In situ conservation, 158–59
In situ populations, 60
In vitro preservation, 142–43
Inadvertent selection, 187
Inbreeding, 17–19, 59–60, 66, 163, 189, 215
  and genetic diversity, 187
Inbreeding coefficient, 18, 191. *See also*
  Ancestry coefficient; Hybridity
  coefficient
Inbreeding depression, 4, 19–25, 38, 44,
  47–48, 89, 96–97, 106, 123, 150, 183,
  185, 189, 190–91. *See also* Outbreeding
  depression
  consequences of, 21
  estimates of, 22–23
  life-cycle stage of expression, 24–25
  relationship of, to environment, 24
  relationship of, to mating system, 22–24
Index species, 159
Integrated conservation strategies, 214
Interdemic selection, 188
Introgression, 181
Ipomopsis aggregata, 26
Island plants, 173

Landscaping, 156
Law of diminishing marginal returns, 217

Layia
  discoidea, 89, 200, 204
  glandulosa, 89, 200
Lethal equivalents, 191
Libocedrus decurrens, 151
Life history, 58–59, 206
Limnanthes
  bakeri, 92
  douglasii, 92
  macounii, 89
Lisianthus, 131
  skinneri, 89
Lobelia cardinalis, 137
Local adaptation, 26, 32–35, 126. *See also*
  Microhabitat adaptation
Local endemics, 64
Lolium multiflorum, 13
Lotus corniculatus, 137
Lupinus texensis, 131
Lycopersicon, 131

Maingaya malayana, 64
Managed populations, loss of genetic
  variation in, 10
Management
  of anthropogenic rarities, 205
  costs, 214
  of existing populations, 205
  mating systems and, 83
  of narrow endemics, 204
Marginal product, 217
Mating systems, 17–18, 21, 58–59, 89–93,
  189
  effects of, 38–39
  pollinator service and, 94–96
  in rare species, 93, 98
  self-fertility, 96
Metapopulation dynamics, 50–51, 55–57, 61
Microeconomic analysis, 212
Microhabitat adaptation, 34–36, 53, 61, 126,
  205, 229–30. *See also* Habitat
  specialization; Local adaptation
Mimulus guttatus, 27
Minimum viable population, 46, 60, 213,
  215. *See also* Population viability
Minisatellite sequences, 133
Mitochondrial DNA, 184, 185
Mixed-mating model, 19
Mixed-mating systems, 21
Morphological variation, 124–26
Mutualism, 58, 63, 68

Narrow endemism. *See* Endemism
Natural catastrophes, 46–47
Natural Diversity Data Base, 196, 212
Natural history, 87
    importance of, 84
Nature Conservancy, 212, 215, 221
Neighborhood size, 52, 65
Neostrearia, 64
Neutral-allele hypothesis, 102–3, 216
Nonrenewable resources, 213
Northern spotted owl, 51–51, 52, 203

Oenothera, 32, 39
    avita subsp. *eurekensis*, 203
    organensis, 93, 201
Offsite breeding. *See* Captive breeding
Opportunity cost, 213
Optimal outcrossing distance, 26
Option value, 223
Opuntia, 37
Orcuttia, 41
Oryctes nevadensis, 197–98
Oroxylum, 64
Orthodox seed storage, 136–38
Outbreeding depression, 4, 25–27, 171, 181,
    189. *See also* Inbreeding depression
Outcrossing rates, 66
Outplanting, 205. *See also* Transplant
    experiments
Overdominance, 21

Pajanelia, 64
Pathogens, 36–37, 201
Pedicularis furbishiae, 42, 49, 56
Pennisetum, 131
Phenotypic plasticity, 34, 54, 61, 124
Phenotypic variation, 124–26
Phlox
    divaricata, 132
    drummondi, 27
Physiological integration, 54
Pinus, 131
    attenuata, 153
    brutia, 156
    contorta, 154
    densiflorus, 156
    monticola, 151–52
    muricata, 152
    ponderosa, 38, 152
    radiata, 152

resinosa, 12, 89, 150
roxburghii, 156
sylvestris, 38
thunbergiana, 156
torreyana, 12, 67, 88, 89, 151
Pithecellobium pedicellare, 66
Plantago cordata, 43, 56, 200, 201,
    205
Pollen dispersal, 52
Pollen preservation, 144–45
Pollinators, 98
Polygenic variation, 124–26
Polymerase chain reaction, 129
Polyploidy, 11
Population bottleneck, 8–17, 84, 89, 203.
    See also Genetic bottleneck
Population decline, 202
Population density, 117
Population expansion, 185
Population size, 89, 101
    historical changes in, 84. *See also* Rarity:
        types of
Population viability, 36–37, 45, 69
    analyses of, 60, 205
    genetic diversity and, 59
    genetic vs, demographic factors in, 203,
        205
    sampling design, 218–19
    systematic threats to, 46
Preserves, 214
Priorities for plant conservation, 157–58,
    198–99, 215–16, 227
    asexual and hybrid taxa, 199
    collection strategies and, 99
    taxonomic rank, 198–99
Pritchardia munroii, 215
Production function, 217, 218
Production-possibility function, 213
Protein electrophoresis. *See* Allozyme
    variation
Prunus maritima var. *gravesii*, 199
Pseudotsuga menziesii, 154

Quantitative genetics, 124–26
Quercus lobata, 157

Rare alleles, 104, 203, 219–20
Rare plants
    abundance of, 204
    number of populations of, 206–7, 229

Rare species
    criteria for listing, 197
    definition of, 196
    procedure for listing, in California, 196
Rare vs. common species, similarities
    between, 200
Rarity
    anthropogenic, 42, 43, 202, 205, 227
    causes of, 158, 227
    consequences of, 6–7
    diffusive, 64
    fitness depression and, 201–2
    geographic range and, 64–65
    of local endemics, 64
    low population density and, 63
    natural, 227
    nonanthropogenic, 42, 202–3
    types of, 5–6, 42, 63–64, 87, 200
    uniform, 64
Recalcitrant seed storage, 138–40
Reciprocal transplants, 26, 109. *See also*
    Transplant experiments
Recombinant DNA, 128
Recordkeeping, importance of in germplasm
    storage, 168
Reestablishment, 28–29, 205
Reforestation, 156, 160
Regional floristic diversity, 118
Reintroduction, 182, 184, 205, 221–23, 228
    and reestablishment, 44
Resource depletion, 223
Restriction fragment length polymorphism,
    129
Ribosomal DNA, 131–33, 176, 179, 184,
    185
Rudbeckia missouriensis, 132

Sabatia angularis, 23, 24, 27
Safe sites, 61
Salicornia, 12–13
Sample size, 216, 232–33
    diminishing returns on, 106
    logarithmic relationship, 104
    minimum per species, 109
    minimum per population, 106, 109–10
    and probability of success, 107
    of seeds, 114–15
Samples, repeated, 207
Sampling, 43, 99, 158–59, 161, 183, 216–20
    allele diversity and, 219–20
    diminishing returns on, 216

of disjunct populations, 220–21
    simple random, 113
    stratified random, 113, 207
    systematic, 113
    theory, 100
Sampling design, 84, 119, 161, 166–68,
    184, 193, 206–7, 220–21, 229–35
    choice of populations, 110–11
    cuttings, 115
    genets, 107
    and mating system, 114
    multiple populations, 110–11
    number of populations, 110–11, 220–21
    rare alleles, 104
    repeated sampling, 112–13
    seed, 114–15
    type of propagule, 113–16
    types of samples, 113
Sampling guidelines, 236–37
Sarracenia, 39
    purpurea, 13
Scarcity models, 213
Seed banks, 206, 216, 221
Seed dispersal, 52
Seed dormancy, 58
Seed longevity, 136
Seed storage, 114, 142, 206
Seed zones, 166
Seeds
    dessication intolerant, 135
    dessication tolerant, 135
    orthodox, 135
    recalcitrant, 135–36
    subfreezing storage of, 137–38
Self-fertilization, 11, 17–18, 19, 59
Self-incompatibility, 93
Shifting balance, 188
Size structure, 57
Small population size
    causes of, 5–7
    effects of, 4–5
Somaclonal variants, 143
Spartina, 34, 35
Speke's gazelle, 59, 185, 188, 190–91, 207
Spotted owl. *See* Northern spotted owl
Spruce grouse, 52
Stabilized hybrid derivative, 176
Stephanomeria
    exiqua ssp. *coronaria,* 92
    malheurensis, 92
Streptanthus niger, 200

Subdivided population, 185, 186, 188, 193.
    See also Metapopulation dynamics
Sustainable harvest, 213
Swallenia alexandrae, 203
Syngameon, 193–94

Tachigali versicolor, 82
Taraxacum, 131
Taxonomic distinctiveness, 199, 228
Threatened species, 215
    definition of, 196
Thuja plicata, 150
Timber harvest, 161
Transition matrices, 49
Transplant experiments, 33, 35–36
Trend information, 212

Trifolium
    repens, 35, 37
    stoloniferum, 200, 205
Tropical forest, 81
Typha, 32

U.S. Fish and Wildlife Service, 211, 215

Variance of reproductive success, 188
Veronica, 34
Viability monitoring, 140–41
Vicia faba, 131

Wild-crop relatives. *See* Genetic resources

Yucca glauca, 55

Zizania texana, 200